Phononic Crystals

Abdelkrim Khelif • Ali Adibi
Editors

Phononic Crystals

Fundamentals and Applications

Springer

Editors
Abdelkrim Khelif
Institut FEMTO-ST
Centre National de la Recherche Scientifique
Besançon Cedex, France

Ali Adibi
School of Electrical Engineering
Georgia Institute of Technology
Atlanta, GA, USA

ISBN 978-1-4614-9392-1 ISBN 978-1-4614-9393-8 (eBook)
DOI 10.1007/978-1-4614-9393-8

Library of Congress Control Number: 2015943375

Springer New York Heidelberg Dordrecht London

Printed on acid-free paper

Springer Science+Business Media LLC New York is part of Springer Science+Business Media (www.springer.com)

Preface

Phononic crystals (PnCs) are novel synthetic periodic materials for controlling and manipulating the propagation of elastic (or acoustic) waves. The periodic nature of PnCs gives them novel properties that cannot be found in bulk materials. For example, PnCs can exhibit acoustic (or phononic) bandgaps, which are frequency ranges in which the propagation of acoustic waves inside the PnC is prohibited. The addition of defects to a perfect PnC with a phononic bandgap allows for the design of devices like waveguides and cavities to control the propagation of acoustic waves inside the bandgap and to enable novel functionalities in a very compact structure.

Imminent impact of PnCs is expected in the near future in applications like wireless communications, sensing, acoustic signal processing, and ultrasound imaging. Novel devices (such as acoustic filters, resonators, sources and lenses) with outstanding performance measures are being enabled by the use of PnCs. In addition, the use of these structures to form acoustic metamaterials can uncover novel effects like negative refraction, acoustic invisibility, or superlensing. This, in turn, can enable researchers to design functional structures with such performance that cannot be obtained with conventional acoustic materials.

While the research in the field of phononic crystals and acoustic metamaterials is at the early stages, their optical counterparts (i.e., photonic crystals) have already been demonstrated to possess unique properties that are not achieved using conventional bulk materials. The properties of photonic crystals have been the subject of intensive investigations in the last decade, and several successful books have been published to address their unique properties and applications. Knowing that the research in PnCs is in its infancy, and more attention is given to this field lately, the field is expected to expand considerably in the next few years.

The purpose of this book is to present a detailed overview of the state of the field from material, device, and application perspectives, and provide the necessary tools for researchers to explore the field. To achieve this goal, this book covers the simulation, fabrication, and characterization methods used to design and experiment with PnCs to the level that is accessible for both the experienced and beginner in the field. The book also reports the most important advances in the field in the last few years.

The idea for this book first came up in summer 2009, where we co-chaired the first International Workshop on Photonic Crystals (Nice, France, 2009), in which all experts in the field were invited. The need for an all-encompassing reference in the field of phononic crystals was recognized in the meeting. After that meeting, we spent an extensive amount of time looking into the needs of the community to form the structure of the chapters in the book and to convince the experts in the field (who were among the participants in the workshop) to write their respective chapters.

The authors of these chapters are among the world leaders in their respective fields with years of experience in performing cutting-edge research and educating young scientists and engineers. In addition to presenting the landscape of the research in this field, we hope that this book can provide interested readers with an in-depth knowledge of the field. The individual chapters are written in such a way that they can be used as the text material for enhancing graduate-level courses in mechanical or electrical engineering disciplines.

At the end of this journey, we would like to thank all those who helped us in forming this book through their discussions, contributions to the book, and reviews of the different sections. We also like to thank the many researchers (students, postdocs, members of technical staff, and professors) whose contributions are covered in this book. Our special thanks go to Dr. Ali A. Eftekhar for his key role in forming the idea of the book, his help in defining different chapters, and his excellent feedback at different stages of forming the book.

Besançon Cedex, France A. Khelif
Atlanta, GA, USA A. Adibi
March 2015

Contents

Chapter 1
Introduction to Acoustics of Phononic Crystals. Homogenization at Low Frequencies

José Sánchez-Dehesa and Arkadii Krokhin

1.1 Introduction

Light and sound are the two most important carriers of information for live organisms. Historically the nature and characteristics of sound have been understood earlier than that of light because the progress in classical mechanics usually advanced the progress in electrodynamics. This is not the case for the recently emerged areas of *photonic* and *phononic* crystals. In the last decade of the twentieth century there was a burst in study of photonic crystals—periodic dielectric structures which may provide much higher efficiency than the traditional optical devices in generation, waveguiding, focusing, splitting, slowing down, etc. of light. The first theoretical publications by Yablonovich [1] predicting suppression of spontaneous emission in semiconductors structures where electron and photon band gaps overlap, and by John [2] predicting easier localization of light near photonic band gap were followed by first experimental realization of 3D photonic crystal with fcc lattice possessing a gap in the microwave region [3]. Regular and intensive study of sound waves in periodic elastic media has retarded by a few years, as compared to similar study of electromagnetic waves. The first phononic band structures for two-dimensional lattices of solid cylinders in solid background were calculated practically simultaneously by Sigalas and Economou [4] and Kushwaha et al. [5].

J. Sánchez-Dehesa (✉)
Grupo de Fenómenos Ondulatorios, Departamento de Ingeniería Electrónica, Universitat Politècnica de València, Camino de Vera s/n, Valencia 46022, Spain
e-mail: jsdehesa@upvnet.upv.es

A. Krokhin
Department of Physics, University of North Texas, 1155 Union Circle #311427, Denton, TX 76203, USA
e-mail: arkady@unt.edu

A. Khelif, A. Adibi (eds.), *Phononic Crystals*, DOI 10.1007/978-1-4614-9393-8_1

Fig. 1.1 This piece of modern art by E. Sempere represents a two-dimensional phononic crystal of steel rods with a diameter of 2.9 cm arranged in a square lattice with a period 10 cm

First experimental measurement of sound attenuation due to periodicity has been performed under ambient conditions using a sculpture by E. Sempere, exhibited in open air in Madrid, see Fig. 1.1.

The transmission coefficient for sound waves propagating perpendicular to the rods exhibited a minimum around frequency of 1.67 kHz [6]. This minimum was originally attributed to a phononic band gap. However, soon the exact band structure calculations [7] have shown that the observed attenuation is due to a dip minimum in the phononic density of states but not to a complete band gap. The origin of the minimum is related to wave diffraction in the lattice, i.e. the waves interference is mostly distractive at the frequency $\nu = 1.67$ kHz, while the waves do not compensate each other completely, giving rise to low (but finite) density of states. Thus, instead of a full gap where the density of states vanishes exactly, the structure in Fig. 1.1 possesses a pseudogap. The full gap opens for the filling fraction of the cylinders—the relative area occupied by metal in the plane of periodicity—exceeding $f = 0.3$ [7]. The filling fraction of the phononic crystal shown in Fig. 1.1 is much lower, $f = 0.066$.

The first phononic crystals possessing a full band gap were reported in 1998. The periodic structures were a square lattice of cylindrical holes drilled in aluminum alloy plate and filled with mercury [8] and a square (and triangular) lattice of metallic rods in air [9]. In the former case the gap for longitudinal elastic waves exists within the frequency range 1–1.12 MHz and in the latter case the gap is within the audible frequencies around 1.5–3 kHz, depending on the filling fraction of the rods. During the last decade a variety of different periodic structures possessing phononic band gaps in a wide range of frequencies have been fabricated. Here we may refer to a comprehensive review by Kushwaha [10] and to more recent topical review by Olsson III and El-Kady [11].

Study of wave propagation in a periodic medium is based on the dynamical equations of motion. Propagating sound wave is oscillating with time and coordinates sequence of material displacements which is accompanied by similar pattern of pressure and/or shear elastic stresses. For a homogeneous and isotropic elastic medium characterized by mass density, ρ, speed of longitudinal, c_l, and transverse, c_t sound, the derivation of the wave equation can be found, e.g., in [11]. In a homogeneous bulk medium the longitudinal and transverse sound waves are decoupled and propagate independently. The displacement vector field $\mathbf{u}$ is potential in a longitudinal wave ($\nabla \times \mathbf{u} = 0$) and it is solenoidal ($\nabla \cdot \mathbf{u} = 0$) in a transverse wave. In presence of a boundary the requirement of continuity of the displacements and stresses leads to mixing of these two modes. As it is known from the theory of surface Rayleigh waves [12], the total elastic displacement is a superposition of potential and solenoidal fields which, thus, cannot be decoupled. In the general case of an arbitrary inhomogeneous elastic medium the longitudinal and transverse displacements also cannot be split and the equation of motion for the components of the displacement vector contains both velocities, c_l and c_t, [4, 5]

$$\rho\frac{\partial^2 u_i}{\partial t^2} = \nabla\cdot\left(\rho c_t^2 \nabla u_i\right) + \nabla\cdot\left(\rho c_t^2 \frac{\partial \mathbf{u}}{\partial x_i}\right) + \frac{\partial}{\partial x_i}\left[\left(\rho c_l^2 - 2\rho c_t^2\right)\nabla\cdot\mathbf{u}\right] \tag{1.1}$$

Here $\rho = \rho(\mathbf{r})$, $c_l = c_l(\mathbf{r})$, and $c_t = c_t(\mathbf{r})$ are arbitrary functions of radius-vector $\mathbf{r} = (x_1, x_2, x_3)$. This complicated equation can be simplified in the particular cases that we will focus on.

In a phononic crystal all the functions characterizing the material properties are periodic in space and can be expanded in the Fourier series over an infinite set of the reciprocal lattice vectors $\mathbf{G}$. For example,

$$\rho(\mathbf{r}) = \sum_{\mathbf{G}} \rho(\mathbf{G})\ \exp(i\mathbf{G}\cdot\mathbf{r}), \tag{1.2}$$

where the Fourier component is defined as follows:

$$\rho(\mathbf{G}) = \frac{1}{V_c}\int_{V_c} \rho(\mathbf{r}) \exp(i\mathbf{G}\cdot\mathbf{r})\ d\mathbf{r}. \tag{1.3}$$

Integral in Eq. (1.3) is taken over the volume of the unit cell V_c, that in particular case of 2D periodicity is replaced by the area of the unit cell A_c, or by the lattice period l_c for 1D superlattices. The displacement vector $\mathbf{u}$, being a solution of the wave Eq. (1.1), satisfies the Bloch theorem and also can be expanded over the reciprocal lattice vectors

$$\mathbf{u}(\mathbf{r}) = \mathbf{u}_{\mathbf{k}}(\mathbf{r}) = \exp(i\mathbf{k}\cdot\mathbf{r}) \sum_{\mathbf{G}} \mathbf{u}_{\mathbf{k}}(\mathbf{G}) \, \exp(i\mathbf{G}\cdot\mathbf{r}). \tag{1.4}$$

The Bloch vector $\mathbf{k}$ plays the role of the phononic momentum; its possible values scan the interior of the irreducible part of the Brillouine zone. Substitution of the Fourier expansions (1.2) and (1.3) (and similar expansions for the elastic constants ρc_t^2 and ρc_l^2) leads to a set of linear homogeneous equations for the coefficients $\mathbf{u}_{\mathbf{k}}(\mathbf{G})$. This set has the nontrivial solutions if the determinant of this set of equations vanishes. As usual, this condition defines the dispersion relation, i.e. the band structure where infinite number of allowed frequencies $\omega_n(\mathbf{k})$ $(n = 1, 2, 3, \ldots)$ correspond to each value of the Bloch vector $\mathbf{k}$. In practice, the size of the determinant, and thus the number of the allowed frequencies (bands), is limited by finite number of plane waves (reciprocal lattice vectors) in the expansions (1.2) and (1.4). This method of calculation of phononic band structures, known as "plane-wave expansion," is the most popular one. Some clear examples of phononic band structure calculations based on the plane-wave expansion method can be found in [5, 7, 8, 13, 14].

1.2 Homogenization: Quasi-Static Limit and the Method of Plane Waves

It is well-known that scattering cross-section decreases with frequency, e.g., for a 3D scatterer it drops as ω^4 (Rayleigh scattering). This means that a wave propagating through an inhomogeneous medium at sufficiently low frequencies suffers very weak multiple scattering. Hence, in the principal approximation it is a plane wave. According to this conclusion the principal contribution to the Fourier expansion Eq. (1.4) comes from the term with $\mathbf{G} = 0$. All other terms with $\mathbf{G} \neq 0$ vanish linearly at $\omega \to 0$. As it will be shown in the next subsections this property is used in calculation of the effective speed of sound

$$c_{\text{eff}} = \lim_{k\to 0} \frac{\omega}{k}. \tag{1.5}$$

The region of frequencies where this formula is valid coincides with the interval of linear dispersion $\omega = c_{\text{eff}} k$ in the lowest allowed band. Here not only the wavelength $2\pi/k$ but also the wavelength in each constituent exceeds the size of the unit cell. Under these conditions all the details of the structure of the phononic crystal cannot be resolved by sound wave. This does not mean, however, that the effective speed of sound Eq. (1.5) is determined solely by average parameters of the structure. It will be shown that the exact formula for the effective speed of sound includes contributions from all reciprocal lattice vectors $\mathbf{G}$.

The value of c_{eff} obtained from Eq. (1.5) corresponds to the so-called quasi-static limit. In this limit the effective medium does not exhibit internal resonances and the dispersion relation is essentially linear. At finite but still low frequencies possible internal resonances in the scattering cross-section may manifest. To take into account the internal resonances the Mie scattering theory has to be applied. Homogenization theory based on this approach is developed in Sect. 1.3 of this chapter.

1.2.1 One-Dimensional Periodicity

The case of one-dimensional phononic crystal, so-called superlattice, is an important particular case when the dispersion relation can be obtained in a closed form. Superlattice is a periodic sequence of two (or more) layers of different elastic materials. We label the characteristics of each material by sub-indices a and b. In this particular case the waves with longitudinal and transverse polarization propagate independently, if the Bloch vector $\mathbf{k}$ is directed along the superlattice axis (axis z). Considering, for example, a longitudinal wave with $u_z \sim \exp(ikz - i\omega t)$ the following dispersion equation is explicitly obtained from direct matching the boundary conditions at the interfaces and applying the Bloch theorem [15]

$$\cos kd = \cos\left(\frac{\omega a}{c_a}\right)\cos\left(\frac{\omega b}{c_b}\right) - \frac{1}{2}\left(\frac{z_a}{z_b} + \frac{z_b}{z_a}\right)\sin\left(\frac{\omega a}{c_a}\right)\sin\left(\frac{\omega b}{c_b}\right). \quad (1.6)$$

Here $d = a + b$ is the period of superlattice with the unit cell containing two layers of width a and b. The elastic materials of the layers are characterized by speed of longitudinal sound c_a and c_b and acoustic impedances $z_a = \rho_a c_a$ and $z_b = \rho_b c_b$. For pure transverse waves the dispersion equation has the same form as Eq. (1.6) with c_a and c_b being the speeds of the transverse acoustic wave [16, 17]. Generalization of Eq. (1.6) for a multi-layered unit cell was done in [18].

For each value of the Bloch vector k lying within the Brillouin zone, $|k| \leq \pi/d$, Eq. (1.6) determines infinite number of frequencies forming the band structure $\omega = \omega_n(k)$. It is easy to see that the absolute value of expression in the rhs of Eq. (1.6) may exceed one within some finite intervals of frequencies. Indeed, for any arbitrary small value of the acoustic contrast between the impedances the sum $(z_a/z_b + z_b/z_a)/2$ exceeds one, therefore the same is true for the rhs as a whole. The intervals of frequencies where the rhs of Eq. (1.6) exceeds one correspond to phononic band gaps. Sound wave with the frequency lying within any of the gaps does not propagate.

In the quasi-static limit when $kd, \omega a/c_a, \omega b/c_b \ll 1$, the trigonometric functions in Eq. (1.6) can be expanded. Keeping quadratic over ω and k terms the following linear relation can be easily obtained:

$$\omega = c_{\mathrm{eff}} k. \quad (1.7)$$

Linear dispersion indicates that the superlattice behaves like a homogeneous elastic medium with speed of sound c_{eff}. The elastic parameters of this effective homogeneous medium are given by the following relations

$$\frac{1}{B_{\mathrm{eff}}} = \frac{f}{B_a} + \frac{1-f}{B_b}, \quad \rho_{\mathrm{eff}} = f\rho_a + (1-f)\,\rho_b, \quad c_{\mathrm{eff}} = \sqrt{\frac{B_{\mathrm{eff}}}{\rho_{\mathrm{eff}}}}. \tag{1.8}$$

Here $f = a/d$ is the filling fraction of the component a. Both the effective bulk constant and the effective mass density are positive and they coincide with their static values. Generalization of Eq. (1.8) to a superlattice with anisotropic layers is given in [19].

Dynamic effects may lead to negative values of ρ_{eff} in the vicinity of internal resonance. In one-dimensional phononic crystal local resonances appear if the unit cell consists of three or more layers of different elastic materials. Original method of homogenization of elastic superlattices near a local resonance has been recently proposed in [20].

1.2.2 Two-Dimensional Periodicity

Here we consider a phononic crystal of infinite rods, all parallel to axis z and arranged periodically in the x–y plane. The wave is assumed to propagate in the x–y plane where the effects of periodicity are the most pronounced. In this geometry the displacement vector $\mathbf{u}$ in Eq. (1.1) is independent of z. Taking the projections of Eq. (1.1) on the x–y plane and axis z, one can see that the obtained equations are uncoupled, i.e. the transverse mode with the displacement $\mathbf{u} = (0, 0, u)$ and the mixed mode with $\mathbf{u} = \left(u_x, u_y, 0\right)$ propagate independently. Here we consider homogenization for the transverse mode [21] which satisfies simple scalar equation

$$\rho\,(\mathbf{r})\,\frac{\partial^2 u}{\partial t^2} = \nabla_t \cdot \left(\tau\,(\mathbf{r})\,\nabla_t u\right). \tag{1.9}$$

Here ∇_t is two-dimensional gradient in the x–y plane and $\tau(\mathbf{r})$ is coordinate-dependent shear modulus. Homogenization of the mixed mode, although follows basically the same procedure [21], requires much longer mathematical treatment. The details of the homogenization procedure for the mixed mode can be found in [22].

Substitution of the Fourier expansions Eqs. (1.2) and (1.4) into Eq. (1.9) leads to a set of linear equations for the coefficients $\mathbf{u_k}(\mathbf{G})$

$$\sum_{\mathbf{G}'} \left[\rho\left(\mathbf{G}-\mathbf{G}'\right)(\mathbf{k}+\mathbf{G})\cdot\left(\mathbf{k}+\mathbf{G}'\right) - \omega^2\tau\left(\mathbf{G}-\mathbf{G}'\right)\right]\mathbf{u_k}\left(\mathbf{G}'\right) = 0. \tag{1.10}$$

In the quasi-static limit modulation of the Bloch wave (1.4) by periodic medium is very weak. Separating the principal term with $G = 0$ from other terms, Bloch wave (1.4) can be rewritten as follows:

$$\mathbf{u}_{\mathbf{k}}(\mathbf{r}) \approx u_0 \exp(i\mathbf{k}\cdot\mathbf{r}) + \sum_{\mathbf{G}\neq 0} \mathbf{u}_{\mathbf{k}}(\mathbf{G}) \exp(i\mathbf{G}\cdot\mathbf{r}). \tag{1.11}$$

The sum in the rhs vanishes linearly with k. Then, Eq. (1.10) contains linear, quadratic, and cubic over k terms. Keeping linear and quadratic terms we come to two coupled equations valid at $k \to 0$,

$$\mathbf{k}\cdot\mathbf{G}\tau(\mathbf{G})\,u_0 + \sum_{\mathbf{G}'\neq 0} \mathbf{G}\cdot\mathbf{G}'\tau(\mathbf{G}-\mathbf{G}')\,u_{\mathbf{k}}(\mathbf{G}') = 0, \tag{1.12}$$

$$\left(k^2\overline{\tau} - \omega^2\overline{\rho}\right)u_0 + \sum_{\mathbf{G}'\neq 0} \mathbf{k}\cdot\mathbf{G}'\tau(-\mathbf{G}')\,u_{\mathbf{k}}(\mathbf{G}') = 0. \tag{1.13}$$

Here $\overline{\tau} = \tau(G=0)$ and $\overline{\rho} = \rho(G=0)$ are the bulk average of the shear modulus and density, respectively. For a binary composite $\overline{\tau} = f\tau_a + (1-f)\tau_b$. Note that the quadratic approximation, Eq. (1.13), contains only the terms with $G = 0$. Eliminating u_0 from Eqs. (1.12) and (1.13) and introducing the effective speed of sound Eq. (1.5), we obtain a homogenized set of equations

$$\begin{aligned}&\left(c_{\text{eff}}^2\overline{\rho} - \overline{\tau}\right)\sum_{\mathbf{G}'\neq 0} \mathbf{G}\cdot\mathbf{G}'\tau(\mathbf{G}-\mathbf{G}')\,u_{\mathbf{k}}(\mathbf{G}') + \\ &\sum_{\mathbf{G}'\neq 0}\left(\widehat{\mathbf{k}}\cdot\mathbf{G}\right)\left(\widehat{\mathbf{k}}\cdot\mathbf{G}'\right)\tau(\mathbf{G})\,\tau(-\mathbf{G}')\,u_{\mathbf{k}}(\mathbf{G}') = 0\,.\end{aligned} \tag{1.14}$$

This set of equations has nontrivial solutions only if the determinant vanishes,

$$\begin{aligned}\det_{\mathbf{G},\mathbf{G}'\neq 0}\Big[&\left(c_{\text{eff}}^2\overline{\rho} - \overline{\tau}\right)\mathbf{G}\cdot\mathbf{G}'\tau(\mathbf{G}-\mathbf{G}') + \\ &\left(\widehat{\mathbf{k}}\cdot\mathbf{G}\right)\left(\widehat{\mathbf{k}}\cdot\mathbf{G}'\right)\tau(\mathbf{G})\,\tau(-\mathbf{G}')\Big] = 0\,.\end{aligned} \tag{1.15}$$

Here $\widehat{\mathbf{k}} = \mathbf{k}/k$ is the unit vector in the direction of propagation.

Although Eq. (1.15) is an infinite-order polynomial equation with respect to $\Lambda = \left(c_{\text{eff}}^2\overline{\rho} - \overline{\tau}\right)$, it turns out that it has only a *single* nonzero solution. It can be easily obtained by multiplying Eq. (1.15) by the determinant of the inverse matrix, $\det\left\{\left[\mathbf{G}\cdot\mathbf{G}'\tau(\mathbf{G}-\mathbf{G}')\right]^{-1}\right\}$,

$$\det_{\mathbf{G},\mathbf{G}'\neq 0}\left[B(\mathbf{G},\mathbf{G}') - \Lambda\delta_{\mathbf{G},\mathbf{G}'}\right] = 0 \tag{1.16}$$

Here

$$B\left(\mathbf{G},\mathbf{G}'\right) = -\widehat{\mathbf{k}}\cdot\mathbf{G}\tau\left(\mathbf{G}\right)\sum_{\mathbf{G}''\neq 0}\widehat{\mathbf{k}}\cdot\mathbf{G}''\tau\left(-\mathbf{G}''\right)\left[\mathbf{G}''\cdot\mathbf{G}'\tau\left(\mathbf{G}''-\mathbf{G}'\right)\right]^{-1}. \quad (1.17)$$

The matrix $B(\mathbf{G},\mathbf{G}')$ represents a projection operator since it is written as a product of two factors, one of which depends only on $\mathbf{G}$ and the other only on $\mathbf{G}'$. Being a projection operator, it has only a single eigenvalue, $\Lambda = \mathrm{Tr}B\left(\mathbf{G},\mathbf{G}'\right)$. This results in the final formula for the effective speed of sound,

$$c_{\mathrm{eff}}^{2}\left(\widehat{\mathbf{k}}\right) = \frac{\overline{\tau}}{\overline{\rho}} - \frac{1}{\overline{\rho}}\sum_{\mathbf{G},\mathbf{G}'\neq 0}\left(\widehat{\mathbf{k}}\cdot\mathbf{G}\right)\left(\widehat{\mathbf{k}}\cdot\mathbf{G}'\right)\tau\left(\mathbf{G}\right)\tau\left(-\mathbf{G}'\right)$$
$$\times\left[\mathbf{G}\cdot\mathbf{G}'\tau\left(\mathbf{G}-\mathbf{G}'\right)\right]^{-1}. \quad (1.18)$$

The effective speed of sound depends on the direction of propagation. Eq. (1.18) is valid for an arbitrary form of the unit cell, geometry of the inclusions, and parameters of the materials forming phononic crystal.

Equation (1.18) can be written in canonical form for a second-order curve,

$$c_{\mathrm{eff}}^{2} = A_{ij}\widehat{k}_i\widehat{k}_j, \quad i,j = x,y, \quad (1.19)$$

where

$$A_{ij} == \frac{\overline{\tau}}{\overline{\rho}}\delta_{ij} - \frac{1}{2\overline{\rho}}\sum_{\mathbf{G},\mathbf{G}'\neq 0}\left(G_iG_j' + G_jG_i'\right)\tau\left(\mathbf{G}\right)\tau\left(-\mathbf{G}'\right)$$
$$\times\left[\mathbf{G}\cdot\mathbf{G}'\tau\left(\mathbf{G}-\mathbf{G}'\right)\right]^{-1}. \quad (1.20)$$

It follows from Eq. (1.20) that radius vector $1/c_{\mathrm{eff}}\left(\widehat{\mathbf{k}}\right)$ sweeps an ellipse with semiaxes $1/A_x$, and $1/A_y$, where A_x and A_y are the principal values of the tensor A_{ij}. If the crystal possesses a third- or higher-order rotational axis z, then any second-rank symmetric tensor such as A_{ij} is reduced to a scalar, $A_{ij} = A\delta_{ik}$. Even in this case of high symmetry the speed of sound is not determined only by the average parameters. The sum over $\mathbf{G}$ and $\mathbf{G}'$ in the rhs of Eq. (1.20) contains information about the structure of the unit cell at all spatial scales. For the case of transverse wave this information enters through modification of the elastic modulus since the dependence of the speed of sound on the densities of the constituents enters to Eq. (1.18) only through $\overline{\rho}$.

1.2.3 Three-Dimensional Periodicity

Here we consider three-dimensional arrangement of inclusions of fluid b in the background of fluid a. The inclusions are of arbitrary shape. The most common case is air bubbles in water or water drops in air. In fluids only longitudinal waves can propagate and Eq. (1.1) after some algebra is reduced to a simple form:

$$\frac{1}{B(\mathbf{r})}\frac{\partial^2 p}{\partial t^2} = \nabla\cdot\left(\frac{\nabla p}{\rho(\mathbf{r})}\right) \tag{1.21}$$

Here $B(\mathbf{r})$ is the elastic bulk modulus. Formally, this equation is a three-dimensional analog of Eq. (1.2). Therefore, the result for the effective speed of sound can be obtained from Eq. (1.18) after the following replacement: $\rho(\mathbf{r})\rightarrow 1/B(\mathbf{r}) = \gamma(\mathbf{r})$, $\tau(\mathbf{r})\rightarrow 1/\rho(\mathbf{r}) = \nu(\mathbf{r})$. Being the speed of longitudinal sound, the resultant formula for c_{eff} supports the widely used Wood's law [23]

$$c_{\text{eff}} = \sqrt{\frac{B_{\text{eff}}}{\rho_{\text{eff}}}}, \tag{1.22}$$

where the effective bulk modulus is obtained by averaging of its inverse value $\gamma(\mathbf{r})$,

$$\frac{1}{B_{\text{eff}}} = \overline{\gamma} = \frac{f}{B_a} + \frac{1-f}{B_b} \tag{1.23}$$

and all the details about the microstructure of the phononic crystal enter through the effective mass density

$$\frac{1}{\rho_{\text{eff}}} = \frac{f}{\rho_a} + \frac{1-f}{\rho_b} - \sum_{\mathbf{G},\mathbf{G}'\neq 0}\left(\widehat{\mathbf{k}}\cdot\mathbf{G}\right)\left(\widehat{\mathbf{k}}\cdot\mathbf{G}'\right)\nu(\mathbf{G})\,\nu\left(-\mathbf{G}'\right)\,. \\ \times\left[\mathbf{G}\cdot\mathbf{G}'\nu\left(\mathbf{G}-\mathbf{G}'\right)\right]^{-1} \tag{1.24}$$

Sums over $\mathbf{G}$ and $\mathbf{G}'$ in Eq. (1.24) run over three-dimensional reciprocal lattice vectors.

An excellent approximation for ρ_{eff} was obtained in [24] for a random distribution of air bubbles in water. In order to compare the results for regular and random distribution of spherical bubbles we plot in Fig. 1.2 c_{eff} vs f in a simple cubic lattice (solid dots). This phononic crystal is isotropic. The solid line shows the dependence obtained in [24] in the coherent potential approximation [25]. One can see that for small and moderate filling fractions ($f < 0.3$) the speed of sound for dilute mixtures ($f < 0.02$) is equally described by the two approaches. Direct numerical solution of the wave equation also gives similar results [26]. For $f > 0.3$ the curve obtained from Eqs. (1.22)–(1.24) grows up rapidly, and at the filling fraction $f_c = \pi/6$ when spheres touch each other c_{eff} reaches the value of $0.18c_0$. It exceeds by almost ten times the value of $c_{\text{eff}} = 0.02c_0$ obtained from the coherent potential approximation [24]. This rapid growth is due to an open air channel in a phononic crystal with overlapping spheres ($f > f_c$) where sound propagates mostly through air. Therefore in a lattice of overlapping spheres c_{eff} is close to the speed of sound in pure air, $c_a = 0.22c_0$. Since for a given structure $c_{\text{eff}}(f)$ is a continuous function, there is a transition region from a low-speed region at small filling fractions ($f < 0.3$) to a

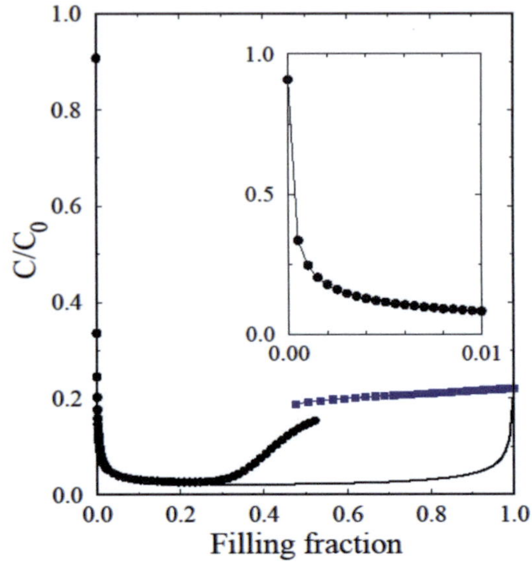

Fig. 1.2 Effective speed of sound calculated from Eqs. (1.22)–(1.24) in a simple cubic lattice of air bubbles, $c_a = 330$ m/s, in water, $c_0 = 1{,}500$ m/s, (*circles*) and in a conjugate lattice of water drops in air (*squares*) vs filling fraction of the air. The number of **G** values (*plane waves*) involved in these calculations is $N = 800$ that guarantees a good convergence of the numerical results. *Solid line* shows the speed of sound obtained in [24]. Insert shows blowup of a region of low concentrations of the air [21]

high-speed region at f close to f_c. Approximate theories fail to explain the region close to f_c. Here the interaction between bubbles becomes very strong, giving rise to a fast increase of the speed of sound.

In a conjugate lattice—water drops in air—the presence of water inclusions does not play a significant role. Squares in Fig. 1.2 show the effective speed of sound in a conjugate lattice. Water drops in air can be considered as rigid spheres where sound does not penetrate. It was demonstrated in [27] that the presence of rigid spheres decreases slightly the speed of sound in a matrix material. This tendency is clearly seen in Fig. 1.2. Note that the curves for the two lattices never meet each other because these two structures are not equivalent. However, for a special case of cubic inclusions the two lattices become geometrically equivalent at $f_c = 0.5$. Therefore at this point the two curves meet each other; see Fig. 1.3. Note that for cubic atoms the transition region starts earlier, at $f \approx 0.12$, and at $f \approx 0.2$ the effective speed of sound practically saturates, reaching the value of c_a. Although the structure with cubic inclusions is not of a real situation, it helps to clarify that the fast increase of the speed of sound in the transition region for the case of noncubic inclusions is a manifestation of the continuity of the curve in Fig. 1.2. The latter is due to an additional symmetry between the direct and conjugate lattices.

Equations (1.22)–(1.24) have been derived for periodic fluids where only longitudinal waves can propagate. However, these results are applicable also for some systems of fluids with periodic solid inclusions. If the acoustic contrast between solid and fluid is so high that sound propagates only through the fluid, the effect of the transverse component of the wave becomes negligible. Then Eqs. (1.22)–(1.24) are valid providing that $\gamma_b = 1/B_b = 0$ and $\nu_b = 1/\rho_b = 0$. In practice, any phononic crystal of metallic inclusions in air can be considered in this

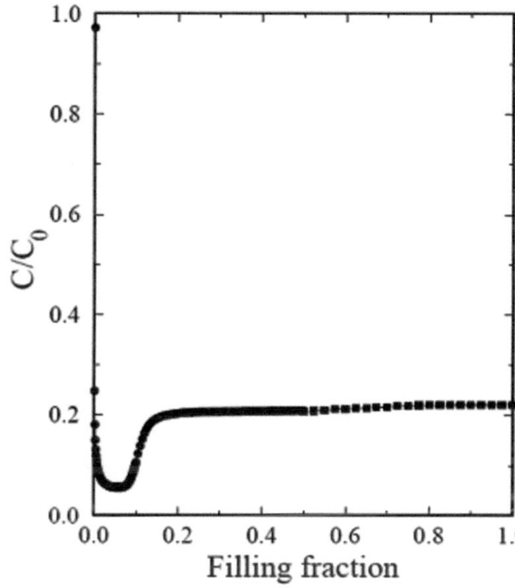

Fig. 1.3 Effective speed of sound in air–water mixture but for cubic form of inclusions [21]

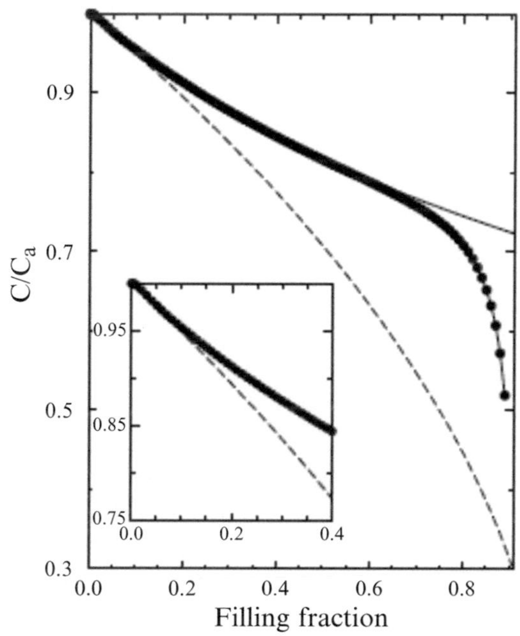

Fig. 1.4 *Circles* represent the speed of sound calculated for phononic crystal of rigid cylinders in air. The unit cell is hexagonal. Acoustic lens of [28]. *Solid* and *dashed lines* show the approximations proposed in [28] and [29], respectively

approximation of rigid inclusion. In Fig. 1.4 we give the dependence of the speed of sound in a phononic crystal of rigid cylinders in air as a function of filling fraction. Two-dimensional version of Eqs. (1.22)–(1.24) was used for these calculations. We assume that the cylinders are arranged in a hexagonal lattice. Because of the presence of the third-order rotational axis of symmetry this phononic crystal is isotropic in the plane of periodicity. In the experiment [28] an acoustic beam was focused by a lens of phononic crystal of aluminum cylinders with the same symmetry of the unit cell. The results shown in Fig. 1.4 are in reasonable agreement with experimental data [28] and with a simple model $c_{\text{eff}} = c_{\text{air}}/\sqrt{1+f}$ proposed there (solid line in Fig. 1.4). Unlike this, the approximation $c_{\text{eff}} = c_{\text{air}}\sqrt{1-f}$ reported in [29] (dashed line in Fig. 1.4) strongly deviates from the exact result. Approximation $c_{\text{eff}} = c_{\text{air}}/\sqrt{1+f}$ will be justified in the next paragraph of this chapter.

1.3 Homogenization: The Multiple-Scattering Method

A periodic distribution of solid cylinders embedded in air, as shown in Fig. 1.1, represents an example of a periodic medium where low-frequency sound travels with speed lower than in free air. This property together with the fact that the reflectance from this medium is much lower than that from its solid constituent has been exploited to design an acoustic lens [28]. It turns out that the refractive properties of this acoustic lens are similar to that of a standard optical lens.

The reduction of speed of sound in a phononic crystal is explained by increase of density (inertia) of the effective medium. When dealing with rigid cylinders in oscillating fluid (i.e., in sound field) we can make use of the reciprocity principle and consider the equivalent problem of oscillating cylinders in air at rest [30]. The air around the oscillating cylinders impedes them from vibrating freely, and this impedance effect results in effective increase of the density of the medium. For a single cylinder, the additional mass per unit length equals the mass of air background displaced by the cylinder, $S\rho_{\text{air}}$, where $S = \pi R^2$ is the cross-section of a cylinder and ρ_{air} is the air density. If N is the number of the cylinders per unit area, then the increase in the effective density is given by $\rho_{\text{eff}} = \rho_{\text{air}} + NS\rho_{\text{air}} = \rho_{\text{air}}(1+f)$. The phase velocity is inversely proportional to the square root of the density ρ_{eff}. Therefore, the effective speed of sound speed within the PC is given by

$$c_{\text{eff}} = \frac{c_{\text{air}}}{\sqrt{1+f}} = \frac{c_{\text{air}}}{n}, \tag{1.25}$$

where $n = \sqrt{1+f}$ is the acoustic refractive index of the effective medium.

Calculation of the effective parameters of the equivalent uniform medium (effective mass density ρ_{eff}, and effective bulk modulus B_{eff}) is a long-standing problem of mathematical theory of homogenization [31, 32]. Approximate formulas for ρ_{eff} as a function of f were reported by Ament [33] and Berryman [34] for the case of random composites. The effects of multiple scattering were neglected. The approach based on the plane wave expansion has been explained in Sect. 1.2, where the exact formulas for $c_{\text{eff}} = \sqrt{B_{\text{eff}}/\rho_{\text{eff}}}$ which take into account the multiple scattering effects and the microstructure of the periodic medium have been obtained.

These calculations are in a good agreement with experimental data [28]. They confirm that a simple model leading to Eq. (1.25) is valid up to rather high values of f, as it is shown in Fig. 1.4.

A different approach to the problem of homogenization is based on the theory of scattering. This approach in application to the case of 2D phononic crystals of rigid cylinders in air has been developed by two different groups in [35–37]. Mei and coworkers [35] have considered an infinite periodic structure and have obtained the result for the effective parameters neglecting the effects of multiple scattering. In the study by Torrent and coworkers [36] and Torrent and Sánchez-Dehesa [37] a finite size cluster of periodically arranged circular cylinders has been analyzed using multiple-scattering theory. The last approach allows simultaneous determination of c_{eff} and ρ_{eff}. Moreover, since these parameters are obtained in the homogenization limit it is of practical interest to formulate the conditions of validity, i.e. the quantitative relation between the cluster size and the wavelengths.

Let us consider a cluster consisting of a large set of N fluid cylinders of equal radius R_a and with the parameters ρ_a and c_a. The background fluid of this cluster is characterized by density ρ_b and speed of sound c_b. Let us also assume that the fluid cylinders are placed at positions corresponding to a "periodic" lattice with hexagonal symmetry. The radius R_{eff} of the effective medium having the same dynamical properties as the inhomogeneous cluster can be estimated by considering that the fraction f of the volume occupied by N cylinders equals the filling fraction of the underlying infinite periodic structure, f_{hex}. For the cluster $f = N(R_a/R_{\text{eff}})^2$ and for the hexagonal lattice $f_{\text{hex}} = \left(2\pi/\sqrt{3}\right)(R_a/R_{\text{eff}})^2$, where a is the lattice period. The condition $f = f_{\text{hex}}$ is true for

$$R_{\text{eff}} = a\sqrt{N\sqrt{3}/2\pi}. \tag{1.26}$$

It is well known that the t-matrix for a circular cylinder is diagonal with the diagonal elements being

$$T_q = -\frac{\chi_q J'_q(k_b R_a) - J_q(k_b R_a)}{\chi_q H'_q(k_b R_a) - H_q(k_b R_a)}, \quad \text{and} \quad \chi_q = \frac{\rho_a c_a}{\rho_b c_b}\frac{J_q(k_a R_a)}{J'_q(k_a R_a)}. \tag{1.27}$$

There are two contributions in the matrix: Bessel and Hankel functions describe the background, while the function χ_q originates from the scatterer.

In the long-wavelength limit the matrix elements can be expanded in powers of the wavenumber $k_b = \omega/c_b$. In the leading order over $k_b R_a \ll 1$, the following expressions are obtained for the first two elements, with q = 0 and q = 1.

$$T_0^a \approx i\frac{\pi R_a^2}{4}\left(\frac{\rho_b c_b^2}{\rho_a c_a^2} - 1\right)k_b^2, \tag{1.28a}$$

$$T_1^a \approx i\frac{\pi R_a^2}{4}\left(\frac{\rho_a - \rho_b}{\rho_a + \rho_b}\right)k_b^2. \tag{1.28b}$$

The constitutive parameters for the effective medium can be derived from the fact that it is also a fluid cylinder with radius R_{eff} and the following diagonal elements of the t-matrix:

$$T_0^{\text{eff}} = i\frac{\pi R_{\text{eff}}^2}{4}\left(\frac{\rho_b c_b^2}{\rho_{\text{eff}} c_{\text{eff}}^2} - 1\right)k_b^2, \tag{1.29a}$$

$$T_1^{\text{eff}} = i\frac{\pi R_{\text{eff}}^2}{4}\left(\frac{\rho_{\text{eff}} - \rho_b}{\rho_{\text{eff}} + \rho_b}\right)k_b^2. \tag{1.29b}$$

The effective bulk modulus $B_{\text{eff}} = \rho_{\text{eff}} c_{\text{eff}}^2$ is obtained from the relation $T_0^{\text{eff}} = NT_0^a$. Then, using Eq. (1.28a) one easily obtains

$$\frac{1}{B_{\text{eff}}} = \frac{f}{B_a} + \frac{1-f}{B_b}, \tag{1.30}$$

which coincides with Eq. (1.23). Similar procedure with the next diagonal term, T_1, leads to the formula for the effective mass density,

$$\frac{\rho_{\text{eff}}}{\rho_b} = \frac{\rho_a(\Delta + f) + \rho_b(\Delta - f)}{\rho_a(\Delta - f) + \rho_b(\Delta + f)}, \tag{1.31}$$

where the term Δ contains contributions from multiple scattering processes at the surfaces of the cylinders [36, 37].

For small values of the filling fraction f multiple scattering interactions are weak, leading to $\Delta \approx 1$. In this case Eq. (1.28) is reduced to

$$\frac{\rho_{\text{eff}}}{\rho_b} = \frac{\rho_b + \rho_a - f(\rho_b - \rho_a)}{\rho_b + \rho_a + f(\rho_b + \rho_a)}. \tag{1.32}$$

In the linear approximation over f this result is the two-dimensional version of the formulas obtained by Ament [33] and Berryman [34] for random distributions of spherical scatterers.

Equations (1.27) and (1.28) can be also applied to extract the effective parameters of a medium consisting of a cluster made of solid cylinders in air background. In this case the cylinders can be considered as rigid scatterers with $B_a, \rho_a = \infty$ and Eqs. (1.27) and (1.28) give

$$\frac{B_{\text{eff}}}{B_b} = \frac{1}{1-f}, \quad \frac{\rho_{\text{eff}}}{\rho_b} = \frac{\Delta + f}{\Delta - f}. \tag{1.33}$$

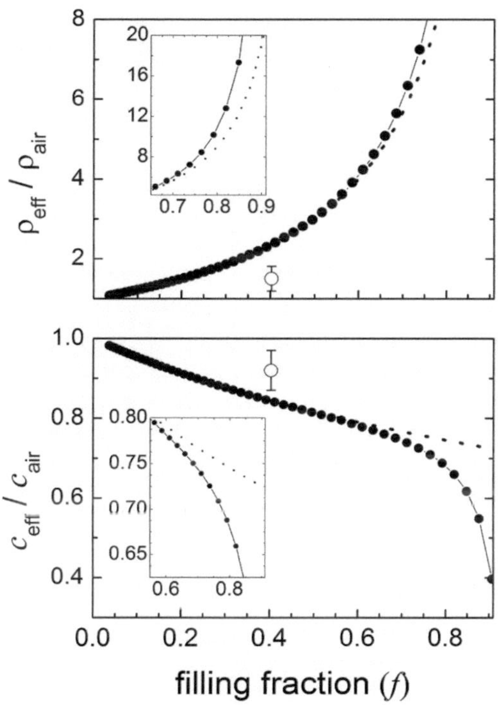

Fig. 1.5 Effective parameters for a circular-shaped cluster made of 151 wooden cylinders embedded in air. The *black circles* defines values obtained from the expression in Eq. (1.31) while the dotted lines were obtained from Eq. (1.34). The *open circles* with error bars represent the experimental results [36]. *Inserts* show the behavior near the values of f where the multiple scattering interaction cannot be neglected

For $f \ll 1$ the multiple scattering interactions are weak (i.e., $\Delta \approx 1$) and we get that the effective parameters depend only on the parameters of the background constituent (air) and the filling fraction

$$\rho_{\text{eff}} = \rho_b \frac{1+f}{1-f}, \quad c_{\text{eff}} = \frac{c_b}{\sqrt{1+f}}. \tag{1.34}$$

These simple expressions have been obtained here using scattering theory. They, of course, reproduce Eq. (1.25) for c_{eff} derived semi-quantitatively. The formula for ρ_{eff} coincides with the result obtained independently by Ament [33] and Berryman [34]. using a phenomenological approach which neglects multiple scattering. The homogenization formulas have been experimentally verified by Torrent and coworkers [36] by studying a circular cluster of wooden cylinders in air. The experiments also demonstrated that the homogenization is valid if the wavelength λ exceeds four the lattice period, i.e., for any $\lambda > 4a$.

The expression for the term Δ is given by

$$\frac{1}{\Delta} = \frac{1}{N} \sum_{\alpha,\beta} \left(\widehat{M}_{\alpha\beta}^{-1} \right)_{11},$$

where the Greek subscripts define the positions of the interacting cylinders in a lattice and the matrix elements contain information about their material parameters and the multiple scattering interactions.

For the clusters containing rigid cylinders ($\rho_a=\infty$) the homogenization procedure described above generates the so-called *magic* clusters [38]. The magic property manifested in unexpectedly homogeneous response from a cluster containing relatively small number of scatterers. In other words, the effective parameters of a magic cluster equal to those of an infinite lattice. For the clusters with hexagonal lattice the magic properties have been observed for the number of cylinders $N=7$, 19, 37, 61, and 85. This property is due to hexagonal symmetry, which produces partial cancellation between double-scattering and triple-scattering events [36]. The existence of magic clusters is of paramount importance because it enlarges the tailoring possibilities of the effective parameters associated with artificial structures consisting of clusters of individual scatterers.

The results discussed so far in this section considered clusters of fluid (or rigid) cylinders in a fluid background. When the cylinders are made of elastic material characterized by their mass density ρ_a together with their longitudinal and transversal velocities, c_ℓ and c_t, respectively, straightforward derivation [39] shows that, in the quasi-static limit, the first diagonal element of the corresponding t-matrix is given by

$$T_0^a \approx i\frac{\pi R_a^2}{4}\left(\frac{\rho_b c_b^2}{\rho_a\left(c_\ell^2 - c_t^2\right)} - 1\right)k^2. \tag{1.36}$$

This element is similar to that of a fluid cylinder with speed of sound

$$c_a = \sqrt{c_\ell^2 - c_t^2}. \tag{1.37}$$

Equation (1.37) is valid only in the long-wavelength limit. It generalizes the expressions obtained for the clusters of fluid cylinders to the case of elastic cylinders in a fluid background and to phononic crystals with unit cell consisting of two types of elastic scatterers, as it is discussed in the next paragraph.

1.3.1 Homogenization of Mixed Lattices

Let us consider a unit cell containing two different elastic cylinders with effective fluid-like parameters ρ_1, B_{a1} and ρ_2, B_{a2}, and radii R_1 and R_2, respectively. If they are distributed, for example, in a square configuration with side a, the parameters of the homogenized medium are

$$\frac{1}{B_{\text{eff}}} = \frac{1-f}{B_b} + \frac{f_1}{B_{a1}} + \frac{f_2}{B_{a2}},$$
$$\rho_{\text{eff}} = \rho_b \frac{1+f_1\eta_1+f_2\eta_2}{1-f_1\eta_1+f_2\eta_2}, \tag{1.38}$$

where $\eta_{1,2} = (\rho_{1,2} - \rho_b)/(\rho_{1,2} + \rho_b)$, $f_{1,2}$ are the partial filling fractions and $f = f_1 + f_2$ is the volume fraction occupied by both cylinders.

Note that by selecting the appropriate pair of materials in a phononic crystal, one can artificially generate acoustic materials with tailored parameters by just changing their relative filling fraction in the total lattice as it is seen in the plots shown in Fig. 1.6. The possibility of tailoring the acoustic refractive index, $n(\mathbf{r})$, locally has been recently employed to design gradient index sonic lenses [40–43].

Another interesting outcome of homogenized 2D phononic crystals results from considering lattices of scatterers possessing rotational axis of symmetry of any order lower than three. If the cylindrical scatterers have circular section, this possibility can be realized in the lattices other than hexagonal or square. The sound

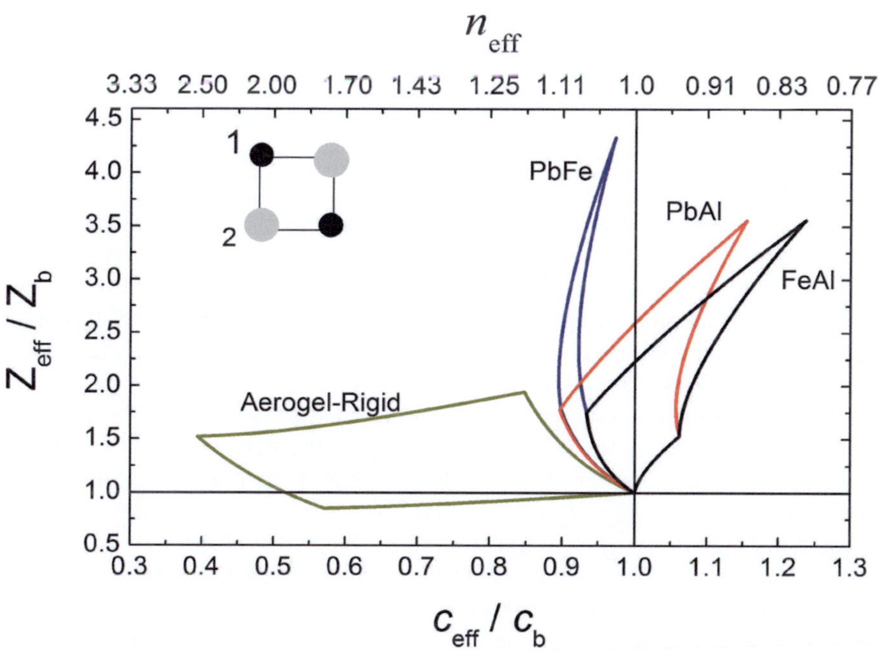

Fig. 1.6 Effective parameters of four phononic crystals with two different solid cylinders in square unit cell (*inset*). In all cases the background is air. By changing the filling fractions, f_1 and f_2, it is possible to obtain the homogenized medium with the parameters lying anywhere within the areas enclosed by the *color lines*. Note that for the case of mixtures of aerogel and rigid cylinders it is possible to obtain the effective media perfectly matching the impedance of air. The *horizontal line* is a guide for the eye defining the condition $Z_{\text{eff}} = Z_b = Z_{\text{air}}$ and the *vertical line* defines the case where $n_{\text{eff}} = 1$

propagation within these lattices becomes anisotropic and the effective medium can be characterized by index ellipsoid, i.e. by the refractive index that depends on the direction of propagation

$$n_{\mathrm{eff}}(\theta) = \frac{c_b}{c_{\mathrm{eff}}(\theta)} \tag{1.39}$$

as it follows from Eq. (1.19). The effective bulk modulus of these lattices is a scalar given by Eq. (1.29). However, the effective mass density is a tensor whose reciprocal is defined as follows:

$$\begin{aligned} \rho_{s+}^{-1} &= \frac{|\Delta|^2 - |\Gamma|^2 - f^2\eta^2}{(\Delta + f\eta)(\Delta^* + f\eta) - |\Gamma|^2}, \\ \rho_{s+}^{-1} &= \frac{2f\eta|\Gamma|\cos\Phi_\Gamma}{(\Delta + f\eta)(\Delta^* + f\eta) - |\Gamma|^2}, \\ \rho_{a+}^{-1} &= \frac{2f\eta|\Gamma|\sin\Phi_\Gamma}{(\Delta + f\eta)(\Delta^* + f\eta) - |\Gamma|^2}. \end{aligned} \tag{1.40}$$

This tensor in Cartesian coordinates can be obtained from

$$\begin{aligned} \rho_{xx}^{-1} &= \rho_{s+}^{-1} - \rho_{s-}^{-1}, \\ \rho_{yy}^{-1} &= \rho_{s+}^{-1} - \rho_{s-}^{-1}, \\ \rho_{yx}^{-1} &= \rho_{xy}^{-1} = \rho_{a+}^{-1}. \end{aligned} \tag{1.41}$$

The complete derivation of these expressions and the physical meanings of the parameters involved are given in [39]. Let us remark that the effective density mainly depends on the lattice structure, its filling fraction and the ratio between the densities of the cylinders and the background. The elastic nature of the cylinders appears in the effective density for high filling fractions, where the higher orders of the t-matrix are present in both the Δ- and the Γ-factors [39].

Artificial fluid-like structures with rectangular lattice have been fabricated and the anisotropy of sound has been measured experimentally [44] in the range 0.5–3 kHz. The elements of the dynamic density matrix ρ_{ik} can be calculated [45] with high accuracy, using Eq. (1.19). This approach gives the results which are undistinguishable from the experimental values obtained in [44].

Artificial structures behaving as fluid-like materials with anisotropic dynamical mass density can be considered as a kind of metamaterial since anisotropy of mass does not exist for natural fluids. Several interesting devices exploring anisotropy of the mass density of the corresponding effective medium (metafluid) have been recently proposed. They are: magnifying hyperlens [46], acoustic cloaks [47], and radial sonic crystals [48]. Here anisotropy plays the paramount role determining the functionality of these devices.

1.3.2 Homogenization of a Medium with Local Resonances

Let us now consider a lattice with "soft" scatterers and "hard" background such that $c_a \ll c_b$. In this case some local resonant effects can be observed even when the wavelength in the background is much longer than the period of the lattice. So, if we consider that the arguments of Bessel and Hankel functions in the expression (1.27) are small ($k_b R_a \ll 1$), we can use their asymptotic expressions to obtain the monopolar and dipolar components of the t-matrix as follows:

$$T_0^a \approx i\frac{\pi R_a^2 k_b^2}{4}\frac{2 + k_b R_a \chi_0}{k_b^2 R_a^2 \ln k_b R_a - k_b R_a \chi_0}, \tag{1.42a}$$

$$T_1^a \approx i\frac{\pi R_a^2 k_b^2}{4}\frac{\chi_1 k_b R_a - \chi_b}{\chi_1 k_b R_a + \chi_b}. \tag{1.42b}$$

Here the logarithmic term cannot be neglected when dealing with materials with local resonances. This term is, however, negligible for structures where the wavelength inside the scatterer and the background are of the same order, $k_a \sim k_b$, as in the case studied in the previous sections.

For the effective medium we expect to obtain a homogeneous scatterer with the elements of the t-matrix obtained from the formulas (1.28). Comparison between Eqs. (1.28a)–(1.28b) and Eqs. (1.42a)–(1.42b) leads to the following *frequency-dependent* bulk modulus and mass density functions:

$$\frac{B_a(\omega)}{B_b} = \frac{k_b^2 R_a^2}{2}\ln k_b R_a - \frac{k_a R_a}{2}\frac{J_0(k_a R_a)}{J_1(k_a R_a)}\frac{B_a}{B_b}, \tag{1.43a}$$

$$\frac{\rho_a(\omega)}{\rho_b} = \frac{1}{k_a R_a}\frac{J_1(k_a R_a)}{J_1'(k_a R_a)}\frac{\rho_a}{\rho_b}, \tag{1.43b}$$

where $k_a = \omega\sqrt{\rho_a/B_a}$. Note that in the limit $k_a \to 0$ we recover the static cylinder parameters.

Negative values of the effective parameters in Eq. (1.43) can be realized in the long-wavelength limit $k_b R_a \ll 1$, while $k_a R_a$ is not necessary a small parameter. The structures exhibiting this dynamical behavior are called acoustic metamaterials. General theory of homogenization of periodic arrangement of "soft" cylinders in "hard" background has been recently presented in [49]. Similar problem for 1D elastic superlattice containing "soft" and "hard" elements and also exhibiting metamaterial behavior has been solved in [20]. It can be concluded that metamaterial behavior is a direct consequence of the complexity of the scatterers employed in the fabrication of the phononic crystal of any dimensionality.

Acknowledgements JSD acknowledges useful discussions with D. Torrent and the support from the ONR (USA) grant N00014-12-1-0216, and the MINECO (Spain) grants #TEC2010-19751 and #CSD2008-66 (CONSOLIDER program). AAK acknowledges support from the DOE grant # DE-FG02-06ER46312.

References

1. E. Yablonovitch, Inhibited spontaneous emission in solid-state physics and electronics. Phys. Rev. Lett. **58**, 2059–2062 (1987)
2. S. John, Strong localization of photons in certain disordered dielectric superlattices. Phys. Rev. Lett. **58**, 2486–2489 (1987)
3. E. Yablonovitch, T.G. Gmitter, K.M. Leung, Photonic band structure: the face-centered-cubic case employing nonspherical atoms. Phys. Rev. Lett. **67**, 2295–2298 (1991)
4. M. Sigalas, E.N. Economou, Band structure of elastic waves in two dimensional systems. Solid State Commun. **86**, 141–143 (1993)
5. M.S. Kushwaha, P. Halevi, L. Dobrzynski, B. Djafari-Rouhani, Acoustic band structure of periodic elastic composites. Phys. Rev. Lett. **71**, 2022–2025 (1993)
6. R. Martínez-Sala, J. Sancho, J.V. Sánchez, V. Gómez, J. Llinarez, F. Meseguer, Sound attenuation by sculpture. Nature **378**, 241 (1995)
7. M.S. Kushwaha, Stop-bands for periodic metallic rods: sculptures that can filter the noise. Appl. Phys. Lett. **70**, 3218–3220 (1997)
8. M.F. de Espinosa, E. Jiménez, M. Torres, Ultrasonic band gap in a periodic two-dimensional composite. Phys. Rev. Lett. **80**, 1208 (1998)
9. J.V. Sánchez-Pérez, D. Caballero, R. Mártinez-Sala, C. Rubio, J. Sánchez-Dehesa, F. Meseguer, J. Llinares, F. Gálvez, Sound attenuation by a two-dimensional array of rigid cylinders. Phys. Rev. Lett. **80**, 5325–5328 (1998)
10. M.S. Kushwaha, Classical band structure of periodic elastic composites. Int. J. Mod. Phys. B **10**, 977–1094 (1996)
11. R.H. Olsson III, I. El-Kady, Microfabricated phononic crystal devices and applications. Meas Sci Technol. **20**, 012002–012015 (2009)
12. L.D. Landau, E.M. Lifshitz, A.M. Kosevich, L.P. Pitaevskii, *Theory of Elasticity* (Pergamon Press, Oxford, 1986)
13. C.G. Poulton, A.B. Movchan, R.C. McPhedran, N.A. Nocorovici, Y.A. Antipov, Eigenvalue problems for doubly periodic structures and phononic band gaps. Proc. R. Soc. A **457**, 2561–2568 (2000)
14. J.O. Vasseur, P.A. Deymier, B. Chenni, B. Djafari-Rouhani, L. Dobrzynski, D. Prevost, Experimental and theoretical evidence for the existence of absolute acoustic band gaps in two-dimensional solid phononic crystals. Phys. Rev. Lett. **86**, 3012–3015 (2001)
15. S.M. Rytov, Acoustic properties of a thinly laminated medium. Sov. Phys. Acoust. **2**, 68 (1956)
16. S. Nemat-Nasser, M. Yamada, Harmonic waves in layered transversely isotropic composites. J. Sound Vibrat. **79**, 161 (1981)
17. R.E. Camley, B. Djafari-Rouhani, L. Dobrzynski, A.A. Maradudin, Transverse elastic waves in periodically layered infinite and semi-infinite media. Phys. Rev. B **27**, 7329 (1983)
18. D. Djafari-Rouhani, L. Dobrzynski, Simple excitations in N-layered superlattices. Solid State Commun. **62**, 609 (1987)
19. M. Grimsditch, Effective elastic constants of superlattices. Phys. Rev. B **31**, 6818 (1985)
20. S. Nemat-Nasser, J.R. Willis, A. Srivastava, A.V. Amirkhizi, Homogenization of periodic composites and locally resonant sonic materials. Phys. Rev. B **83**, 104103 (2011)
21. A.A. Krokhin, J. Arriaga, L.N. Gumen, Speed of sound in periodic elastic composites. Phys. Rev. Lett. **91**, 264302 (2003)
22. Q. Ni, J. Cheng, Anisotropy of effective velocity for elastic wave propagation in two-dimensional phononic crystals at low frequencies. Phys. Rev. B **72**, 014305 (2005)

23. A.W. Wood, *Textbook of Sound* (Macmillan, New York, 1941)
24. M. Kafesaki, R.S. Penciu, E.N. Economou, Air bubbles in water: a strongly multiple scattering medium for acoustic waves. Phys. Rev. Lett. **84**, 6050 (2000)
25. M. Kafesaki, E.N. Economou, Multiple-scattering theory for three-dimensional periodic acoustic composites. Phys. Rev. B **60**, 11993 (1999)
26. A.A. Ruffa, Acoustic wave propagation through periodic bubbly liquids. J. Acoust. Soc. Am. **91**, 1 (1992)
27. D. Bai, J.B. Keller, Sound waves in a periodic medium containing rigid spheres. J. Acoust. Soc. Am. **82**, 1436 (1987)
28. F. Cervera, L. Sanchis, J.V. Sánchez-Pérez, R. Martínez-Sala, C. Rubio, F. Meseguer, C. López, D. Caballero, J. Sánchez-Dehesa, Refractive acoustic devices for airborne sound. Phys. Rev. Lett. **88**, 023902 (2002)
29. B.C. Gupta, Z. Ye, Theoretical analysis of the focusing of acoustic waves by two-dimensional sonic crystals. Phys. Rev. E **67**, 036603 (2003)
30. E. Meyer, E.G. Neumann, *Physical and Applied Acoustics* (Academic Press, New York, 1972)
31. A. Bensoussan, J.-L. Lions, G. Papanicolau, *Asymptotic Analysis for Periodic Structures* (North-Holland, Amsterdam, 1978)
32. N.S. Bakhvalov, G.P. Panasenko, *Homogenization. Averaging Process in Periodic Media. Mathematical Problems in the Mechanics of Composite Materials* (Kluwer, New York, 1989)
33. W.S. Amont, Sound propagation in gross mixtures. J. Acoust. Soc. Am. **25**, 638–641 (1953)
34. J.G. Berryman, Long-wavelength propagation in composite elastic media I. Spherical inclusions. J. Acoust. Soc. Am. **68**, 1809–1819 (1980)
35. J. Mei, Z. Liu, W. Wen, P. Sheng, Effective mass density of fluid-solid composites. Phys. Rev. Lett. **96**, 024301 (2006)
36. D. Torrent, A. Hakansson, F. Cervera, J. Sánchez-Dehesa, Homogenization of two-dimensional clusters of rigid rods in air. Phys. Rev. Lett. **96**, 204302 (2006)
37. D. Torrent, J. Sánchez-Dehesa, Effective parameters of clusters of cylinders embedded in a non-viscous fluid or gas. Phys. Rev. B **74**, 224305 (2006)
38. D. Torrent, J. Sánchez-Dehesa, F. Cervera, Evidence of two-dimensional magic clusters in the scattering of sound. Phys. Rev. B (RC) **75**, 241404 (2006)
39. D. Torrent, J. Sánchez-Dehesa, Anisotropic mass density by two-dimensional acoustic metamaterials. New J. Phys. **10**, 023004 (2008)
40. D. Torrent, J. Sánchez-Dehesa, Acoustic metamaterial for new two-dimensional sonic devices. New J. Phys. **9**, 323 (2007)
41. A. Climente, D. Torrent, J. Sánchez-Dehesa, Sound focusing by gradient index sonic lenses. Appl. Phys. Lett. **97**, 104103 (2010)
42. T.P. Martin, M. Nicholas, G. Orris, L.W. Cai, D. Torrent, J. Sánchez-Dehesa, Sonic gradient index lens for aqueous applications. Appl. Phys. Lett. **97**, 113503 (2010)
43. L. Zigoneanu, B.-I. Popa, S.A. Cummer, Design and measurements of a broadband two-dimensional acoustic lens. Phys. Rev. B **84**, 024305 (2011)
44. L. Zigoneanu, B.-I. Popa, A.F. Starr, S.A. Cummer, Design and measurements of a broadband two-dimensional acoustic metamaterial with anisotropic effective mass density. J. Appl. Phys. **109**, 054906 (2011)
45. L.N. Gumen, J. Arriaga, A.A. Krokhin, Metafluid with anisotropic dynamic mass. Low Temp. Phys. **37**, 1221–1224 (2011)
46. J. Li, L. Fok, X. Yin, G. Bartal, X. Zhang, Experimental demonstration of an acoustic magnifying hyperlens. Nat. Mater. **8**, 931–934 (2009)
47. S.A. Cummer, D. Schurig, One path to acoustic cloaking. New J. Phys. **9**, 45 (2007)
48. D. Torrent, J. Sánchez-Dehesa, Radial wave crystals: Radially periodic structures from metamaterials for engineering acoustic or electromagnetic waves. Phys. Rev. Lett. **103**, 064301 (2009).
49. D. Torrent, J. Sánchez-Dehesa, Multiple scattering formulation of two-dimensional acoustic and electromagnetic metamaterials. New J. Phys. **13**, 093018 (2011)

Chapter 2
Fundamental Properties of Phononic Crystal

Yan Pennec and Bahram Djafari-Rouhani

2.1 Introduction to the Concept of Phononic Crystals and Their Band Structures

The control and manipulation of acoustic/elastic waves is a fundamental problem with many potential applications, especially in the field of information and communication technologies. For instance, confinement, guiding, and filtering phenomena at the scale of the wavelength are useful for signal processing, advanced nanoscale sensors, and acousto-optic on-chip devices; acoustic metamaterials, working in particular in the sub-wavelength regime can be used for efficient and broadband sound isolation as well as for imaging and super-resolution.

Phononic crystals, which are artificial materials constituted by a periodic repetition of inclusions in a matrix, are proposed to achieve these objectives via the possibility of engineering their band structures. The elastic properties, shape, and arrangement of the scatterers modify strongly the propagation of the acoustic/elastic waves in the structure. The phononic band structure and dispersion curves can then be tailored with appropriate choices of materials, crystal lattices, and topology of inclusions.

Similarly to any periodic structure, the propagation of acoustic waves in a phononic crystal is governed by the Bloch [1] or Floquet theorem from which one can derive the band structure in the corresponding Brillouin zone. The periodicity of the structures, that defines the Brillouin zone, may be in one (1D), two (2D), or three dimensions (3D). The propagation of acoustic waves in layered periodic materials or superlattices which are now being considered as 1D phononic crystals has been extensively studied [2] since the early paper of Rytov [3]. However, the

Y. Pennec • B. Djafari-Rouhani (✉)
Institut d'Electronique, de Microélectronique et de Nanotechnologie,
Université de Lille 1, IMEN Bat P5, Villeneuve d'Ascq 59655, France
e-mail: yan.pennec@univ-lille1.fr; bahram.djafari-rouhani@univ-lille1.fr

A. Khelif, A. Adibi (eds.), *Phononic Crystals*, DOI 10.1007/978-1-4614-9393-8_2

concept of phononic crystal was introduced only two decades ago in relation with 2D [4–6] and 3D [7] periodic media, especially to seek for the possibility of the so-called absolute band gaps [8–10]. Indeed, the dispersion curves exhibit band gaps in which the propagation of waves is prohibited. Such gaps may occur for particular directions of the wave vector, but they can also span the whole 2D or 3D Brillouin zone where the propagation of elastic waves becomes forbidden for any polarization and any incident angle. Then, the structure behaves like a perfect mirror for any incidence angle, thus prohibiting the transmission of sound waves.

The concept of phononic crystal followed by a few years the analogous concept of photonic crystals [11, 12] for the propagation of electromagnetic waves. The existence of band gaps is especially well-known in solid state physics in the field of electronic band structure of crystalline materials. In particular, the properties of semiconductors, such as electronic, conduction, and optical properties, are dominated by the band gap separating the valence and conduction bands and, moreover, these properties can be drastically modified and tailored by introducing defects into the semiconductor due to the emergence of new states inside the band gaps (the so-called localized modes associated with the defects which have a decaying wave function far from the defect position). Similarly, the introduction of defects such as waveguides and cavities in phononic or photonic crystals are at the origin of many of their potential applications for confinement, guiding, filtering, and multiplexing of acoustic waves at the level of the wavelength [10] and pave the way for the realization of advanced sensors and acousto-optic devices.

The progress in the field of phononic crystals goes in parallel with their photonic counterpart, although they involve a larger variety of materials as concerns the possibility of high contrast among the elastic properties, large acoustic absorption and the solid or fluid nature of the constituents. Since the band structure is scalable with the dimensions of the structure (as far as the linear elasticity theory applies), a great deal of works has been devoted to macroscopic structures in the range of sonic (kHz) and ultrasonic (MHz) frequencies where the proof of concepts of band gaps and manipulation of sound (such as wave guiding, confinement, sharp bending) have been established with simple demonstrators. Yet, there is a continuous interest in the engineering of band structures with new structures and materials as well as the technological fabrication of sub-micron scale structures working in the hypersonic (GHz) regime.

The general mechanism for the opening of a gap is based on the destructive interference of the scattered waves by the inclusions and therefore requires a high contrast between the elastic properties of the materials. In periodic structures, this is called the Bragg mechanism and the first band gap generally occurs at a frequency which is about a fraction of c/a, where c is a typical velocity of sound, and a the period of the structure. However, when the propagating waves in the embedding medium are strongly scattered by the internal resonances of the individual inclusions, one may obtain a so-called hybridization gap which results from the coupling between the propagating waves of the matrix and the localized mode of the scatterers [13, 14]. Such a gap is less sensitive to the periodicity and can persist even in presence of some disorder in the structure [15, 16]. For common materials, it may happen that both types of gaps arise in the same frequency range since the internal resonances of the inclusions would be of the order of c/d where d is the typical diameter or size of the inclusion. In such cases, the combination of the two effects can widen the actual band gap. It is also worthwhile mentioning the concept of locally resonant sonic materials (LRSM) introduced by Ping Sheng et al. [17] which later developed into the field of acoustic metamaterials. In the latter work, the coating of hard inclusions by a very soft rubber produced a very low frequency resonance gap situated two orders of magnitude below the Bragg gap, thus allowing the sound isolation below kHz by a sample with a thickness of a few centimeters only.

Point or linear defects [18] such as cavities or waveguides [19] can be introduced into the phononic crystal by removing or modifying one, a few or a row of inclusions. Depending on their geometries and constitutions, such defects can give rise to new modes inside the band gap of the phononic crystal that correspond to localized or evanescent waves with a decaying displacement field far from the defect [20–22]. Therefore, they can be used for confinement and guiding [23, 24] of the acoustic waves and the coupling between a waveguide and cavities provide the possibility of filtering devices [25, 26, 10].

In this preliminary chapter, we limit ourselves to a basic presentation of the trends on the dispersion curves and band gaps in different types of phononic crystals with solid or fluid constituents. For the sake of simplicity, we consider only the case of 2D crystals constituted by a periodic array of infinitely long bars in a matrix background. Then, we review the localized modes associated with some simple defects and their applications in filtering and multiplexing phenomena. In a final section, we briefly summarize further developments in the field of phononic crystals.

2.2 Dispersion Curves and Band Gaps in 2D Phononic Crystals

2.2.1 Origin of the Band Gaps: Bragg Gaps and Local Resonances

An absolute phononic gap, if one exists, can be a Bragg type gap, which appears at about an angular frequency ω of the order of c/a where c is a typical velocity of sound in the structure and a the lattice parameter. The existence of absolute band gaps was predicted theoretically [4–8] prior to being demonstrated experimentally in various phononic crystals constituted of solid components [27, 28] or mixed solid/fluid components [29]. It has been shown that the existence and bandwidth of the gaps depend strongly on the nature of the constituent materials (solid or fluid), the contrast between the physical characteristics (density and elastic constants) of the inclusions and the matrix, the geometry of the array of inclusions, the inclusion shape and the filling factor.

It can be also a resonance type gap, which can appear at frequencies below the Bragg limit. In the latter case, it is possible to obtain absolute gaps at frequencies one to two orders of magnitude lower than the Bragg diffraction threshold, without increasing the size of the unit cell in the crystal. Such gaps can be realized in the so-called LRSM, whose building units exhibit localized resonant modes at specific frequencies [17, 30]. Forming a phononic crystal from such components, the resonances interact and give rise either to flat bands or to resonance gaps about the corresponding eigenfrequencies. As these localized resonances depend on the properties of the individual scatterers, their position in frequency can be tuned by properly choosing the properties (elastic or geometric) of the scatterer. These materials could found several potential applications, in particular in the field of sound isolation or in the realization of vibrationless environment for high precision mechanical systems, negative refraction or cloaking acoustic metamaterials.

2.2.2 *Behavior of the Band Gaps as a Function of the Geometrical and Physical Parameters*

Phononic crystals are heterogeneous elastic media composed of a periodic array of inclusions embedded in a matrix. The main characteristic of such composite media is to exhibit stop bands in their transmission spectra, in which the propagation of waves is forbidden. Three classes of phononic crystal can be defined, which differ from each other by the physical nature of the inclusions and the matrix. One can thus define solid–solid, fluid–fluid, and mixed solid–fluid composite phononic crystals. The opening of wide acoustic band gaps requires two main conditions. The first one is to have a large physical contrast, such as density and speed of sound, between the inclusion and the matrix. The second condition is to present a sufficient filling factor of the inclusion in the matrix unit cell. One can note that the forbidden band gap occurs in a frequency domain given by the ratio of an effective sound velocity in the composite material to the value of the lattice parameter of the periodic array of inclusions. In two-dimensional solid–solid phononic crystal, the modes of vibration can be decoupled between the in-plane propagation where the elastic displacement is perpendicular to the cylinders and the out-of-plane propagation where the elastic displacement is parallel to the cylinders axis. In fluid–fluid phononic crystals, only longitudinal modes are allowed. In mixed phononic crystals, complex modes of vibration can exist, ranging from longitudinal in the fluid to longitudinal and transverse in the solid part. In this section, we will give a few examples of the three classes of phononic crystals, dealing with the nature, the composition, and the geometry of the constituents. Most of the following calculations have been performed using improved methods such as plane wave expansion (PWE), finite difference time domain (FDTD), well-known in the field of photonic crystals, and finite element method (FEM).

Table 2.1 Mass density ρ and elastic constants C_{11}, C_{44}, and C_{12} of silicon and epoxy. $c_l = \sqrt{\frac{C_{11}}{\rho}}$ and $c_t = \sqrt{\frac{C_{44}}{\rho}}$ represent, respectively, the longitudinal and transverse speed of sound

Material	ρ (kg/m^3)	C_{11} ($\times 10^{11}$ dyn/cm^2)	C_{44} ($\times 10^{11}$ dyn/cm^2)	C_{12} ($\times 10^{11}$ dyn/cm^2)	c_l (m/s)	c_t (m/s)
Silicon	2,331	16.57	7.962	6.39	8,430	5,844
Epoxy	1,180	0.761	0.159	0.443	2,540	1,161

2.2.3 Solid–Solid Phononic Crystal

The elastic band structure of two-dimensional solid–solid composite materials has been investigated independently in a few works by Sigalas and Economou [4, 7, 31] and Kushwaha et al. [5, 6]. These authors demonstrate the existence of absolute phononic band gaps in the first irreducible Brillouin zone. The dependence of the band gap on the composition of the material and on the physical parameters of the constituents was investigated in [6, 8, 27]. In the following, we propose to examine in detail the elastic band structures and existence of absolute band gaps in phononic crystals made of two common materials, silicon and epoxy. Silicon is considered to be a cubic material with a crystallographic axis [001] parallel to the direction of propagation whereas epoxy is isotropic. The physical parameters of the two materials are reported in Table 2.1. These materials present a strong contrast between both their densities and elastic constants, meaning that silicon is the hard material while epoxy is the soft one. The first general requirement corresponding to the existence of absolute band gap is then respected.

The purpose of the section is to investigate three lattices of periodic structures, i.e. square, hexagonal, and honeycomb, as depicted in Fig. 2.1. The two-dimensional cross section of the three investigated arrays is represented, in which a is the lattice parameter. The corresponding Brillouin zone is also represented where (Γ, X, M) (resp. (Γ, J, X)) are the high symmetry points of the first irreducible Brillouin zone for the square (resp. hexagonal and honeycomb) array.

We first deal with hard material inclusions inside a soft matrix. Figure 2.2a shows one example of the dispersion curves for a square array of silicon cylinders in epoxy matrix, the filling factor defined by β being equal to 0.68. In the range of frequency of Fig. 2.2a, two complete band gaps are found for the in-plane and out-of-plane polarizations of the modes. The choice of the filling factor $\beta = 0.68$ produces almost the largest complete band gap.

Indeed, in Fig. 2.2b, the evolution of the band gap widths (white area) is presented as a function of the filling factor. The first complete band gap is the largest one and is open over a large range of filling factor, above 0.2. We note that the largest width of the band gap ($\frac{\Delta(fa)}{(fa)_{\max}} = 28$ % at $\beta = 0.74$) is open for very high filling fraction which can be a limitation for technological fabrication. A second smaller band gap opens for $\beta > 0.55$. The central frequencies of both band gaps increase with increasing the filling factor.

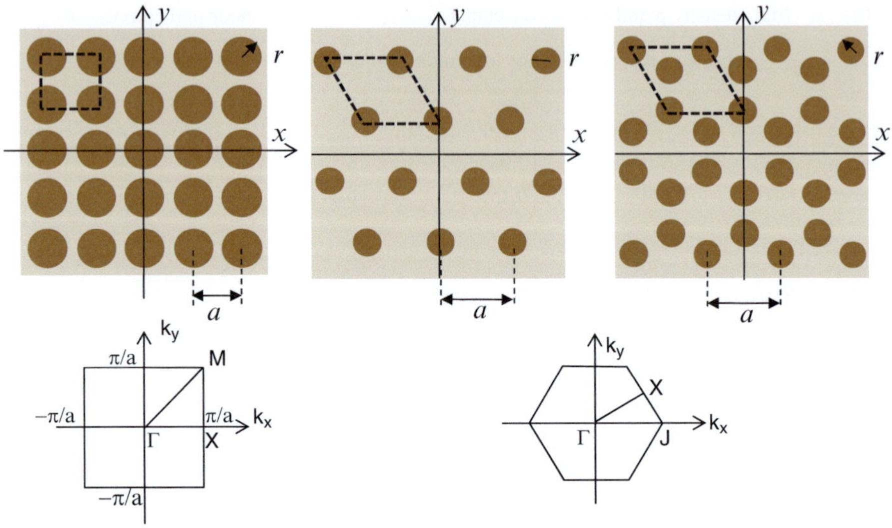

Fig. 2.1 Two-dimensional cross sections of square, hexagonal, and honeycomb lattices with the corresponding Brillouin zone. The *dashed lines* represent the elementary unit cell of lattice parameter *a*. *r* is the radius of the inclusions

In Fig. 2.2c, we present the evolution of the band gaps for the hexagonal lattice. We obtain now three band gaps where the largest opens up for a filling fraction of $\beta > 0.36$, with a maximum width ($\frac{\Delta(fa)}{(fa)_{\max}} = 37\ \%$) around $\beta = 0.80$.

Finally, for the honeycomb lattice, (Fig. 2.2d), a large and complete band gap opens at higher frequencies and for filling fraction in the range $0.24 < \beta < 0.44$. In this composite system, the gap width ($\frac{\Delta(fa)}{(fa)_{\max}} = 8\ \%$ at $\beta = 0.34$) is much lower than those obtained for the two preceding geometries. As a conclusion, for hard inclusions in a soft matrix the largest band gaps are obtained for the hexagonal and square lattices and the former allows lower filling fractions.

It is worth noticing that the band gaps are also dependent upon the shape of the inclusions. For example, we have shown [8] that their positions and widths can be changed if the circular inclusions are replaced by squares. Moreover, by rotating the squares with respect to the axes of the photonic crystals, one can also tune the band gaps.

In the opposite situation of soft epoxy inclusions in a silicon matrix, the square and hexagonal lattices display absolute band gaps only for very high filling fraction which may be not interesting from a fabrication point of view. On the contrary, for the honeycomb lattice (Fig. 2.3), one can observe the opening of an absolute band gap as far as the filling fraction exceeds $\beta = 0.34$. Moreover, the band gap width increases strongly and reaches the larger value of $\frac{\Delta(fa)}{(fa)_{\max}} = 78\ \%$ at $\beta = 0.60$.

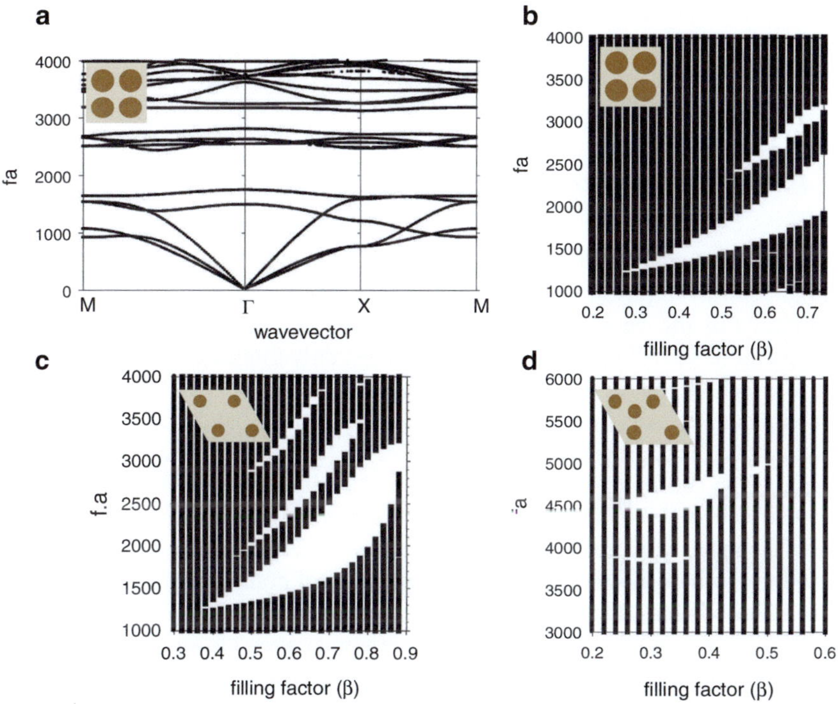

Fig. 2.2 Band gap existence in phononic crystal made of hard silicon inclusion in soft epoxy matrix. (**a**) Example of dispersion curve for the square array of symmetry with filling factor $\beta = 0.68$. Band gap maps for (**b**) square, (**c**) hexagonal, and (**d**) honeycomb arrays as a function of the filling factors

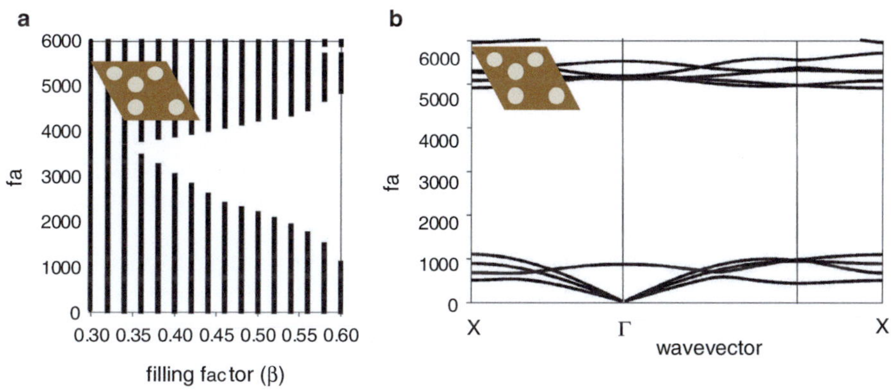

Fig. 2.3 (**a**) Band gap map for the honeycomb array of soft epoxy inclusions in hard silicon matrix. (**b**) Example of dispersion curves for the honeycomb structure with filling factor $\beta = 0.60$

Table 2.2 Mass density ρ and speed of sound of steel and water

Material	ρ (kg/m^3)	c_l (m/s)	c_t (m/s)
Steel	7,780	5,825	3,227
Water	1,000	1,490	–

2.2.4 *Solid–Fluid Phononic Crystal*

We now turn to solid–fluid periodic structures known as mixed phononic crystals. A large contrast in physical properties between the two materials is often satisfied particularly in the case of solid/gas combinations. The mixed systems present complex modes of vibration as the liquid allows only longitudinal modes while the solid allows both longitudinal and transverse modes. Due to this difficulty, the PWE method generally fails to predict accurately the acoustic band structures for such a mixed system. This drawback can be alleviated by imposing the condition of elastic rigidity to the solid inclusions [32, 33] which is satisfactory to describe the sound propagation in a phononic crystal made of solid inclusions in air. Nevertheless, this difficulty can be overcome by making the band structure calculations with the FDTD method [34] which allows defining the real nature of both solid and liquid [28, 35]. In the mixed systems, the fluid can be either a condensed liquid [28, 36, 37] or a gas [38–40]. We propose here to consider two different cases, i.e. when solid inclusions are inserted in a liquid matrix and the opposite situation.

We first investigate the case of a phononic crystal made of steel cylinders in a water matrix. The density and elastic constants of the two materials are given in Table 2.2. We present the calculations for the in-plane vibrations.

The left diagram of Fig. 2.4a shows the band structure along the direction ΓX of the irreducible Brillouin zone, calculated for a square array of steel cylinders of radius $r/a = 0.38$. The right part of Fig. 2.4a shows the transmission coefficient using the FDTD calculation. The incoming wave is a longitudinal pulse, uniform along the X direction and with a Gaussian profile along the Y axis. The transmitted signal is recorded as a function of time over the cross section of the waveguide, and finally Fourier transformed to obtain the transmission coefficient versus frequency. The spectrum is normalized with respect to the signal obtained without the phononic crystal sample. For the ΓX direction, the band diagram shows a large band gap from 500 m/s to almost 1,000 m/s.

We have then computed similarly the dispersion curves and transmission spectrum for hollow steel cylinder filled with water [22, 41]. The inner radius r_i/a is chosen equal to 0.22 with the same outer radius $r/a = 0.38$ as in the previous case. One can see that the insertion of the hollow tubes widens the stop band, with the upper edge of the band gap moving towards higher frequencies. But looking at the transmission spectrum, the most remarkable feature is the existence of a narrow pass band localized inside the band gap of the hollow cylindrical phononic crystal, at $fa = 780$ m/s. As the inner radius increases from 0.2 to 0.25, the frequency of the narrow pass band decreases.

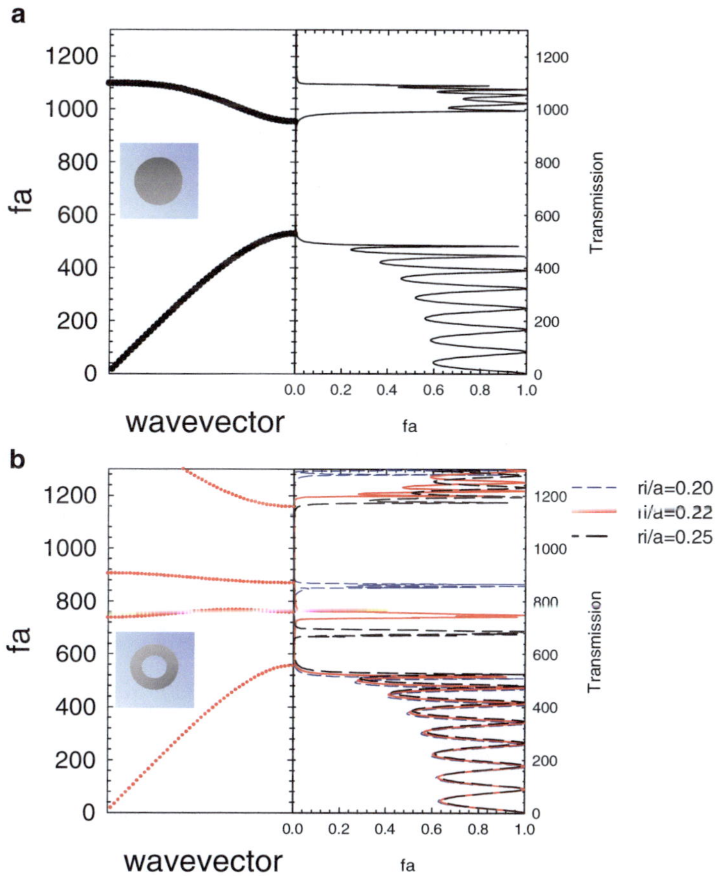

Fig. 2.4 (**a**) Dispersion (*left*) and transmission (*right*) curves of the phononic crystal composed of steel cylinders of radius $r/a = 0.45$ in water matrix. (**b**) (*Left*) Dispersion curve for hollow cylinders of inner radius $r_i/a = 0.22$ and filled with water. (*Right*) Transmission curves for hollow cylinders of variable inner radius

The dispersion curves, calculated for the inner radius $r/a = 0.22$, present two flat bands inside the band gap. The lower one, at $fa = 780$ m/s, fits perfectly the narrow pass band observed in the transmission spectrum. The upper one, at 900 m/s, does not contribute to the transmission. Such a band is named a "deaf band" because it cannot be excited due to symmetry reason [42]. A detailed analysis of the eigenvectors associated with these vibration modes is reported in [43]. We have also shown that the nearly flat transmitted branch does not correspond to a mode localized in the water-filled cavities inside the hollow cylinders but to a propagative branch with very slow group velocity.

We now consider the opposite situation where the two-dimensional phononic crystal is made of water cylinders in silicon. The conclusion can be also extended to

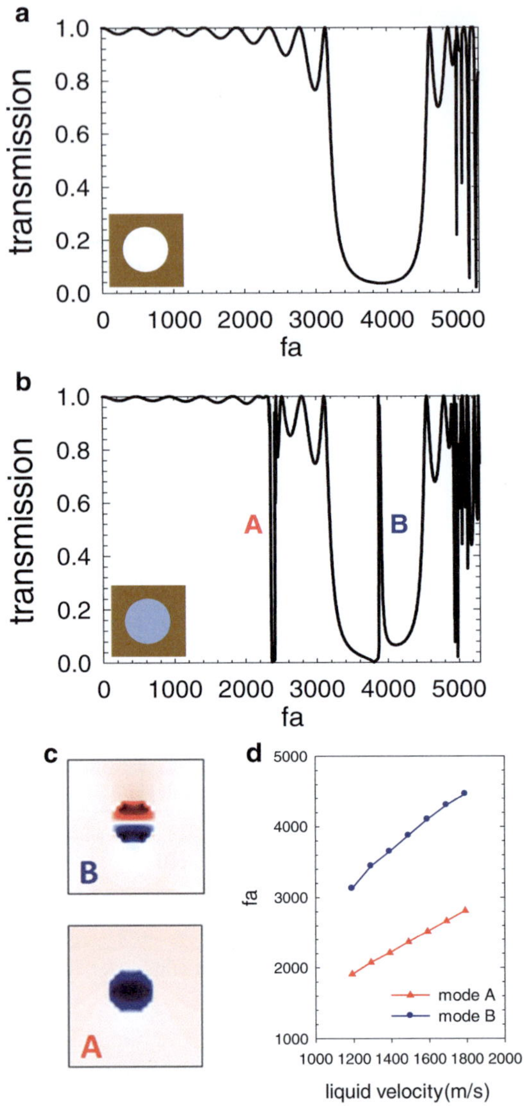

Fig. 2.5 Transmission curve through a 2D square lattice phononic crystal made of periodic holes of radius $r/a = 0.18$ in a silicon substrate when (**a**) the holes are empty and (**b**) the holes are filled with water. (**c**) Map of the displacement field at the dip *A* and the peak *B*. (**d**) Evolution of the frequencies of the resonant modes A and B as a function of the velocity of the liquid inside the holes

the case where the cylinders are filled with a liquid polymer [44]. Figure 2.5a shows as a reference the calculation of the transmission curve when the phononic crystal is made of air holes with radius $r/a = 0.18$ inside the silicon matrix. One can see that the spectrum presents a large pass band below 3,000 m/s then a band gap between 3,000 and 4,200 m/s.

When the holes are filled with water (Fig. 2.5b), the transmission curve exhibits two new features labeled A and B, which appear as a dip in the transmitted branch and a peak in the band gap. To give a deeper insight of the two features A and B, we calculated (Fig. 2.5c) their corresponding maps of the displacement field. The dip A and the peak B are associated with a high confinement of the field inside the water holes. Due to the large contrast between the acoustic velocities and impedances of water and silicon, these modes can be considered as cavity resonances inside the holes surrounded by an almost rigid material. Therefore, their frequencies are very close to the solution of the equation $J'_m\left(\omega r/c_{\text{liq}}\right) = 0$ where J'_m is the derivative of the Bessel function of order m, ω the frequency, r the radius of the cylinder, and c_{liq} the velocity of sound in water. In the transmission curve of Fig. 2.5b, it appears that the resonant modes of the cavity give rise, respectively, to a dip or a peak as far as their corresponding frequencies fall inside a pass band or a band gap of the phononic crystal. In Fig. 2.5d, we give the evolution of the features A and B when changing the longitudinal acoustic velocity c_{liq} of the liquid filling the holes with respect to the water. The frequencies of the resonant modes increase by increasing the sound velocity of the liquid and in both cases the relative shift in frequency ($\Delta(fa)/(fa) = 20\ \%$) has almost the same order of magnitude as the relative shift of the sound velocity ($\Delta c_{\text{liq}}/c_{\text{liq}} = 24\ \%$).

One interest of such mixed structure is to present a new way to sense the sound velocity of bio-chemical liquids [45, 46]. To make a phononic sensor, the well-defined features should display a high quality factor, be very sensitive to the acoustic velocity of the liquid, and remain relatively isolated in frequency from each other in order to allow the sensing of the probed parameter on a sufficiently broad range. Such ultra-compact structure can be shown as label-free, affinity-based acoustic nanosensor, useful for bio-sensing applications in which the amount of analyte is often limited.

2.2.5 Fluid–Fluid Phononic Crystal

In this section we assume that the materials constituting the phononic crystal are made of two different fluids. An interesting example is provided by air cylinders (in 2D) or air bubbles (in 3D) in a water matrix. Indeed, these structures display giant sonic stop bands resulting from a combination of Bragg and resonance scatterings that can be obtained whatever the symmetry of the lattice [47–49]. Figure 2.6a illustrates the transmission coefficient for a square lattice of air cylinders in water background for a filling factor $\beta = 20\ \%$. The lattice parameter taken equal to $a = 20$ mm in order to fall in the audible frequency range. One can see a large stop band extending from 0.5 to 20 kHz, followed by a sharp peak. Thus this system would have the property of preventing the propagation of sound in a large frequency domain, with the period of the sonic crystal being much smaller than the acoustic wavelength in air. The few peaks of transmission below 0.5 kHz come from the lowest dispersion curve. The peak A, at 20 kHz, corresponds to a mode localized inside the air cylinder (a resonance of the air cylinder), as can be seen in Fig. 2.6a. Such localization is possible due to the huge density and compressibility contrasts between air and water.

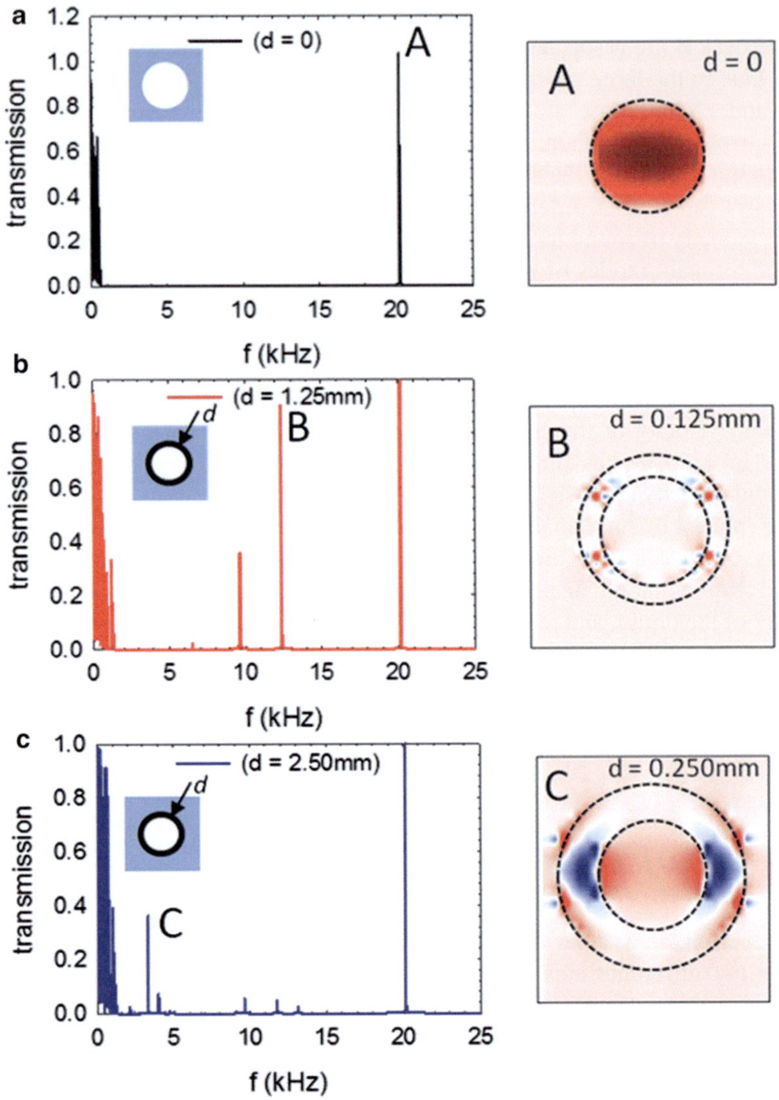

Fig. 2.6 Spectral transmission coefficient for three values of the polymer thickness: (**a**) $d = 0$, (**b**) $d = 1.25$ mm, and (**c**) $d = 2.50$ mm. The lattice parameter is $a = 20$ mm and the inner radius of the tube (air cylinder) is 5 mm. The maps of displacement field close to each diagram correspond to one example of the transmitted peaks

In the following, we consider the more practical system where air inside cylinders is surrounded by a thin polymer shell immersed in water. The transmission calculations are presented for different thicknesses of the polymer shell (Fig. 2.6b, c). The density and elastic constants of the two materials are reported in Table 2.3.

Table 2.3 Mass density ρ and speed of sound of air and polymer [50]

Material	ρ (kg/m^3)	c_l (m/s)	c_t (m/s)
Soft polymer	995	1,000	20
Air	1,000	340	–

In contrast to [34, 17] where the velocities of sound in the polymer were assumed very low and especially unrealistic as concerns the longitudinal velocity, here we chose realistic values of longitudinal (1,000 m/s) and low transverse (20 m/s) velocities as reported in [50]. However, the physical conclusions which are much dependent upon the transverse velocity of the polymer will remain very similar to those presented in [34].

In Fig. 2.6b (resp. c), the thickness of the polymer shell is $d=1.25$ mm (resp. 2.50 mm), while keeping the air cylinder at $r=5$ mm. A large and low frequency stop band is still observed but now starting at 1.2 kHz. Moreover, while the peak at 20 kHz is still present, some new ones appear in the transmission as B or C, mainly localized inside the polymer layer of the inclusion as seen in the map of the displacement fields of Fig. 2.6b, c. As a conclusion, it has been shown that hollow cylinders made of an elastically soft polymer containing air inside and arranged on a square lattice in water can still give rise to very large acoustic band gaps at low frequencies. In the opposite case of water cylinders in an air background, large band gaps can be obtained with a honeycomb lattice with a very high filling fraction (touching cylinders) [34].

2.2.6 *Locally Resonant Phononic Crystal*

As introduced previously, an absolute phononic gap, if one exists, can be a Bragg type gap or a resonance type gap, which can appear at frequencies well below the Bragg limit. Such structure, known as acoustic metamaterials, presents an important issue for phononic crystals related to their property of perfectly reflecting mirror for the purpose of sound isolation, negative refraction, and sub-wavelength imaging. The objective consists of finding structures that attenuate the propagation of sound over a sample whose thickness remains smaller than the wavelength in air. Most of the recent studies have been directed towards a new class of phononic crystals, the so-called locally resonant materials [17]. These structures essentially consist of a hard core, such as a metal, surrounded by a soft coating (silicone rubber) and immersed in a polymer such as epoxy. Due to the local resonances associated with the soft coating material, dips can appear in the transmission coefficient at very low frequencies situated about two orders of magnitudes below the Bragg frequency. Such behaviors have been obtained in both 3 and 2D locally resonant phononic crystals.

In this section, we present for a 2D phononic crystal, a generalization of the preceding structure to a multilayer cylindrical core constituted by two or several coaxial shells surrounding the internal hard core [30]. The structural unit of the phononic crystal consists of an infinitely long cylinder, composed of multicoaxial shells, embedded in a water matrix. The inner (core) cylinder is made of steel. This core is coated by alternate shells constituted, respectively, by a thin layer of an elastically soft material and a thin layer of a hard material (steel). In this calculation, the soft polymer is chosen to have very small transverse velocity $c_t = 19$ m/s with a longitudinal velocity of $c_l = 55$ m/s. In the following, we fix the outer radius of the cylinder equal to 8.4 mm and the thickness of each layer in the coating equal to 1.6 mm. The filling fraction of the whole cylinder, taken to be $\beta = 55$ %, will be kept constant. Finally, the sonic crystal is constituted by five rows of elementary units arranged on a square lattice, with a lattice parameter of $a = 20$ mm, embedded in water. The whole size of the sonic crystal is therefore 10 cm.

Figure 2.7a reports the transmission through a phononic crystal made of a bi-layer inclusion constituted by a steel core coated with one polymer and one steel layer. At very low frequency, a sharp dip appears in the transmission spectrum ($f = 1.45$ kHz) for which the displacement field (Fig. 2.7b) shows an elastic field localized inside the inclusion. The displacement can be understood as a motion of the core and the outer steel layer in phase opposite each other, while the polymer acts as a spring (see the schematic representation). This behavior can be interpreted as the appearance of a dynamic negative effective mass density in the frequency range of the dip [17].

We investigate now the case of a multicoaxial cylinder containing an even number of shells, 4 and 6, and we consider that the uttermost shell in contact with water is made of steel. In this way, we obtain an alternation of hard and soft materials with a solid core. Figure 2.8a presents the low frequency transmission curves in which the number of low frequency dips evolves in relation with the number of shells. A number of 2 (resp. 3) bi-layers give rise to 2 (resp. 3) low frequency peaks.

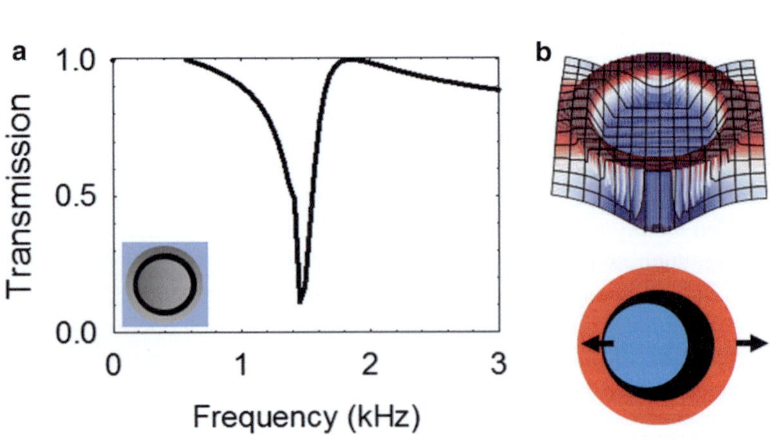

Fig. 2.7 (**a**) Transmission curve through a low resonant phononic crystal made of a steel core coated with a polymer and a steel layer, embedded in water. (**b**) Displacement field calculation at the frequency of the dip and corresponding schematic representation on the motion of the mode

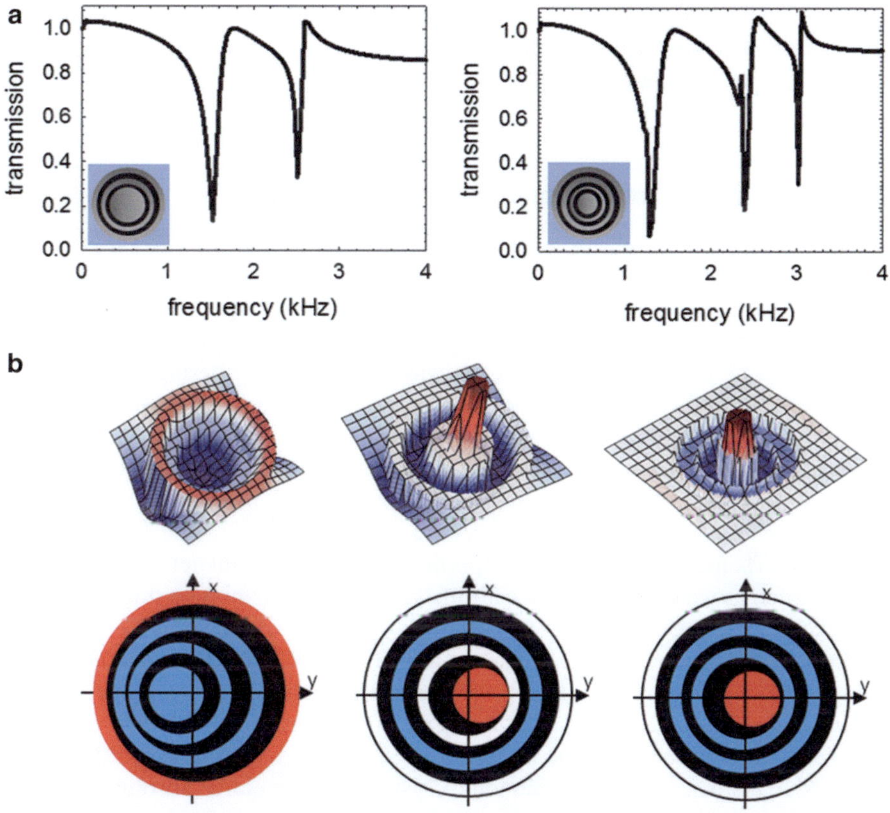

Fig. 2.8 (**a**) Transmission curves through a low resonant phononic crystal made of a steel core coated with 2 (*left*) and 3 (*right*) bi-layers constituted of polymer and steel, embedded in water. (**b**) Displacement field at the frequency of the dips for 3 bi-layers and the corresponding schematic representation of the rigid motions of the steel core and shells

Figure 2.8b gives an illustration with $N = 3$ of the displacement fields of the three resonance modes. For each frequency, we give the component of the displacement along the direction of propagation, as well as a schematic view of the vibrations. The common feature to all these three modes is the fact that the hard parts of the inclusion, namely the inner core and the three steel cylindrical shells, vibrate as rigid bodies linked together through the polymer shells that act as springs. In the lowest mode, occurring at $f = 1.61$ kHz, the inner core and the two following steel shells vibrate in phase along the propagation direction, while the outer steel shell moves with the opposite phase. The displacement fields of the second ($f = 3.0$ kHz) and third ($f = 3.77$ kHz) resonant modes correspond to other vibrational states of four rigid bodies linked together through the polymer shells. Therefore, we show the possibility of obtaining several dips in the transmission coefficient in a given frequency range. By combining two or more phononic crystals of different parameters, we could also show that it is possible to overlap some dips and obtain a widening of the frequency gaps [30].

2.3 Localized Modes Associated with Defects

2.3.1 Guiding

The existence of band gaps in phononic crystals may be useful for the purpose of introducing functionalities such as waveguiding and filtering in integrated structures. The ability to tailor the acoustic properties of phononic crystals and more specifically of their waveguides makes them particularly suitable for a wide range of applications from transducer technology to filtering and guidance of acoustic waves. They can operate at the frequencies of telecommunications (about 1 GHz) when the lattice parameter of the phononic crystal is in the micron range. This section is dealing with some examples of the properties of linear and point defects in phononic crystals such as wave bending and splitting [24, 52] or transmission through perfect or defect-containing waveguides [19, 23, 53].

As a basic structure, we consider a mixed (solid/fluid) 2D phononic crystal composed of steel cylinders in a water matrix. The inclusions are arranged periodically on a square lattice. Throughout this section, we assume the lattice parameter $a = 3$ mm and the radius of the inclusion $r = 1.25$ mm resulting in a filling factor $\beta = 0.55$. This insures that the phononic crystal displays a large absolute band gap of Bragg type in the ultrasonic range, extending from 250 to 325 kHz. All numerical simulations are based on the finite difference time-domain (FDTD) method.

We first investigate the properties of the phononic crystal containing a simple straight waveguide obtained by removing one row of cylinders along the direction of propagation (Fig. 2.9a). We have calculated the transmission through the guide as a function of the frequency. As seen in Fig. 2.9a, the guide exhibits a full transmission band in the frequency range (270–300 kHz) that covers a large part of the phononic crystal stop band. The map of the displacement field corresponding to the frequency 290 kHz shows that the transmission can be associated with a high confinement of the field inside the waveguide.

One can also demonstrate the bending of acoustic wave constructed by removing holes over a large frequency range inside the absolute band gap of the perfect phononic crystal [24]. Figure 2.9b shows the transmission curve obtained through the bending waveguide formed by two sharp corners with 90° angle. We show that most of the linear guided modes are transmitted except a transmission dips at 275 kHz. Figure 2.9b shows a numerical illustration of the propagation of the wave at 290 kHz through the bending waveguide in which the incident wave propagates along the first straight waveguide, couples successfully with the perpendicular one, then reaches to the last horizontal one.

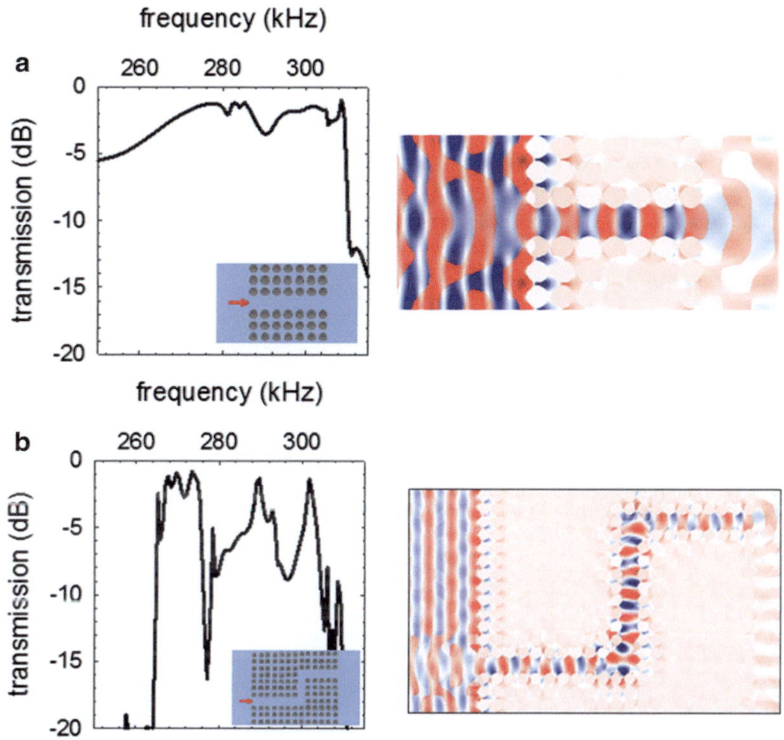

Fig. 2.9 Calculated transmission spectra in the frequency range of the band gap and displacement fields at $f = 290$ kHz through (**a**) a straight and (**b**) a bent waveguide

2.3.2 *Filtering*

We now turn to the behavior of the phononic crystal where a point defect is inserted inside the waveguide. A resonant cavity (or stub) of nominal length and width equal to one period is simply obtained by removing one cylindrical inclusion attached to the guide as sketched in the insert of Fig. 2.10a. As compared to Fig. 2.9a, the transmission remains almost unchanged except for one narrow dip occurring at the frequency of 290 kHz where the transmission becomes very small. It clearly appears that in the presence of a stub the transmission through the waveguide can be significantly altered due to the interference phenomena. In Fig. 2.10a we have represented the map of the displacement field at the frequency of the dip. One can see the wave entering the guide, penetrates into the stub, reflects at the end of the stub, and then returns back to the entrance of the guide while the transmission towards the end of the guide remains negligible. The eigenmodes of the cavity have been used advantageously to induce a very narrow stopping band in the pass band of the waveguide.

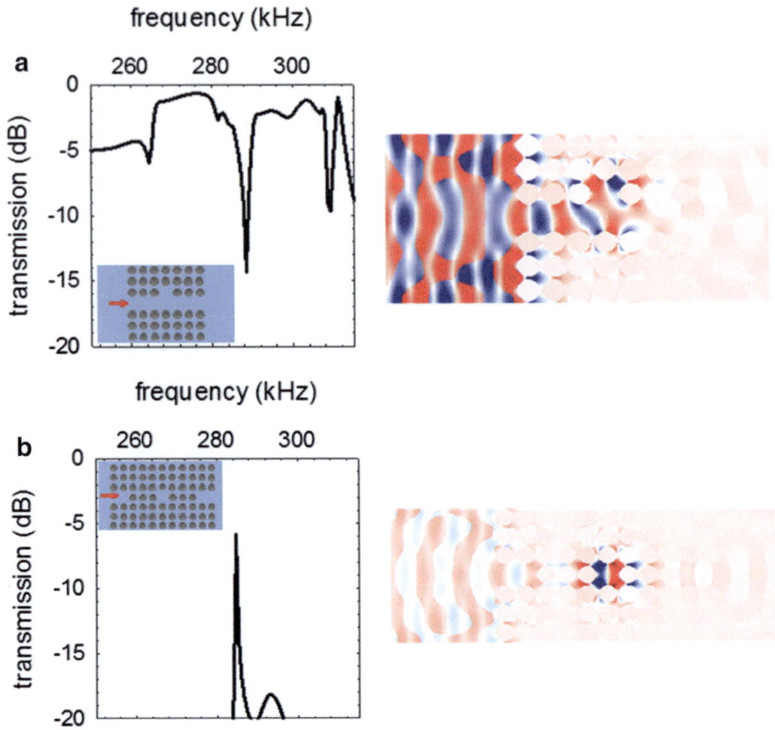

Fig. 2.10 Calculated transmission spectra in the frequency range of the band gap and displacement fields at the frequency of the dip (resp. peak) when a cavity is inserted (**a**) at the side of the guide or (**b**) inside the waveguide

In Fig. 2.10b, we have considered the same cavity incorporated inside the waveguide. The cavity is isolated from the entrance and the exit of the waveguide by three steel cylinders. Nevertheless, the transmission spectrum exhibits a peak which occurs at the resonance frequency of the cavity. This transmission is due to a coupling between the cavity modes and the waveguide one, via tunneling effects. Indeed, a single cavity incorporated into the waveguide limits the transmission mainly to the frequencies situated in the neighborhood of the eigenfrequencies of the cavity.

So, the same cavity can have two opposite effects depending on whether it is incorporated inside or at the side of the waveguide, leading, respectively, to applications as transmitted selective or rejective filters.

2.3.3 Demultiplexing

Based on the previous results, we have studied an acoustic channel drop tunneling in a phononic crystal, i.e., the possibility of transferring one particular acoustic wavelength between two parallel waveguides coupled through an appropriate coupling element which is composed of two coupled cavities interacting with stubs located at the sides of the two parallel guides (see Fig. 2.11a) [26]. The incoming wave is a longitudinal pulse with a Gaussian profile which only covers the entrance of port 1 (black arrow), leaving port 4 essentially unaffected. The transmitted signals, displayed in Fig. 2.11b, are recorded at ports 2 (blue arrow) and 3 (red arrow). It can be observed that the direct transmission at port 2 drops almost to zero at the frequency of 290 kHz. At the same time, a significant peak of transmission occurs at port 3, with a magnitude comparable to the loss at port 2. This means that, at this frequency, the incoming signal is essentially transferred to the second wave

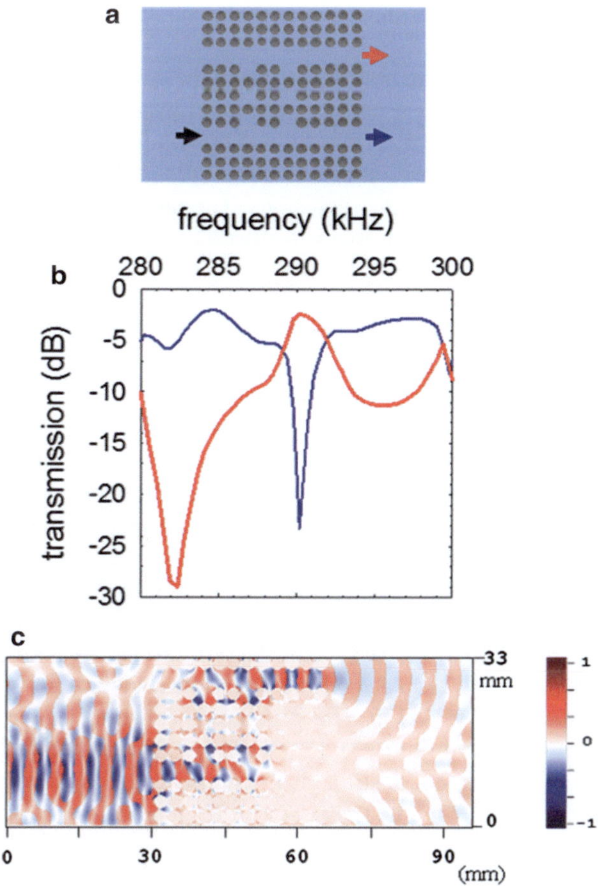

Fig. 2.11 (**a**) Schematic view of the phononic crystal with two waveguides coupled through an element constituted by two vacancies. Stubs along the guides ensure the efficiency of the coupling. The *black*, *red*, and *blue arrows* indicate the entrance and exit signal. (**b**) Calculated transmission spectra at the output ports for an input Gaussian signal coming from port 1. At the frequency of 290 kHz, the incident wave drops from the first to the second waveguide. (**c**) Calculated displacement field along the direction of propagation at a frequency of 290 kHz, averaged over one period of oscillation. The *red color* (*blue*) corresponds to the highest (lowest) value of the displacement field given in arbitrary units

guide towards port 3, leaving all other exits of the structure unaffected. In other words, the input signal tunneled through the coupling element and dropped inside the second wave guide.

To obtain a direct confirmation of the demultiplexing phenomenon, the FDTD computation was used to simulate a monochromatic source at the frequency of 290 kHz. The computed displacement field along the direction of propagation is displayed in Fig. 2.11c. The transfer of the input signal from port 1 to port 3 is clearly apparent together with an absence of signal at port 2.

2.3.4 Tunability

The tunability can be achieved by a modification of the geometrical parameters, the nature of the constituents [22, 41] or by an external physical stimulus applied to the phononic crystal. The purpose of the tunability is to modify some specific properties of the phononic crystal such as the band gap width or the position of singular features.

In Sect. 2.4, we have investigated the case of a two-dimensional phononic crystal that incorporates a narrow pass band inside a band gap. They are constituted of a periodic repetition of hollow cylinders filled and immersed in water. As seen in Fig. 2.4b, the position of the narrow pass band is quite sensible to the value of the inner radius, r_i of the inclusions. The modification of the frequency of the narrow pass band has been also investigated when the hollow cylinders are filled with a fluid other than water. We chose mercury because of the large contrast between its physical parameters and those of water. In Fig. 2.12a, we report the transmission coefficient through the phononic crystal with hollow cylinders with inner radius $r_i/a = 0.22$, filled with water or mercury and we clearly notice the shift of the frequency of the narrow pass band.

Figure 2.12b summarizes the values of the frequencies for a set of inner radius. One can conclude that the narrow pass band can be tuned and may offer a mean for selective transmission. The value of the frequency can be adjusted both by changing the inner radius of the cylinders or the nature of the fluid that fills them. In the latter case, the tuning of the frequency can be made either in a passive way or actively by injecting and flushing the fluids contained in the interior of the cylinders.

We can also combine tunability and guiding by design phononic crystals waveguides with narrow pass band. We will discuss the multiplexing and demultiplexing properties of Y-shaped waveguides constituted of hollow cylinders. Let us consider sketched I Fig. 2.13a. The structure is constituted by a heteroradii waveguide with alternating radii $r_i/a = 0.24$ and $r_i/a = 0.20$ then divided at its end into two branches. Each branch is constituted of hollow cylinders designed for the propagation of waves with only one frequency corresponding to one specific narrow pass band.

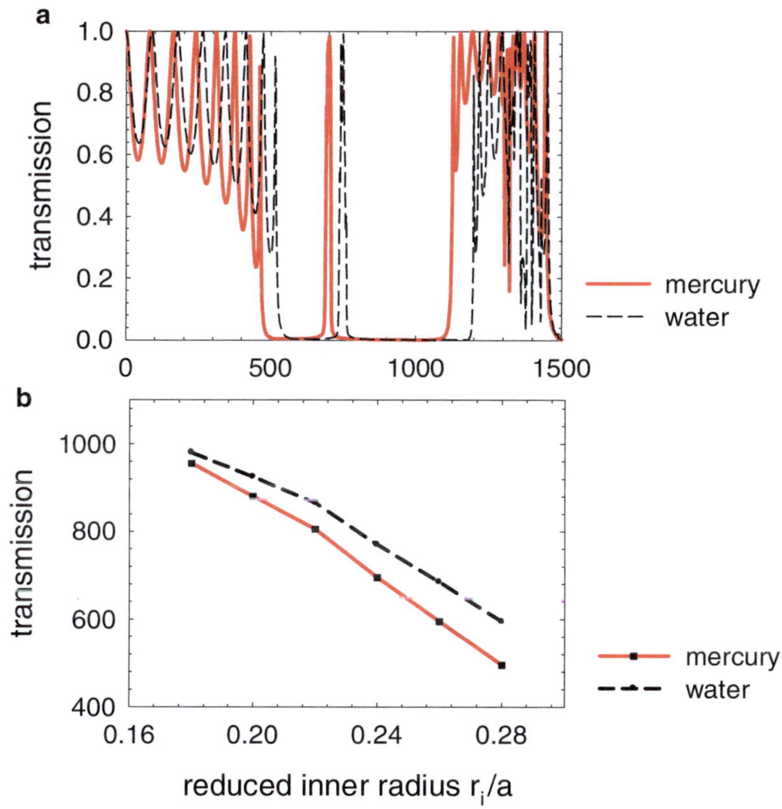

Fig. 2.12 (**a**) Transmission spectra calculated for a phononic crystal composed of hollow cylinders with inner radius $r_i/a = 0.22$ containing mercury (*solid red line*) or water (*dashed black lines*). (**b**) Values of the narrow pass band centered frequencies for phononic crystals of hollow cylinders with different inner radius containing mercury or water

We launch at the entrance of the structure a monochromatic excitation at the value of frequency of the narrow pass band corresponding to $r_i/a = 0.24$ and $r_i/a = 0.20$. One can see on the displacement field calculations of Fig. 2.13 that each frequency is guided through the heteroradii waveguide and then directed towards the branch corresponding to the respective inner radius. It means that, when an initial broad band signal is sent from the right of the system, each branch of the Y-shaped waveguide will select its own narrow pass band. These two signals are then superimposed into the heteroradii waveguide. As a result, the transmitted spectrum will contain two peaks corresponding to both narrow pass band and selectively transmitted. Finally, similar conclusions hold if the Y-shaped waveguide contains hollow cylinders of same inner radius filled with two different fluids [41].

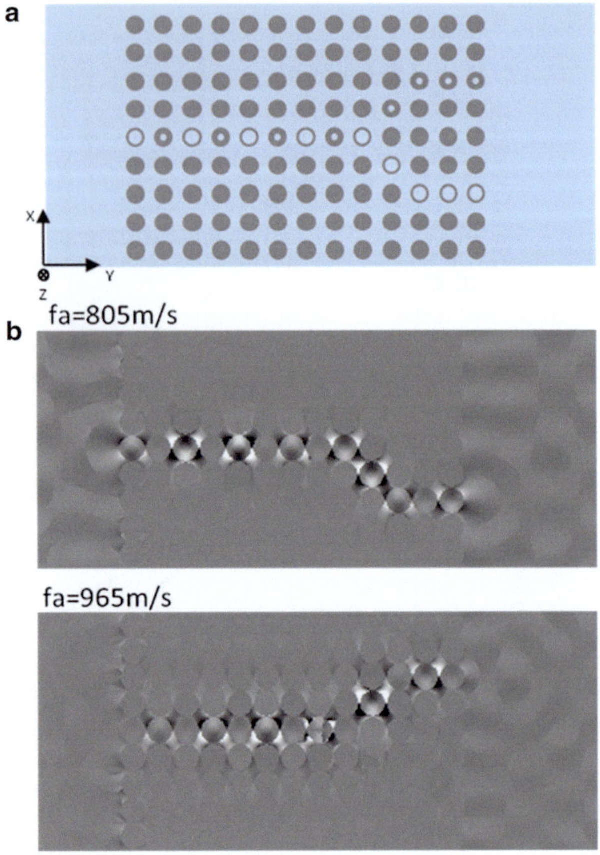

Fig. 2.13 (**a**) Schematic representation of the Y-shaped waveguide. The left part of the waveguide contains two types of cylinders with inner radii $r_i/a = 0.24$ and $r_i/a = 0.20$, in alternation. Each branch of the Y contains one type of cylinder to permit the separation of the two narrow pass bands. (**b**) Representation of the displacement field for a Y-shaped waveguide at the two frequencies of $fa = 805$ and 965 m/s

2.4 Concluding Remarks and Further Developments in the Field of Phononic Crystals

The main object of this chapter was to present the basic results about the trends of the dispersion curves and band gaps in phononic crystals, as well as the emergence of localized modes associated with cavities and waveguides and their functionalities in acoustic devices. Another type of localized modes which will be described in detail in a next chapter concerns the surface acoustic modes when the phononic crystal is cut along a plane. Besides the surface modes of lamellar materials (or superlattices) that have been widely studied [54, 2], an early paper presented the Rayleigh waves and their folding when a superlattice is cut normal to the laminations [55]. Later, the surface modes of a 2D phononic crystal cut perpendicular to the cylinders were calculated [56, 57] and then observed experimentally a few years after [58–61]. The possibility of an absolute gap in the band structure of surface waves was also demonstrated [62, 59]. Other works studied the surface waves of a 2D crystal cut parallel to the cylinders [63] or of a 3D crystal composed of spheres in a matrix [64].

Phononic crystals of finite thickness, such as a periodic array of holes in a plate or a periodic array of pillars on a membrane, started to be studied during the last decade. It was demonstrated that they can also exhibit absolute band gaps, thus providing the same functionalities associated with defects as in infinite phononic crystals. In the case of periodic holes in a plate [65, 66], the existence of an absolute band gap requires having a thickness of the slab about half of the period. In the case of periodic pillars on a membrane [67–70], besides the possibility of wide Bragg gaps, a low frequency gap exhibiting metamaterial type behavior can be obtained with an appropriate choice of the geometrical parameters [67, 70], in particular a small thickness of the membrane. With the advancements of nanotechnologies, there is a great deal of interest on nanophononics [71, 72], in particular phononic circuits with waveguides and cavities inside sub-micron phononic membranes working at a few GHz.

In this paper, we briefly mentioned examples of tunable phononic crystal where the band structure can be modified by changing the geometrical parameters (for instance, rotating square shape inclusions [8, 51]) or material parameters (for instance, filling hollow inclusions [41]). More generally, such modifications can be induced dynamically by the application of external stimuli, for instance an electric or magnetic field with piezoelectric or magnetoelastic materials [73–76], a stress in elastomeric structures [77], or the change of temperature [78, 79] (for example, the phase transitions of a polymer infiltrating the holes of a phononic crystal).

A new emerging topic concerns the search of dual phononic and photonic band gap materials in which the phonon–photon interaction can be drastically enhanced with the simultaneous confinement of both electromagnetic and acoustic waves [80–82]. For instance, stimulated Brillouin scattering can be expected over a short distance in a so-called phoxonic membrane while in general it happens in fibers which are several meters long. The optomechanic interaction between phonons and photons can take place through either the photoelastic or the interface deformation mechanisms. The latter has been investigated intensively during the last few years to cool or amplify the mechanical vibrations of a resonator via its coupling to the light. Optomechanical effect at the quantum level may be expected in micro or nanoscale systems hosting both optical and mechanical degrees of freedom. During the last few years, dual phononic–photonic membranes and strip waveguides are proposed to sustain such effects [83, 84]. Finally, more advanced dual phononic–photonic sensors allowing a simultaneous determination of the index of refraction and the acoustic velocity of an embedding liquid can be envisaged [85, 86]. The transmission spectra for each type of waves should display narrow peaks that are sensitive to the corresponding property of the liquid.

Besides the topics related to the existence of absolute band gaps, there is a continuous growing interest on refractive properties of phononic crystals, in particular: negative refraction phenomena and their applications in imaging and sub-wavelength focusing in phononic crystals [87–91], self-collimation and beam-splitting in relation with the shape of the equifrequency surfaces [92], control of the sound propagation with metamaterials with emphasis on cloaking and hyperlens phenomena.

Thermal transport in nonmetallic nanostructured materials can be strongly affected by the specific phonon dispersion curves as well as by different scattering mechanisms such as phonon–phonon interaction and phonon-boundary scattering which become very important at THz frequencies [93]. The existence of band gaps and flat dispersion curves can reduce the thermal transport and be useful for thermoelectric applications. Different frequency domains of phonons can be involved depending on the temperature and on the wavelength dependent mean free paths. Insights into the latter can be derived from molecular dynamic calculations.

In conclusion, one can expect that the field of phononic crystals will acknowledge a continuous growth in relation with the fundamental understanding of the wave phenomena in these heterogeneous materials and with their numerous expected technological applications. The latter cover a broad range of frequencies from the sonic regime for sound isolation and metamaterial behaviors, to GHz for telecommunications and phonon–photon interaction, and to terahertz for thermal transport phenomena.

References

1. F. Bloch, Z. Phys. **52**, 555 (1928)
2. For a review, see E.H. El Boudouti, B. Djafari Rouhani, A. Akjouj, L. Dobrzynski, Acoustic waves in solids and fluid layered materials. Surf. Sci. Rep. **64**, 471 (2009)
3. S.M. Rytov, Sov. Phys. Acoust. **2**, 6880 (1956)
4. M.M. Sigalas, E.N. Economou, Band structure of elastic waves in two dimensional systems. Solid State Commun. **86**, 141 (1993)
5. M.S. Kushwaha, P. Halevi, L. Dobrzynski, B. Djafari-Rouhani, Acoustic band structure of periodic elastic composites. Phys. Rev. Lett. **71**, 2022 (1993)
6. M.S. Kushwaha, P. Halevi, L. Dobrzynski, B. Djafari-Rouhani, Theory of acoustic band structure of periodic elastic composites. Phys. Rev. B **49**, 2313 (1994)
7. M.M. Sigalas, E.N. Economou, Elastic and acoustic wave band structure. J. Sound Vib. **158**, 377 (1992)
8. J.O. Vasseur, B. Djafari-Rouhani, L. Dobrzynski, M.S. Kushwaha, P. Halevi, Complete acoustic band gaps in periodic fibre reinforced composite materials: the carbon/epoxy and some metallic systems. J. Phys.: Condens. Matter **7**, 8759–8770 (1994)
9. For a review, see M. Sigalas, M.S. Kushwaha, E.N. Economou, M. Kafesaki, I.E. Psarobas, W. Steurer, Classical vibrational modes in phononic lattices: theory and experiment. Z. Kristallogr. **220**, 765–809 (2005)
10. For a recent review, see Y. Pennec, J. Vasseur, B. Djafari Rouhani, L. Dobrzynski, P.A. Deymier, Two-dimensional phononic crystals: examples and applications. Surf. Sci. Rep. **65**, 229 (2010)
11. E. Yablonovitch, Inhibited spontaneous emission in solid-state physics and electronics. Phys. Rev. Lett. **58**, 2059–2062 (1987)
12. J.D. Joannopoulos, R.D. Meade, J.N. Winn, *Molding the Flow of Light* (Princeton University Press, Princeton, 1995)

13. I.E. Psarobas, A. Modinos, R. Sainidou, N. Stefanou, Acoustic properties of colloidal crystals. Phys. Rev. B **65**, 064307 (2002)
14. R. Sainidou, N. Stefanou, A. Modinos, Formation of absolute frequency gaps in three-dimensional solid phononic crystals. Phys. Rev. B **66**, 212301 (2002)
15. T. Still, W. Cheng, M. Retsch, R. Sainidou, J. Wang, U. Jonas, N. Stefanou, G. Fytas, Simultaneous occurrence of structure-directed and particle-resonance-induced phononic gaps in colloidal films. Phys. Rev. Lett. **100**, 194301 (2008)
16. C. Croënne, E.J.S. Lee, H. Hu, J.H. Page, Band gaps in phononic crystals: generation mechanisms and interaction effects. AIP Adv. **1**, 041401 (2011)
17. Z. Liu, X. Zhang, Y. Mao, Y.Y. Zhu, Z. Yang, C.T. Chan, P. Sheng, Locally resonant sonic materials. Science **289**, 1734–1736 (2000)
18. M. Torres, F.R. Montero de Espinosa, D. Garcia-Pablos, N. Garcia, Sonic band gaps in finite elastic media: surface states and localization phenomena in linear and point defects. Phys. Rev. Lett. **82**, 3054 (1999)
19. M. Kafesaki, M.M. Sigalas, N. Garcia, Frequency modulation in the transmittivity of wave guides in elastic-wave band-gap materials. Phys. Rev. Lett. **85**, 4044 (2000)
20. A. Khelif, B. Djafari-Rouhani, J.O. Vasseur, P.A. Deymier, P. Lambin, L. Dobrzynski, Transmittivity through straight and stublike waveguides in a two-dimensional phononic crystal. Phys. Rev. B **65**, 174308 (2002)
21. A. Khelif, B. Djafari-Rouhani, J.O. Vasseur, P.A. Deymier, Transmission and dispersion relations of perfect and defect-contained waveguide structures in phononic band gap materials. Phys. Rev. B **68**, 024302 (2003)
22. A. Khelif, B. Djafari-Rouhani, V. Laude, M. Solal, Coupling characteristics of localized phonons in photonic crystal fibers. J. Appl. Phys. **94**, 7944–7946 (2003)
23. A. Khelif, A. Choujaa, B. Djafari-Rouhani, M. Wilm, S. Ballandras, V. Laude, Trapping and guiding of acoustic waves by defect modes in a full band-gap ultrasonic crystal. Phys. Rev. B **68**, 214301 (2003)
24. A. Khelif, A. Choujaa, S. Benchabane, B. Djafari-Rouhani, V. Laude, Guiding and bending of acoustic waves in highly confined phononic crystal waveguides. Appl. Phys. Lett. **84**, 4400 (2004)
25. S. Benchabane, A. Khelif, A. Choujaa, B. Djafari-Rouhani, V. Laude, Interaction of waveguide and localized modes in a phononic crystal. Europhys. Lett. **71**, 570 (2005)
26. Y. Pennec, B. Djafari Rouhani, J.O. Vasseur, H. Larabi, A. Khelif, A. Choujaa, S. Benchabane, V. Laude, Acoustic channel drop tunneling in a phononic crystal. Appl. Phys. Lett. **87**, 261912 (2005)
27. J.O. Vasseur, P.A. Deymier, G. Frantziskonis, G. Hong, B. Djafari Rouhani, L. Dobrzyński, Experimental evidence for the existence of absolute acoustic band gaps in two-dimensional periodic composite media. J. Phys.: Condens. Matter **10**, 6051 (1998)
28. J.O. Vasseur, P.A. Deymier, B. Chenni, B. Djafari-Rouhani, L. Dobrzynski, D. Prevost, Experimental and theoretical evidence for the existence of absolute acoustic band gaps in two-dimensional solid phononic crystals. Phys. Rev. Lett. **86**, 3012 (2001)
29. F.R. Montero de Espinosa, E. Jimenez, M. Torres, Ultrasonic band gap in a periodic two-dimensional composite. Phys. Rev. Lett. **80**, 1208 (1998)
30. H. Larabi, Y. Pennec, B. Djafari-Rouhani, J.O. Vasseur, Multicoaxial cylindrical inclusions in locally resonant phononic crystals. Phys. Rev. E **75**, 066601 (2007)
31. E.N. Economou, M.M. Sigalas, Classical wave propagation in periodic structures: cermet versus network topology. Phys. Rev. B **48**, 13434 (1993)
32. M.M. Sigalas, E.N. Economou, Attenuation of multiple-scattered sound. Europhys. Lett. **36**, 241 (1996)
33. M.S. Kushwaha, Stop-bands for periodic metallic rods: sculptures that can filter the noise. Appl. Phys. Lett. **70**, 3218 (1997)
34. P. Lambin, A. Khelif, J.O. Vasseur, L. Dobrzynski, B. Djafari-Rouhani, Stopping of acoustic waves by sonic polymer-fluid composites. Phys. Rev. E **63**, 066605 (2001)

35. J.O. Vasseur, P.A. Deymier, A. Khelif, P. Lambin, B. Djafari-Rouhani, A. Akjouj, L. Dobrzynski, N. Fettouhi, J. Zemmouri, Phononic crystal with low filling fraction and absolute acoustic band gap in the audible frequency range: a theoretical and experimental study. Phys. Rev. E **65**, 056608 (2002)
36. Z. Liu, C.T. Chan, P. Sheng, A.L. Goertzen, J.H. Page, Elastic wave scattering by periodic structures of spherical objects: theory and experiment. Phys. Rev. B **62**, 2446 (2000)
37. J.H. Page, A.L. Goertzen, S. Yang, Z. Liu, C.T. Chan, P. Sheng, in *Photonic Crystals and Light Localization in the 21st Century*, ed. by C.M. Soukoulis (Kluwer, Dordrecht, 2001), p. 59
38. M.S. Kushwaha, B. Djafari-Rouhani, Complete acoustic stop bands for cubic arrays of spherical liquid balloons. J. Appl. Phys. **80**, 3191 (1996)
39. R. Martinez-Sala, J. Sancho, J.V. Sanchez, V. Gomez, J. Llinares, F. Meseguer, Sound attenuation by sculpture. Nature **378**, 241 (1995)
40. D. Caballero, J. Sanchez-Dehesa, C. Rubio, R. Martinez-Sala, J.V. Sanchez-Perez, F. Meseguer, J. Llinares, Large two-dimensional sonic band gaps. Phys. Rev. E **60**, 6316 (1999)
41. Y. Pennec, B. Djafari-Rouhani, J. Vasseur, P.A. Deymier, A. Khelif, Tunable filtering and demultiplexing in phononic crystals with hollow cylinders. Phys. Rev. E **69**, 046608 (2004)
42. J.V. Sanchez-Pérez, D. Caballero, R. Mártinez-Sala, C. Rubio, J. Sánchez-Dehesa, F. Meseguer, J. Llinares, F. Gálvez, Sound attenuation by a two-dimensional array of rigid cylinders. Phys. Rev. Lett. **80**, 5325 (1998)
43. Y. Pennec, B. Djafari-Rouhani, J.O. Vasseur, P.A. Deymier, A. Khelif, Transmission and dispersion modes in phononic crystals with hollow cylinders: application to waveguide structure. Phys. Stat. Sol. (c) **1**(11), 2711–2715 (2004)
44. A. Sato, W. Knoll, Y. Pennec, B. Djafari-Rouhani, G. Fytas, M. Steinhart, Anisotropic propagation and confinement of high frequency phonons in nanocomposites. J. Chem. Phys. **130**, 111102 (2009)
45. R. Lucklum, J. Li, Phononic crystals for liquid sensor applications. Meas. Sci. Technol. **20**, 124014 (2009)
46. M. Ke, M. Zubtsov, R. Lucklum, Sub-wavelength phononic crystal liquid sensor. J. Appl. Phys. **110**, 026101 (2011)
47. M.S. Kushwaha, B. Djafari-Rouhani, Giant sonic stop bands in two-dimensional periodic system of fluids. J. Appl. Phys. **84**, 4677 (1998)
48. M.S. Kushwaha, B. Djafari Rouhani, L. Dobrzynski, Sound isolation from cubic arrays of air bubbles in water. Phys. Lett. A **248**, 252–256 (1998)
49. V. Leroy, A. Bretagne, M. Fink, H. Willaime, P. Tabeling, A. Tourin, Design and characterization of bubble phononic crystals. Appl. Phys. Lett. **95**, 171904 (2009)
50. T. Still, M. Oudich, G.K. Auernhammer, D. Vlassopoulos, B. Djafari-Rhouani, G. Fytas, P. Sheng, Soft silicone rubber in phononic structures: correct elastic moduli. Phys. Rev. B **88**, 094102 (2013)
51. C. Goffaux, J.P. Vigeron, Theoretical study of a tunable phononic band gap system. Phys. Rev. B **64**, 075118 (2001)
52. M. Torres, F.R. Montero de Espinosa, J.L. Aragon, Ultrasonic wedges for elastic wave bending and splitting without requiring a full band gap. Phys. Rev. Lett. **86**, 4282 (2001)
53. J.O. Vasseur, P.A. Deymier, M. Beaugeois, Y. Pennec, B. Djafari-Rouhani, D. Prevost, Experimental observation of resonant filtering in a two-dimensional phononic crystal waveguide. Z. Kristallogr. **220**, 824 (2005)
54. B. Djafari Rouhani, L. Dobrzynski, O. Hardouin Duparc, R.E. Camley, A.A. Maradudin, Sagittal elastic waves in infinite and semi-infinite superlattices. Phys. Rev. B **28**, 1711 (1983)
55. B. Djafari Rouhani, A.A. Maradudin, R.F. Wallis, Rayleigh waves on a superlattice stratified normal to the surface. Phys. Rev. B **29**, 6454 (1984)
56. Y. Tanaka, S. Tamura, Surface acoustic waves in two-dimensional periodic elastic structures. Phys. Rev. B **58**, 7958 (1998) and Acoustic stop bands of surface and bulk modes in two-dimensional phononic lattices consisting of aluminum and a polymer. Phys. Rev. B **60**, 13294 (1999)

57. T.-T. Wu, Z.-G. Huang, S. Lin, Surface and bulk acoustic waves in two-dimensional phononic crystals consisting of materials with general anisotropy. Phys. Rev. B **69**, 094301 (2004)
58. T.-T. Wu, L.-C. Wu, Z.-G. Huang, Frequency band-gap measurement of two-dimensional air/silicon phononic crystals using layered slanted finger interdigital transducers. J. Appl. Phys. **97**(094916) (2005)
59. S. Benchabane, A. Khelif, J.-Y. Rauch, L. Robert, V. Laude, Evidence for complete surface wave band gap in a piezoelectric phononic crystal. Phys. Rev. E **73**, 065601 (2006)
60. B. Bonello, C. Charles, F. Ganot, Velocity of a SAW propagating in a 2D phononic crystal. Ultrasonics **44**, 1259–1263 (2006)
61. S. Benchabane, O. Gaiffe, R. Salut, G. Ulliac, Y. Achaoui, V. Laude, Observation of surface-guided waves in holey hypersonic phononic crystal. Appl. Phys. Lett. **98**, 171908 (2011)
62. V. Laude, M. Wilm, S. Benchabane, A. Khelif, Full band gap for surface acoustic waves in a piezoelectric phononic crystal. Phys. Rev. E **71**, 036607 (2005)
63. B. Manzanares-Martinez, F. Ramos-Mendieta, Surface elastic waves in solid composites of two-dimensional periodicity. Phys. Rev. B **68**, 134303 (2003)
64. R. Sainidou, B. Djafari Rouhani, J.O. Vasseur, Surface acoustic waves in finite slabs of three-dimensional phononic crystals. Phys. Rev. B **77**, 094304 (2007)
65. A. Khelif, B. Aoubiza, S. Mohammadi, A. Adibi, V. Laude, Complete band gaps in two-dimensional phononic crystal slabs. Phys. Rev. E **74**, 046610 (2006)
66. J.O. Vasseur, P. Deymier, B. Djafari Rouhani, Y. Pennec, A.C. Hladky-Hennion, Absolute forbidden bands and waveguiding in two-dimensional phononic crystal plates. Phys. Rev. B **77**, 085415 (2008)
67. Y. Pennec, B. Djafari Rouhani, H. Larabi, J. Vasseur, A.C. Hladky-Hennion, Low frequency gaps in a phononic crystal constituted of cylindrical dots deposited on a thin homogeneous plate. Phys. Rev. B **78**, 104105 (2008)
68. T.T. Wu, Z.G. Huang, T.-C. Tsai, T.C. Wu, Evidence of complete band gap and resonances in a plate with periodic stubbed surface. Appl. Phys. Lett. **93**, 111902 (2008)
69. T.C. Wu, T.T. Wu, J.C. Hsu, Waveguiding and frequency selection of Lamb waves in a plate with a periodic stubbed surface. Phys. Rev. B **79**, 104306 (2009)
70. Y. Pennec, B. Djafari Rouhani, H. Larabi, A. Akjouj, J.N. Gillet, J.O. Vasseur, G. Thabet, Phonon transport and waveguiding in a phononic crystal made up of cylindrical dots on a this homogeneous plate. Phys. Rev. B **80**, 144302 (2009)
71. S. Mohammadi, A.A. Eftekhar, W.D. Hunt, A. Adibi, High-Q micromechanical resonators in a two-dimensional phononic crystal slab. Appl. Phys. Lett. **94**, 051906 (2009)
72. C.M. Reinke, M.F. Su, R.H. Olsson, I. El-Kady, Realization of optimal bandgaps in solid–solid, solid–air, and hybrid solid–air–solid phononic crystal slabs. Appl. Phys. Lett. **98**, 061912 (2011)
73. Z. Hou, F. Wu, Y. Liu, Phononic crystals containing piezoelectric material. Solid State Commun. **130**(11), 745–749 (2004)
74. X.-Y. Zou, Q. Chen, B. Liang, J.-C. Cheng, Control of the elastic wave bandgaps in two-dimensional piezoelectric periodic structures. Smart Mater. Struct. **17**, 015008 (2008)
75. Y.-Z. Wang, F.-M. Li, K. Kishimoto, Y.-S. Wang, W.-H. Huang, Elastic wave band gaps in magnetoelectrostatic phononic crystals. Wave Motion **46**(47) (2009)
76. J.-F. Robillard, O. Bou Matar, J.O. Vasseur, P.A. Deymier, M. Stippinger, A.C. Hladky-Hennion, Y. Pennec, B. Djafari-Rouhani, Tunable magnetoelastic phononic crystals. Appl. Phys. Lett. **95**, 124104 (2009)
77. K. Bertoldi, M.C. Boyce, Mechanically triggered transformations of phononic band gaps in periodic elastomeric structures. Phys. Rev. B **77**, 052105 (2008)
78. Z.-G. Huang, T.-T. Wu, Temperature effect on the bandgaps of surface and bulk acoustic waves in two-dimensional phononic crystals. IEEE Trans. Ultrason. Ferroelectr. Freq. Contr. **52**, 365 (2005)
79. A. Sato, Y. Pennec, N. Shingne, T. Thurn-Albrecht, W. Knoll, M. Steinhart, B. Djafari-Rouhani, G. Fytas, Tuning and switching the hypersonic phononic properties of elastic impedance contrast nanocomposites. ACS Nano **4**, 3471 (2010)

80. M. Maldovan, E.L. Thomas, Simultaneous localization of photons and phonons in two-dimensional periodic structures. Appl. Phys. Lett. **88**, 251907 (2006)
81. S. Mohammadi, A.A. Eftekhar, A. Khelif, A. Adibi, Simultaneous two-dimensional phononic and photonic band gaps in opto-mechanical crystal slabs. Opt. Express **18**, 9164 (2010)
82. Y. Pennec, B. Djafari-Rouhani, E.H. El Boudouti, C. Li, Y. El Hassouani, J.O. Vasseur, N. Papanikolaou, S. Benchabane, V. Laude, A. Martinez, Simultaneous existence of phononic and photonic band gaps in periodic crystal slabs. Opt. Express **18**, 14301 (2010)
83. M. Eichenfield, J. Chan, R.M. Camacho, K.J. Vahala, O. Painter, Optomechanical crystals. Nature **462**, 78 (2009)
84. Y. Pennec, B. Djafari-Rouhani, C. Li, J.M. Escalante, A. Martinez, S. Benchabane, V. Laude, N. Papanikolaou, Band gaps and cavity modes in dual phononic and photonic strip waveguides. AIP Adv. **1**, 041901 (2011)
85. S. Amoudache, Y. Pennec, B. Djafari Rouhani, A. Khater, R. Lucklum and R. Tigrine, Simultaneous sensing of light and sound velocities of fluids in a two-dimensional phoXonic crystal with defects, J. Appl. Phys. **115**, 134503 (2014)
86. R. Lucklum, Y. Pennec, A. Kraych, M. Zubstov, B. Djafari Rouhani, in *Phoxonic Crystal Sensor*, SPIE Photonics Europe, Photonic Crystal Materials and Devices X, Brussels, Belgium, April 16–19, 2012, Proc. SPIE-Int. Soc. Opt. Eng., 8425 (2012) 84250N-1-8, ISBN 978-0-8194-9117-6. doi:10.1117/12.922553
87. S. Yang, J.H. Page, Z. Liu, M.L. Cowan, C.T. Chan, P. Sheng, Focusing of sound in a 3D phononic crystal. Phys. Rev. Lett. **93**, 024301 (2004)
88. K. Imamura, S. Tamura, Negative refraction of phonons and acoustic lensing effect of a crystalline slab. Phys. Rev. B **70**, 174308 (2004)
89. X. Zhang, Z. Liu, Negative refraction of acoustic waves in two-dimensional phononic crystals. Appl. Phys. Lett. **85**, 341 (2004)
90. A. Sukhovich, L. Jing, J.H. Page, Negative refraction and focusing of ultrasound in two-dimensional phononic crystals. Phys. Rev. B **77**, 014301 (2008)
91. A. Sukhovich, B. Merheb, K. Muralidharan, J.O. Vasseur, Y. Pennec, P.A. Deymier, J.H. Page, Experimental and theoretical evidence for subwavelength imaging in phononic crystals. Phys. Rev. Lett. **102**, 154301 (2009)
92. J. Bucay, E. Roussel, J.O. Vasseur, P.A. Deymier, A.-C. Hladky-Hennion, Y. Pennec, K. Muralidharan, B. Djafari-Rouhani, B. Dubus, Positive, negative, zero refraction, and beam splitting in a solid/air phononic crystal: theoretical and experimental study. Phys. Rev. B **79**(214305) (2009)
93. P.E. Hopkins, C.M. Reinke, M.F. Su, R.H. Olsson III, E.A. Shaner, Z.C. Leseman, J.R. Serrano, L.M. Phinney, I. El-Kady, Reduction in the thermal conductivity of single crystalline silicon by phononic crystal patterning. Nano Lett. **11**, 107 (2011)

Chapter 3
The Three-Dimensional Phononic Crystals

Badreddine Assouar, Rebecca Sainidou, and Ioannis Psarobas

3.1 Introduction

The conceptual impact of crystalline matter in condensed matter physics has set the framework of understanding the morphological laws of crystals in terms of atomic dynamics. A definite state of atomic equilibrium for an atomic ensemble will necessitate an atomic arrangement in a regular system of points so that the nature of atoms composing the crystal determines under given external conditions their metric disposition, which is summed up in the 230 groups of symmetry. The crystals occurring in nature display the possible types of symmetry enclosed in an abundance of different forms as a result of constitution and environment. The dynamics of the crystal lattice is also responsible for the crystal's physical behavior initiating a connection of outmost importance between quantum mechanics and symmetry. Symmetry plays a great role in ordering the atomic and molecular spectra, for the understanding of which the principles of quantum physics provide the key. An exact classical analogue of a natural crystal is a three-dimensional (3D) phononic

B. Assouar (✉)
International Joint Laboratory (CNRS - Georgia Institute of Technology),
777 Atlantic Dr. NW, Atlanta, GA 30332-0250, USA

Institut Jean Lamour, UMR 7198, CNRS - Nancy University,
Boulevard des Aiguillettes, 54506 Vandoeuvre lès Nancy, France
e-mail: Badredine.Assouar@gatech.edu

R. Sainidou
Laboratoire Ondes et Milieux Complexes UMR CNRS 6294, Université du Havre, 75 rue Bellot,
76600 Le Havre, France
e-mail: sainidor@univ-lehavre.fr

I. Psarobas
Section of Solid State Physics, University of Athens, Panepistimioupolis,
GR-157 84 Athens, Greece
e-mail: ipsarob@phys.uoa.gr

A. Khelif, A. Adibi (eds.), *Phononic Crystals*, DOI 10.1007/978-1-4614-9393-8_3

crystal [28, 35], a composite structure of identical macroscopic scatterers arranged in a 3D crystal lattice points embedded in a homogeneous medium enclosure. Its vibrational spectra is determined by means of multiple elastic-acoustic scattering, a mechanism triggered by interaction with propagating elastic waves of wavelengths comparable to the characteristic size of the crystal (i.e., its lattice constants or else the size of its structured scatterers). It is then evident that by analogy to phononic crystals of lower dimensions, 3D phononic crystals will exhibit as well frequency spectral gaps and in particular omnidirectional elastic wave cutoff behavior [53]. In such a case a well-established topology argument can be stated:

Omnidirectional phononic frequency gaps are favored by cermet topology (a structure of non-touching periodic scatterer arrangement) as opposed to photonic crystals which favor network topology [53].

Since there are striking analogies between the propagation of electrons in crystalline matter and phononic crystals [28], a great variety of multiple-scattering methods (MS) as well as other traditional methods originally developed for electronic-structure calculations have been transferred to the field of acoustics. Thus, the frequency dispersion along the symmetry points in k-space of a phononic crystal, or else known as its phononic band structure, can be calculated by the plane-wave method (PW) [19, 52] and by a more efficient MS method [16] coming from the traditional Korringa–Kohn–Rostoker (KKR) method. On the other hand, well-established purely numerical methods such as the finite-difference-time domain (FDTD) [10], appeared in order to add to the band structure calculations, the computation of the scattering properties (transmission, reflection, and absorption) of finite slabs of a 3D phononic crystal, although it appears to be more effective in 2D problems [62]. In 3D phononic problems, however, the PW method or the FDTD and other finite difference techniques are insufficient to gain physical insight, while in many situations they are computationally cumbersome. The layer-multiple-scattering (LMS) method [37, 46] is a very powerful computational methodology for studying three-dimensional (3D) phononic crystals, a method similar to the layer-KKR developed for low energy electron diffraction (LEED) [27, 28, 33]. An important aspect of the method is that, contrary to traditional band-structure or time-domain methods, it solves the wave equation of elasticity in frequency domain and thus one can allow the elastic coefficients of any of the constituent materials to depend on the frequency, including also absorptive losses of any kind. Besides the complex phononic band structure of an infinite crystal, associated with a given crystallographic plane, the LMS method can also provide the transmission, reflection, and absorption coefficients of an elastic wave of any polarization incident at a given angle on a finite slab of the crystal and, therefore, it can describe an actual transmission experiment. A further advantage of the method is that it does not require periodicity in the direction perpendicular to the layers, which must only have the same 2D periodicity. Therefore a number of interesting configurations, such as planar defects [38], heterostructures [36], stacking disorder [47], phononic crystal slabs on homogeneous plates and semi-infinite substrates [41] can be treated in a more or less straightforward manner. Finally, the LMS method can incorporate multiple-scattering Green's-function techniques [44] that allow one to calculate the (local) density of states of the elastic field, as well as to treat defects [49], disorder [47], and the system response to an external perturbation [40], in general. The LMS method has been proven very efficient for phononic crystals of spherical particles. It applies to any material type combination (fluid or solid) and with the same efficiency, it can treat, also, coated spheres and generally spherical particles consisting of an arbitrary number of concentric spherical shells by a powerful recursive algorithm [46]. Recently, the method has been extended to scatterers of arbitrary shape with the addition of the extended-boundary-condition (EBC) technique [9, 15]. Thus, in addition to the physical clarity and computational efficiency, the generalized LMS method would offer an efficient and versatile alternative to treat a variety of complex phononic structures. Most aspects of the method will therefore emerge throughout this chapter as the main tool to unfold the physics and applications related to 3D phononic crystals.

We close this part by noting a few words about experiments concerning 3D phononic structures. Although their fabrication is, in general, more difficult and delicate [12, 18, 26] than in the corresponding 2D case, new chemical techniques together with non-destructive measurement methods such as Brillouin light scattering, allowed recently the study of phononic nanostructures operating in hypersonic (GHz) regime, including the case of complex in structure or composition spherical scatterers [5, 57–60]. On the other hand, 3D phononic structures on the macroscale continue to be fabricated and studied experimentally being in very good agreement with LMS theoretical results [25, 30, 61, 65].

3.2 Phononic Lattices

3.2.1 Multiple Scattering and the LMS Method

The essence of any multiple-scattering theory consists in taking into account, explicitly and in an exact and self-consistent manner, all scattering processes occurring in an assembly of scatterers centered on sites $\mathbf{R}_n$ in a homogeneous medium. An outgoing wave from the n'th scatterer can reach the nth scatterer directly, or indirectly after scattering any number of times by any number of scatterers (including those at $\mathbf{R}_n$ and $\mathbf{R}_{n'}$). We can then generalize the scattering process at $\mathbf{R}_n$ by treating it in two distinct stages. The first stage is described by the individual scattering properties of each scatterer via its scattering matrix, $\mathbf{T}$, while the second expresses the sum of the contributions to the coefficient of an incident wave on $\mathbf{R}_n$ from all possible scattering paths originating from the outgoing

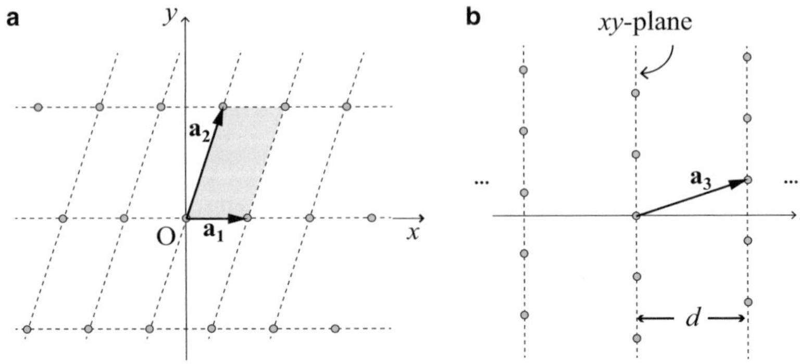

Fig. 3.1 (**a**) A layer of the crystal consisting of a 2D array of scatterers on *xy*-plane. (**b**) The phononic crystal viewed as a succession of identical layers, along *z*-direction

waves from the other scatterers. The latter is achieved through the introduction of structural Green's functions elements, $\mathbf{\Omega}$. The reader may consult [28] for an extended presentation on the matter. Some elements are also given in Sect. 3.3.2.1.

When scatterers are arranged in 3D periodic lattices an efficient and convenient way to perform the multiple scattering is by using the LMS technique. The 3D crystal is viewed as a succession of identical characteristic planes of spheres (layers) of a given crystallographic direction (let's denote it by *z*-axis direction); each layer has its scatterers arranged in a 2D lattice defined by lattice vectors $\mathbf{a}_1$, $\mathbf{a}_2$ (see Fig. 3.1a), the same for every layer, and is connected to its neighboring layers by a translation lattice vector $\mathbf{a}_3$ (see Fig. 3.1b), with $d = a_{3z}$ being the distance between consecutive planes of spheres, i.e. the thickness of the layer. By combining several such layers (let's denote their number by N) a finite part of a 3D phononic crystal (slab), of thickness Nd, is formed. We assume a plane wave of angular frequency ω, incident on the slab, with its wavevector component parallel to xy plane, $\mathbf{k}_{\|}$, being a conserved quantity. The LMS finds the exact elastic wave solution layer by layer through the calculation of the four transmission and reflection matrices $\mathbf{Q}$ [37, 46] for each individual layer, by performing the multiple scattering on it; then, it matches the different layers together by combining their Q-matrices to find the corresponding matrices of the whole composite system. Their knowledge permits the evaluation of the transmittance, reflectance, and absorbance of the slab.

An example of the assembly of spherical scatterers of radius S filling the sites $\mathbf{R}_n$ of a 2D lattice, in particular a square lattice of lattice constant a_0, is given in the left panel of Fig. 3.2 [42]. Parts (a)–(d) compare theory with experiment [12, 26] to a more than a satisfactory level and thus prove one of the important assets of the LMS method. In Fig. 3.2e, f one can see the evolution of the transmittance through a finite slab of the crystal consisting of N such layers for a longitudinal or transverse wave incident normally on it, with the thickness Nd of the slab. The thicker the slab is, the clearer the regions of frequency where the transmittance practically vanishes, become. For sufficiently thick samples, these regions coincide with the frequency band gaps of the corresponding infinite crystal.

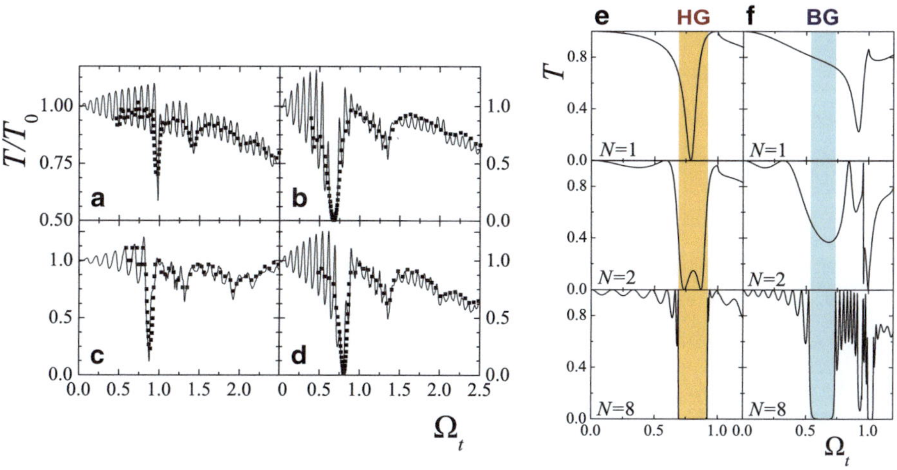

Fig. 3.2 After [42], normalized transmittance for a longitudinal elastic wave incident normally on a square lattice of (**a**) glass spheres (S=0.56 mm, a_0=2.63 mm), (**b**) lead spheres (S=0.60 mm, a_0=2.63 mm), (**c**) steel spheres (S=0.585 mm, a_0=3.95 mm), (**d**) steel spheres (S=0.585 mm, a_0=2.63 mm), in the middle of a polyester slab, 7 cm thick, immersed in water. *Solid lines*: theoretical results; *squares*: experimental data. In the *right panel*, we present the transmittance for a longitudinal (**e**) and a transverse (**f**) elastic wave incident normally on a slab of N layers of an fcc crystal ($a = a_0\sqrt{2}$ 3.72 mm), parallel to the (001) surface, of steel spheres (S=0.585 mm) in an infinite, nonabsorbing polyester matrix. Here $\Omega_t = \omega a_0/(2\pi c_t)$ and the *shaded areas* denote the gaps of the corresponding infinite crystal, discussed in Fig. 3.3

The complex band structure can be calculated by viewing the infinite crystal as a sequence of layers, extending over all space from $z = -\infty$ to $z = \infty$. After imposing periodic boundary conditions (Bloch's theorem), for a given $\mathbf{k}_\parallel$ we determine $k_z(\omega)$ of the 3D wavevector $\mathbf{k} = [\mathbf{k}_\parallel,\ k_z(\omega)]$. This is presented, following Fig. 3.2, in Fig. 3.3a. In general, two types of frequency gaps exist:

(a) the *Bragg gap* (BG), opening up at the Brillouin zone (BZ) boundaries ($k_z d/\pi = 0, \pm 1$) and originating from the destructive interference of the waves scattered by the periodically arranged layers along a given crystallographic direction. In Fig. 3.3a a BG gap for transverse bands appears about $\omega a/c_t \sim 2\pi$, as shown in Fig. 3.3b. Here and throughout this chapter $c_{l(t)}$ denotes the longitudinal (transverse) elastic velocity in the host matrix, unless otherwise stated.

(b) the *hybridization gap* (HG), opening up when two bands of the same symmetry cross each other; usually, the most narrow of them originates from virtual bound states localized in the building units (spheres) of the crystal. In Fig. 3.3a the frequency gap for longitudinal bands is a HG, as explained in Fig. 3.3c, d. When a single steel sphere is embedded in polyester, there exists, at $\Omega_t = 0.51$, a virtual bound state (a resonant state of finite lifetime) of the dipole-type displacement field, localized in its interior, as confirmed by calculating

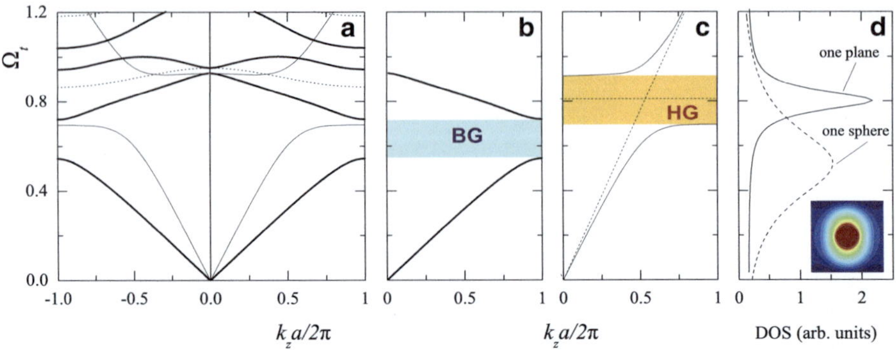

Fig. 3.3 (**a**) After [42]: The phononic frequency band structure normal to the (001) surface of an fcc crystal (a=3.72 mm) of steel spheres (S=0.585 mm) in nonabsorbing polyester. The *thin (thick) solid lines* refer to longitudinal (transverse) bands and the *dotted lines* refer to deaf bands. For the system in (**a**), we give a schematic representation of (**b**) a Bragg and (**c**) a hybridization-induced gap with *solid (dotted) lines* showing the hybridized (unhybridized) bands after (before) interaction. In (**d**) the DOS for one sphere (*dashed line*) and one fcc (001) plane of spheres (*solid line*) for the system under study. In the inset the eigenmode map for a single sphere at resonance, showing strong localization (*red color*) in its interior

the density of states (DOS) by $\frac{1}{\pi}\partial_\omega[\mathrm{ImTr}\ \ln(\mathbf{I}+\mathbf{T})]$ [44]. Next, putting the spheres together to form a layer results in a collective virtual bound state of the longitudinal displacement field, for $\mathbf{k}_\parallel$=0, which peaks about the said plane but falls to a much lower value away from it, originating from the interaction of the individual resonant states of the spheres. The corresponding DOS for one layer (an fcc (001) plane of spheres) given now by the formula $\frac{1}{\pi}\partial_\omega\,\{\mathrm{Im}\,[\mathrm{Tr}\ \ln(\mathbf{I}+\mathbf{T})-\mathrm{Tr}\ \ln(\mathbf{I}-\mathbf{T}\boldsymbol{\Omega})]\}$ [44] confirms its existence at $\Omega_t=0.80$. Therefore, virtual bound states on neighbor planes of spheres will couple weakly with each other, resulting in a relatively flat band as shown schematically in Fig. 3.3c.

It is worthnoting that a HG manifests itself as a dip in the transmission spectrum of a slab even for very thin slabs (monolayers) (see Fig. 3.2e) and does not require any periodicity to exist [57], while a BG appears in the spectrum as a dip progressively by increasing the thickness Nd of the slab, and becomes evident only for sufficiently thick slabs (see Fig. 3.2f).

3.2.2 Formation of Omnidirectional Frequency Gaps, Attenuation, and Tunneling

One of the merits of the LMS method is the reduced **k** zone scheme, a surface Brillouin zone (SBZ) which is completely equivalent to the commonly used bulk Brillouin zone (BZ), in the sense that a point in one of them lies also in the other

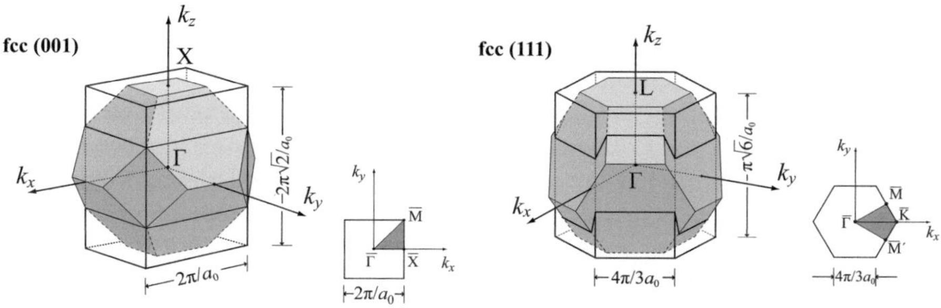

Fig. 3.4 The reduced **k** zones associated with the fcc (001) and (111) crystallographic planes and the corresponding SBZ. The bulk BZ is also shown for comparison

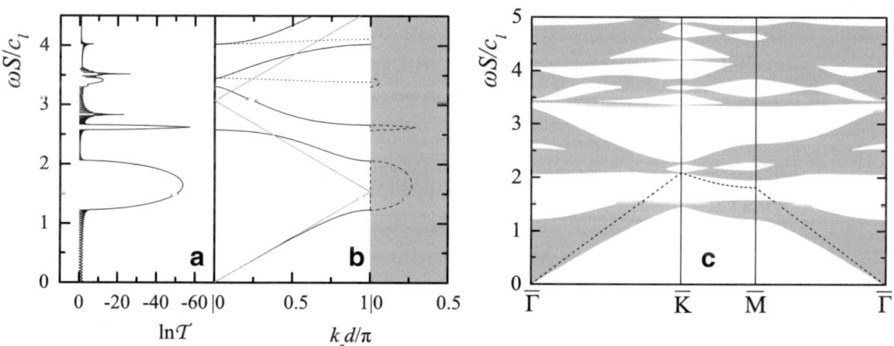

Fig. 3.5 After [45]: (**a**) Transmittance of a slab consisting of $N = 32$ fcc (111) planes of touching silica spheres in air, at normal incidence. (**b**) The complex phononic band structure normal to the (111) plane of the corresponding infinite crystal. The *black solid/dotted lines* are bands with Λ_1/Λ_3 symmetry, respectively. The *gray lines* correspond to the effective-medium approximation. Over the frequency gaps, we show by *broken lines* the bands with the smallest imaginary part. The imaginary part is shown in the *shaded region*. (**c**) Projection of the phononic band structure of an fcc crystal of touching silica spheres in air (volume filling fraction $f = 74\%$) on the SBZ of the fcc (111) surface (see Fig. 3.4), along high symmetry lines. Propagating waves in the air about a slab of the crystal exist for frequencies above a threshold value (a function of $\mathbf{k}_\parallel$) $\omega_{\text{inf}} = c_l|\mathbf{k}_\parallel|$ denoted by the *dashed line*

or differs by a vector of the 3D reciprocal lattice (see Fig. 3.4). A full multiple-scattering calculation by means of the LMS method for an fcc crystal of touching silica spheres in air is depicted in Fig. 3.5, where transmission, complex band structure, and the projection on the SBZ of a (111) surface reveal an omnidirectional phononic frequency gap. In this case, together with an extended study of phononic colloidal crystals [39], one can discover interesting symmetry properties of the bands, as well as a clear physical explanation on the origin of hybridization gaps [45, 60].

Absolute spectral gaps do not occur easily in 3D solid phononic crystals. Elastic waves in a 3D solid phononic crystal have longitudinal and transverse components, and corresponding gaps must overlap, for any direction of propagation, if an absolute gap is to appear. One may think that this is not likely in view of the different propagation velocities of longitudinal and transverse waves. However, under specific conditions, absolute gaps do occur in 3D solid phononic crystals. It has been established that periodic binary composites consisting of nonoverlapping high-density spheres in a low-density host material [17], as well as three-component phononic crystals consisting of coated spheres inside an embedding medium [25], exhibit absolute phononic gaps. The physical origin of the widest of these gaps lies beyond the Bragg gaps at the BZ boundaries: it is a HG. In reality, there are also bands originating from the resonant elastic modes of the individual scatterers: resonant states on neighboring scatterers that couple weakly with each other, resulting in relatively narrow bands, as explained in Fig. 3.3. These bands hybridize with continuum bands of an almost free propagation in an effective homogeneous elastic medium [37, 39, 43]. The opening of the hybridization gap is favored by an increased volume filling fraction, but at the same time it is compromised by the widening of the resonance bands which are also favored by an increased value of the same property of the crystal. This compromising effect will be seized by introducing a mild dissipation on the scatterers of the crystal [34]. In Fig. 3.6, we scan for absolute frequency gaps in a phononic crystal consisting of steel spheres in a polyester matrix with different cubic, tetragonal and in between cases of family arrangements. We show the ratio of the gap width to its respective midgap frequency, $\Delta\omega_G/\overline{\omega}_G$, against the volume fraction f occupied by the spheres. The filling fractions of the above investigation maximize the omnidirectional gap width, at high values of f but below $f_{\max}$. If we add to the picture a compact arrangement such as the hcp, which has the same $f_{\max} = 74\,\%$ as the fcc, and the Diamond structure which is not a compact structure, we can draw the following conclusion:

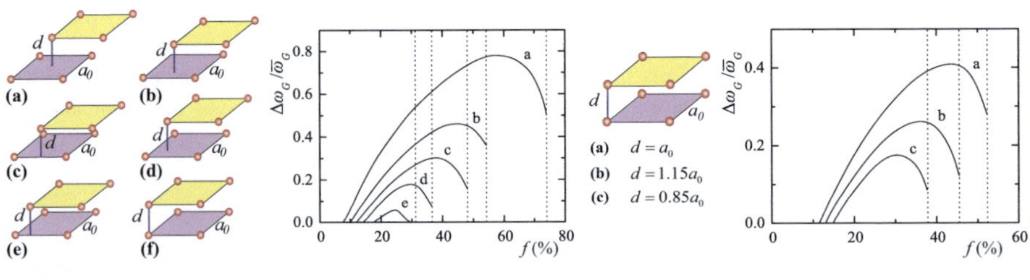

Fig. 3.6 After [43], on the *left panel*, absolute frequency gaps in phononic crystals of steel spheres, centered at the sites of various Bravais lattices specified by the primitive vectors $\mathbf{a}_1 = a_0\,(1, 0, 0)$, $\mathbf{a}_2 = a_0\,(0, 1, 0)$, and $\mathbf{a}_3 = a_0\,(\alpha, \beta, \sqrt{2}/2)$, in a polyester matrix. (**a**) $\alpha = 0.5,\ \beta = 0.5$ (fcc lattice). (**b**) $\alpha = 0.5,\ \beta = 0.25$. (**c**) $\alpha = 0.5,\ \beta = 0$. (**d**) $\alpha = 0.5,\ \beta = 0.5$. (**e**) $\alpha = 0.25,\ \beta = 0.5$. The *dotted lines* (normal on the abscissa) indicate the values of $f_{\max}$ corresponding to these structures. For $\alpha = \beta = 0$ (st lattice, $f_{\max} = 26.2\,\%$) no absolute gap appears. On the right, we present a family of various tetragonal lattices with different interplanar distances d. Note that the sc lattice is the special case when $d = a_0$

Solid–Solid 3D phononic crystals of large density contrast with respect to their embedding medium favor compact structures for the formation of wide absolute spectral gaps. Thus, the geometry of the structure is the most important issue for the formation of omnidirectional phononic gaps in 3D [43].

Damping is an interesting tool in further engineering phononic structures [14] to comply with specific needs in complex applications. An interesting example is the case of an fcc crystal of close-packed rubber spheres in air (see Fig. 3.7). In such a system [34] the induced resonant absorption on the scatterers, and not the host, serves as a cutoff to disturbing resonant states on the frequency spectrum of the phononic crystal, which spoil the formation of clear omnidirectional and partial stop-gaps. This might be extremely useful in relation to waveguiding applications.

Finally, an extended study of finite slabs of 3D phononic crystals with tungsten carbide (or steel) beads embedded in epoxy [30] provided interesting evidence of ultrasonic tunneling and anomalous dispersion of ultrasound near the band edges. Conclusively, in the absence of absorption, the ultrasound tunneling time was found to be independent of the phononic crystal thickness, with $\Delta\omega_{gap}\, t_{tun} \sim 1$ in the middle of the gap (where t_{tun} is the tunneling time and $\Delta\omega_{gap}$ is the width of the phononic band gap).

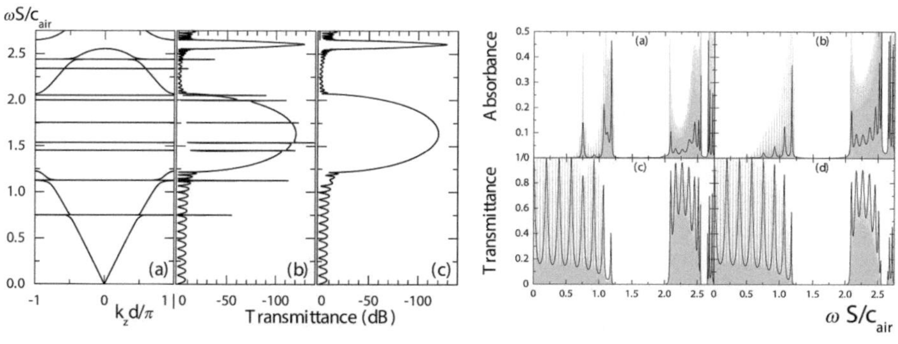

Fig. 3.7 On the *left*: The sonic band structure at the center of the SBZ of a (111) surface of an fcc crystal (of lattice constant a) of close-packed lossless rubber spheres in air. The corresponding transmittance curve of a slab of 16 layers parallel to the same surface is given in the middle. The third graph from the *left* is the same transmittance curve but with spheres of a low viscous level. d is the distance between successive (111) planes of the fcc crystal under consideration. On the *right*: Absorbance and transmittance curves of slabs of the rubber sonic crystal consisting of 8 (*left column*) and 32 (*right column*) planes of spheres, respectively. The *black line* (*shaded curve*) corresponds to the low (high) viscous level. (After [34])

3.2.3 Next-Generation 3D Phononic Structures

The LMS method has been recently extended to nonspherical particles, which are versatile building units for designing functional phononic structures for polarization-selective applications as well as novel acousto-optic devices [9]. A detailed analysis of transmission spectra of such finite phononic crystal slabs by reference to corresponding complex-band-structure and density-of-states diagrams corroborates that nonspherical particles provide an additional degree of freedom for tailoring the modes of the elastic field. There are many applications expected from the effect of distortion of spherical scatterers, an example being single-mode acoustic waveguides based on weakly coupled defects in a phononic band gap crystal. The high degeneracy of the modes of a single spherical defect would imply a large number of bands in the same frequency region within the gap for a chain of such defects. By deforming the spherical shape, a single band could split off, thus ensuring single-mode operation in a given range of frequencies. In Fig. 3.8 we show an example of model phononic crystal consisting of PMMA oblate spheroidal particles.

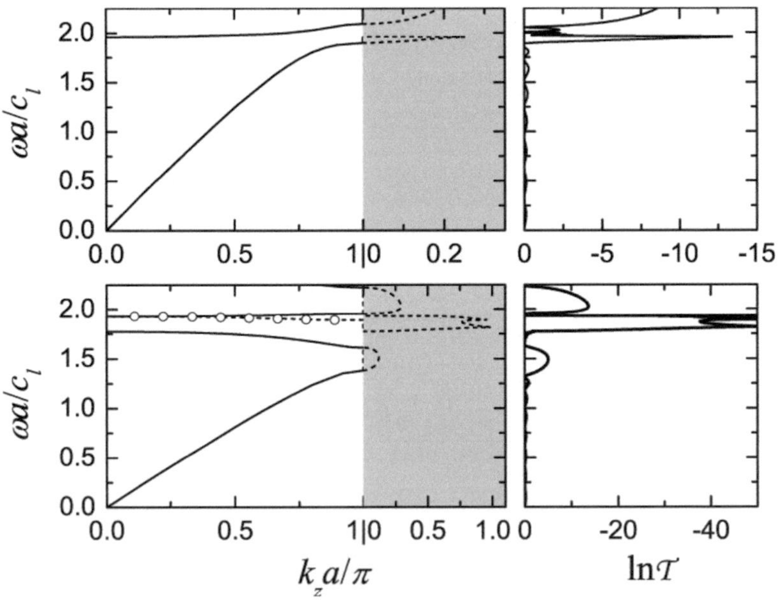

Fig. 3.8 After [9]: (*Left panels*) The phonic band structure of an infinite sc crystal, with lattice constant a, of PMMA oblate spheroidal particles, with A = 0.88a and B = 3A/4, in silicon, along its [001] direction. Over the frequency gaps, we show by *broken lines* the corresponding complex bands with the smallest in magnitude imaginary part of k_z. The imaginary part is shown in the *shaded region*. (*Right panels*) Transmittance of a slab of eight (001) planes of the above crystal at normal incidence for longitudinal (*top*) and transverse (*bottom*) incident waves

Classical wave transport in periodic media can provide the means to control light, sound or both with development of phoXonic crystals [31], a special class of dual spectral-gap materials which can integrate the management of sound, light and heat in a versatile manner. Such next-generation structures can operate as dual sensors and they can modulate localized photonic states by hypersound [32, 40] (acousto-optic interactions) or/and create phononic states through optomechanics [6]. In particular, the acousto-optic interaction realized in the merging fields of nanophoTonics and nanophoNonics could lead to unprecedented control of light and sound in very small regions of space [32, 40]. In the regime of inelastic light scattering by sound, one can have phonon-assisted light emission, control of light speed (delay-storage) by stimulated Brillouin scattering and the realization of highly sensitive dual phoXonic sensors. Although it is quite an easy task to realize 1D and 2D phoXonic crystals with omnidirectional spectral gaps, 3D phoXonic crystals were highly unlike to construct with dual functionality a few years ago, because of the different topology arguments that favor photonic and phononic structures, respectively. Nevertheless, in Fig. 3.9 we present an example of a 3D phoXonic crystal with predicted dual functionality. The phoXonic crystal consists of Au nanospheres in epoxy and for the sc structure at filling fractions above 40 % exhibits an absolute photonic spectral gap of relative width 15 % around the telecom 1.55 μm, as well as an absolute phononic gap of 53 % around the 2 GHz hypersonic frequency.

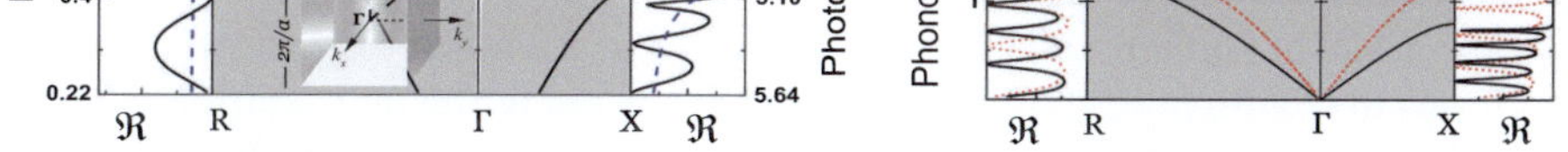

Fig. 3.9 After [31]: (*Left panel*) Complex photonic band structure and reflectance, ℜ, of an *sc* crystal with lattice constant $a = 480$ nm of Au spheres with diameter 460 nm in an epoxy matrix, along the ΓX and ΓR directions (see *inset*). The *white lines* show the imaginary part of the wave vector. The calculation was done with the experimental dielectric function of Au, which includes absorption. Next to the band diagrams we display the corresponding reflection spectra of (001) and (111) slabs of the crystal, five-layers thick, together with the reflectivity of the semi-infinite crystal (*dashed lines*). The extent of the absolute gap is marked by the *horizontal lines*. (*Right panel*) Phononic band structure and reflectance, ℜ, of the same crystal along the ΓX and ΓR directions. The *solid and dashed lines* represent doubly degenerate (shear) and non-degenerate (compressional or deaf) bands, respectively. Next to the band diagrams we display the corresponding reflection spectra of (001) and (111) slabs of the crystal, five-layers thick. Shear (compressional) waves correspond to *solid (dashed) lines*. The extent of the absolute gaps is marked by the *horizontal lines*

3.3 Imperfect Phononic Structures: From Periodicity to Disorder

Deviations from periodicity are responsible for several interesting phenomena already encountered in solid state physics and electromagnetic theory, such as Anderson localization of electrons in solids [2], waveguiding, introduction of impurity states within gaps and tunneling, etc. Based on the same principles they are expected, by analogy, to be also found in the case of elastic waves, although the full vectorial character of the elastic problem will add some complexity in the solution of the system.

3.3.1 Layered Non-periodic Heterostructures

As already stated (see Sect. 3.2.1), a perfectly periodic 3D phononic crystal can always be seen as a succession of identical characteristic planes of spheres (layers) of a given crystallographic direction (z-axis), each layer having its scatterers arranged in a 2D lattice (see Fig. 3.1), the same for every layer. Therefore, the simplest way to deviate from periodicity is by letting each layer have its own characteristics: for instance, the size or the material of the spheres and/or the layer thickness, d.

Periodicity can then be destroyed along z-direction (a) *locally* by just introducing some planar (layer) defects [38], (b) *progressively* by changing the layers' characteristics in a smooth manner (following a gradient law [36], or (c) even completely *randomly* by applying a random distribution of size disorder or position disorder (with the fcc stacking faults being the most famous example) [47]. In all these cases periodicity still remains on the 2D lattices parallel to xy plane. We will present them in detail in the next sections.

The LMS is perfectly adapted for the study of such heterostructures, since it performs the multiple scattering on each layer separately, the only restriction being that all layers must have the same 2D periodicity. Therefore, the evaluation of the transmittance, reflectance or absorbance of the composite slab, periodic or not, along z-direction, can be carried out, using the computer code of [46] (all the results shown in this section are obtained by this code).

3.3.1.1 Planar Defects

We take as an example a pure (perfect) crystal consisting of nonoverlapping lead spheres of radius $S = 0.25a$ centered on the sites of an fcc lattice of lattice constant a, embedded in epoxy. An absolute frequency gap extends between $\omega a/c_t = 5.29$ and 7.38. We will show how a slight deviation from periodicity (one impurity plane, i.e. a planar defect, different from the rest) changes dramatically the transmittance

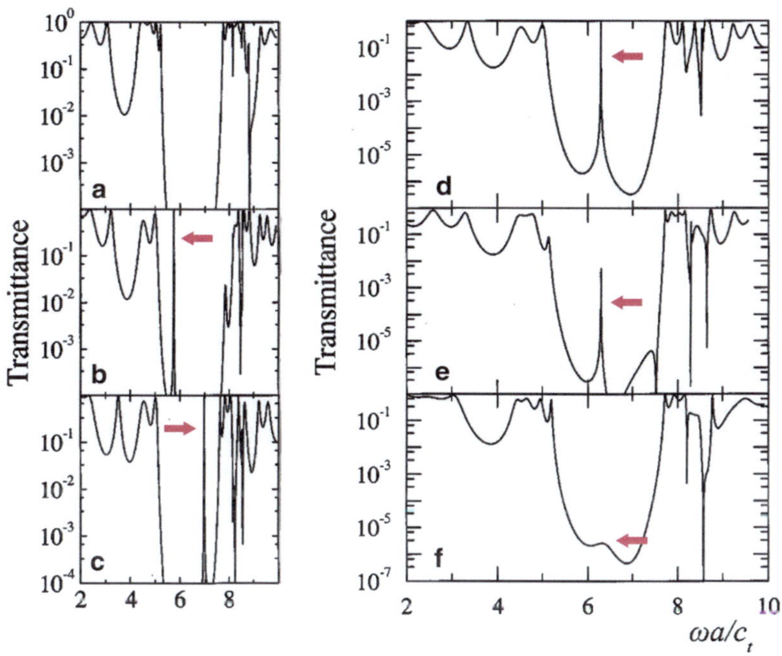

Fig. 3.10 After [38]: Transmittance of transverse elastic waves incident normally on a slab of an fcc crystal of lead spheres in epoxy. The slab consists of five planes of spheres parallel to the (001) surface. *Left*: The spheres have a radius $S = 0.25a$, except those of the middle plane which have a different radius S_i [(**a**): $S_i = S$, (**b**): $S_i = 0.8S$, (**c**): $S_i = 0.6S$]. *Right*: The spheres have a radius $S = 0.25a$, except those of (**d**) the middle plane, or of (**e**) the second plane from the surface, or of (**f**) the surface of the slab, which have $S_i = 0.7S$. *Red arrows* show the transmission resonance due to impurity planes

of elastic waves incident on a slab of the crystal. A complete study can be found in [38]. When a plane of impurity spheres of radius S_i, bigger or smaller than those of the spheres of the other planes, replaces one of the five (001) fcc planes of a finite slab of such a crystal (see Fig. 3.10a), modes of vibration of the elastic field, localized on this plane, are introduced at frequencies within the frequency gap of the corresponding pure phononic crystal. These show up as sharp resonances in the transmittance of elastic waves incident on a slab of the crystal (see Fig. 3.10b, c). The transmission resonance that now appears at a frequency within the gap signifies the existence of a state of the elastic field centered on the impurity plane: a mode of vibration of the elastic field that extends to infinity parallel to the surface of the slab (in the manner of a Bloch wave), but decays rapidly normal to the impurity plane on either side of it.

The transmission coefficient equals unity at the resonance frequency, when the impurity plane lies in the middle of the slab. When the impurity plane moves away from the middle plane, the transmission coefficient at the resonance frequency becomes less than unity and the resonance disappears altogether when the impurity plane is removed to the surface of the slab, as demonstrated in Fig. 3.10d–f. Further detailed analysis for phononic-crystal slabs with surface-located impurity planes can be found in [50].

Moreover, a periodic arrangement of impurity planes along a given direction creates narrow impurity bands with a width which depends on the position of these bands within the frequency gap of the pure crystal and on the separation between the impurity planes.

3.3.1.2 Gradient Layered Structures

An example of gradient layered heterostructures is given in [36]. The base phononic crystal is made of mercury spheres embedded in aluminum host and arranged in an fcc lattice of lattice constant a. In such a system absolute gaps appear for filling ratios f between 5.2 % and 12.1 % and the maximum absolute gap is 2.52 % at $f = 8.2\,\%$, as shown in Fig. 3.11a depicting the width of the gap over the midgap frequency. The full-band-gap feature of the system under study can be enhanced by using a heterostructure containing slabs corresponding to different f (spheres of different size). More importantly, by stacking together seven slabs of different spheres sizes, having consequently overlapping full gaps, one can obtain a resulting $\Delta\omega_G/\omega_{MG} = 14.4\,\%$ as shown in Fig. 3.11b. In particular, the filling ratios used are 5.9 %, 6.3 %, 6.8 %, 7.5 %, 8.23 %, 8.93 %, and 9.6 % (denoted by the vertical lines in Fig. 3.11b). It should be noted that slabs of at least eight planes of spheres are needed in order to preserve with accuracy the full-band-gap feature of the crystal. Therefore a minimum width of $4a$ per slab should be appropriate for stacking and a width of $28a$ for the above heterostructure is recommended at least. We note here that all consecutive (001) fcc slabs have the same 2D periodicity on xy-plane.

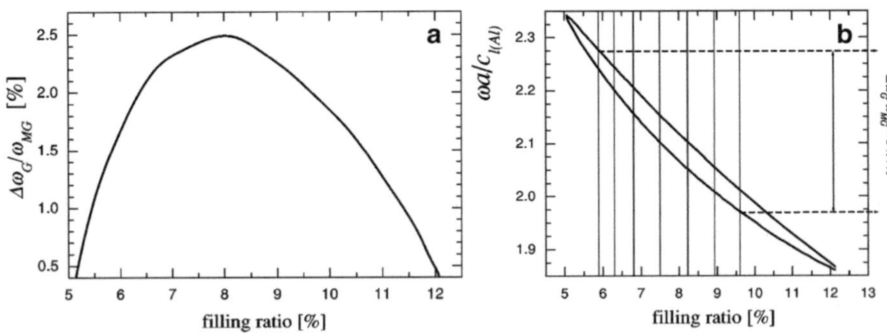

Fig. 3.11 After [36]: (**a**) The width of the gap over midgap frequency for the fcc crystal of mercury spheres in aluminum as a function of the filling ratio. (**b**) The edges of the full gap for this crystal as a function of the filling ratio

3.3.1.3 Planar Disorder

We next introduce a random distribution of defect planes along z-direction, having different positions or different sphere sizes with respect to the reference (periodic) phononic crystal.

We first choose for the latter an fcc phononic crystal of steel spheres of radius $S = 0.31a_0$ in a polyester matrix (ordered system I), where a_0 is the nearest-neighbor distance of the crystal and the subindexes l and t refer to longitudinal and transverse wave, respectively. This crystal has an absolute phononic gap extending from $\omega a_0/c_l = 2.32$ to $\omega a_0/c_l = 3.23$ [43]. In Fig. 3.12a we show the frequency band structure normal the (111) surface of the crystal about the region of this gap. Only non-degenerate (longitudinal) bands, denoted by thin lines, are excited when a longitudinal wave is incident normally on a finite slab of the material parallel to the (111) surface; this is not the case for doubly degenerate and deaf bands (thick and dotted lines, respectively). Such a slab of the (ordered) crystal, consisting of a number N of (111) fcc planes of spheres (layers), has thickness Nd (d is the distance between two consecutive planes). When immersed in water and assumed sufficiently thick ($N = 128$ layers), this slab has, as expected, a transmission coefficient (solid line in Fig. 3.12b) practically equal to one over the extension of longitudinal band and vanishing elsewhere, for a longitudinal wave incident normally on it.

Position disorder is next introduced, assuming the simplest case: fcc stacking faults. An fcc slab grown along the [111] direction (let this be the z-direction) consists, in the absence of disorder, of parallel planes of spheres in a sequence: $\cdots$ABCABC$\cdots$. The layers A, B, C are identical, having the same 2D periodicity in the xy plane, and each one is displaced relative to the one preceding by $\mathbf{a_3} = \mathbf{a_0}(\mathbf{1/2}, \sqrt{\mathbf{3}}/\mathbf{6}, \sqrt{\mathbf{6}}/\mathbf{3})$ (see Fig. 3.12d). Every time this sequence is destroyed as in $\cdots$ABC<u>B</u>ABC$\cdots$, we have a stacking fault <u>B</u> corresponding to a displacement by $\tilde{\mathbf{a}}_3 = \mathbf{a_0}(-\mathbf{1/2}, -\sqrt{\mathbf{3}}/\mathbf{6}, \sqrt{\mathbf{6}}/\mathbf{3})$. The dashed line in Fig. 3.12b gives the ensemble-averaged logarithm of the transmittance, $\langle \ln \mathscr{T} \rangle$, of a normally incident longitudinal wave through a slab, the same as that of the solid line, except that it is disordered by stacking faults with a 20 % probability for a stacking fault to appear at each point of the sequence; $\langle \ln \mathscr{T} \rangle$ means averaging over an ensemble of 100 different realizations of random sequences of the given number of layers with a 20 % disorder probability. We observe a reduction in $\langle \ln \mathscr{T} \rangle$, relative to the ordered structure at frequencies above and below the edges of the gap, but not a dramatic one. This is due to Anderson localization and to be certain that this occurs, the quantity

$$l/d = -2N/\langle \ln \mathscr{T} \rangle, \tag{3.1}$$

must be independent of N, for sufficiently large values of N. Only then we can refer to l as the localization length. As seen in Fig. 3.12c where l/d is shown for $N = 128$, and $N = 2,048$, l has converged for certain frequencies above the gap (about the edges of the longitudinal band of Fig. 3.12a) which confirms the existence of Anderson localization for these frequencies, for very large samples ($N = 2,048$), without however producing any dramatic difference in the transmission gap.

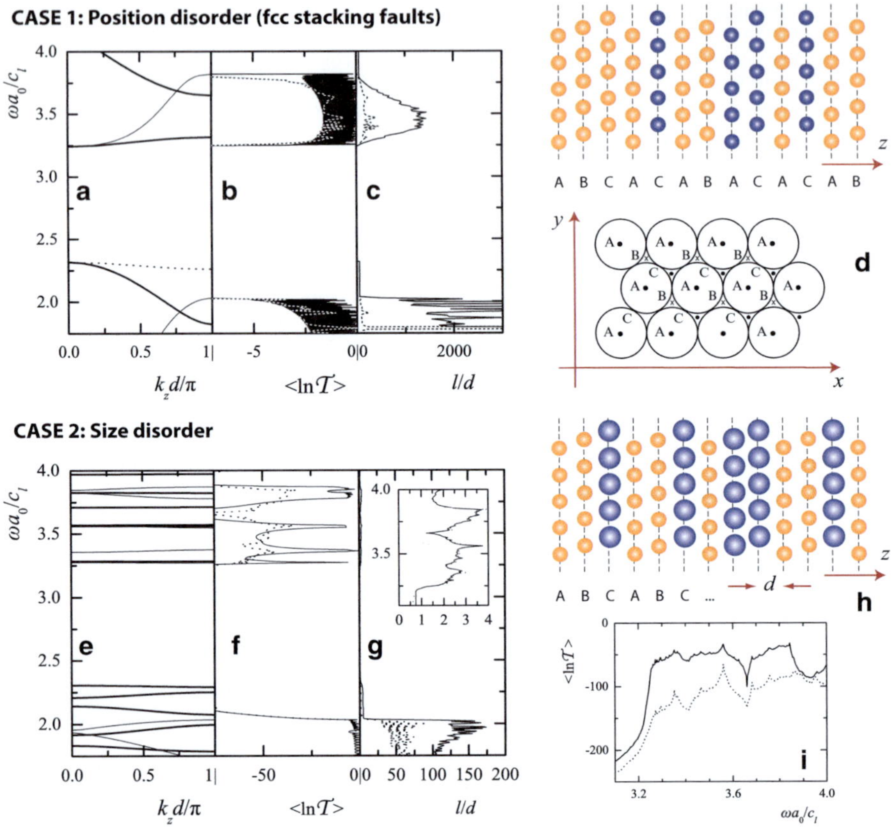

Fig. 3.12 After [47]: *Top panel*: Stacking faults. (**a**) The frequency band structure along the [111] direction for the ordered system I. *Thin/dotted/thick lines* refer to bands of $\Lambda_1/\Lambda_2/\Lambda_3$ symmetry. (**b**) $\langle \ln \mathcal{T} \rangle$ of a longitudinal wave incident normally on a slab of the material immersed in water. The slab is 128-layers thick and contains 0 % (*solid line*) and 20 % (*dashed line*) stacking faults. (**c**) l/d for slabs with 20 % stacking faults (*dashed line*: $N = 128$, *solid line*: $N = 2{,}048$). (**d**) Schematic view of disorder due to fcc stacking faults. *Bottom panel*: Size disorder. (**e**) The frequency band structure along the [111] direction for the ordered system II. *Thin/thick lines* refer to bands of Λ_1/Λ_3 symmetry. (**f**) $\langle \ln \mathcal{T} \rangle$ of a longitudinal wave incident normally on a slab of the material immersed in water. The slab is 65-layers thick, with 0 % (*solid line*) and 20 % (*dashed line*) size disorder. (**g**) l/d for slabs with 20 % size disorder (*dashed line*: $N = 65$, *solid line*: $N = 256$). (**h**) Schematic view of size disorder along [111] direction. (**i**) $\langle \ln \mathcal{T} \rangle$ at normal incidence for a slab of system II 64-layers thick, immersed in water, with 20 % (*solid line*) and 40 % (*dotted line*) size disorder

When size disorder is considered strong Anderson localization occurs leading to considerable widening of the absolute phononic gap. Now the reference crystal (ordered system II) is built as a sequence of fcc (111) planes of spheres as before, except that the last plane in a period of five planes has spheres with a radius $S' = 1.2S$ different from that of the spheres in the rest of the planes ($S = 0.31a_0$).

System II has a period five times larger than that of system I, thus the relatively wide bands about the gap of system I have now to be folded into a reduced zone five times smaller than the original one. This results in a number of narrow bands in system II together with opening of Bragg gaps between them (Fig. 3.12e); the frequency gap of Fig. 3.12a is retained. Comparison of the logarithm of the transmittance of a longitudinal wave incident normally on a slab of this crystal consisting of 65 planes embedded in water (solid line in Fig. 3.12f) to the corresponding $\langle \ln \mathscr{T} \rangle$ of a disordered slab (dashed line in Fig. 3.12f) with 20 % size disorder probability for each plane of the slab to have the larger spheres (Fig. 3.12h) shows a dramatic reduction of the latter over a region of frequency extending well above the frequency gap of the (ordered) crystal. Of course this is due to Anderson localization, as confirmed by the results of Fig. 3.12g: l has converged to a value independent of the thickness for slabs with 65 layers for frequencies above the gap, and more importantly, because the localization length is relatively small (less than the thickness of five layers), slabs of reasonable thickness will now produce the desired effect. At frequencies within the gap, the transmission through the disordered slab remains negligibly small.

Widening of the transmission gap, at least double that of the band gap, due to Anderson localization in such size disordered structures, is favored by the presence of the narrow bands which become easily localized when disorder is introduced. It occurs not only at normal incidence and persists even if water is replaced by polyester on either side of the slab, so that transverse waves are included, i.e. whatever the direction or polarization of the incident wave [47]. Increasing the size disorder (from 20 % to 40 %) leads to further reduction of the transmittance, exponentially at most frequencies (Fig. 3.12i).

3.3.2 Linear Defects

All non-periodic structures presented up to here have retained their periodicity parallel to their characteristic surface, and consequently the LMS technique [37, 46] is perfectly adapted for their theoretical description. However, for point or linear defects or when disorder needs to be extended in 3D, this technique has to be used in its primitive form, i.e. the multiple scattering (MS) has to be performed in the whole space where scatterers exist.

In this section, we will not be extended ourselves to completely random systems. Some work has already been done in bulk (volume) disordered bubbly liquids using multiple scattering techniques (see, for instance, [67]). Recently, random systems composed by solid spheres in a fluid host have been studied experimentally, in ultrasound [11] or hypersound [57] regime, assuming, respectively, position disorder or position and size disorder simultaneously. We note in passing that the case of 3D phononic quasicrystals has been also considered experimentally: they exhibit a "hybrid" behavior between periodic and random structures [61] concerning persistence of both Bragg and hybridization gaps together with reduction of transmission levels due to localization. A general conclusion could be made here: although the prominent feature of perfectly periodic phononic structures is the appearance of frequency gaps, periodicity is not a precondition for them to exist: hybridization gaps originating from individual resonances localized in the structural units of the system will be still present even when periodicity is removed [57, 61].

We will focus next on the case of linear defects introduced to phononic crystals, directly related to waveguiding. A general description of the multiple scattering in the real space, which of course can be also applied in random media, will be presented first.

3.3.2.1 Multiple Scattering in the Real Space

We will briefly give the principle of the MS in the general case: the inhomogeneities (spheres in our case) are distributed randomly in the space and do not need to be the same. We know that for a homogeneous space a wave of angular frequency ω is transferred from a site n' of position vector $\mathbf{R}_{\mathbf{n}'}$ to another site n of position vector $\mathbf{R}_{\mathbf{n}}$ through the so-called propagators of the free space, $\boldsymbol{\Omega}_{nn'}$ (see Fig. 3.13a). On the other hand, each time a wave is incident on a sphere at position $\mathbf{R}_{\mathbf{n}}$ it is diffused by it and the scattering process is described by the scattering-matrix, $\mathbf{T}_n^0$, which gives the amplitude of the outgoing wave from that sphere (see Fig. 3.13b). Let's now assume

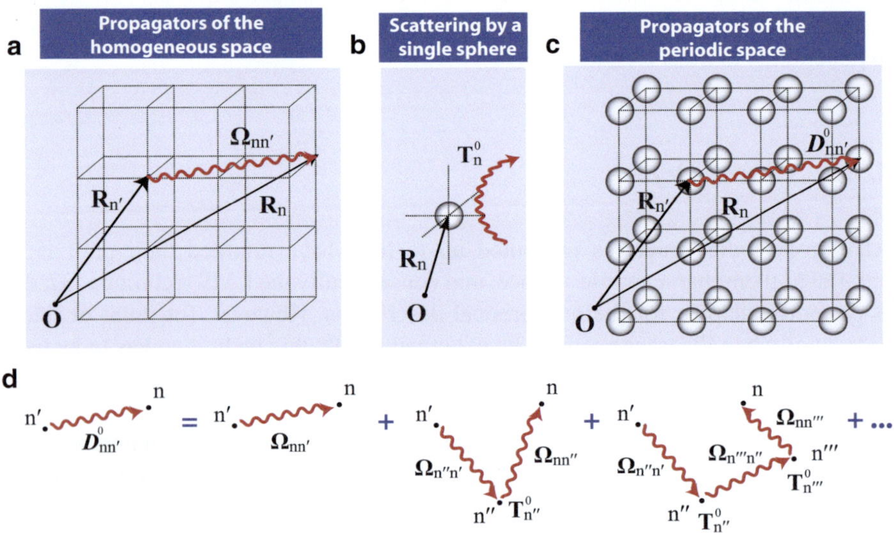

Fig. 3.13 Principle of the multiple scattering in an inhomogeneous medium. (**a**) Free-host propagator functions describing the propagation of a wave between two sites in the homogeneous space. (**b**) T-matrix describing the scattering of a wave by an inhomogeneity (*sphere*). (**c**) Propagator functions describing the propagation of a wave between two sites for a periodic (unperturbed) system. (**d**) Schematic physical process of propagators based on the multiple-scattering approach

a periodic (unperturbed) system (see Fig. 3.13c) for the purpose we need it later, although what we will present is valid for any type of disorder (in size, material, or position). These two quantities, $\mathbf{\Omega}_{nn'}$ and $\mathbf{T}_n^0$ whose explicit expressions can be found elsewhere [44, 46], are the key for the exact description of the propagation of the wave in that complex system from a site n' to another site n. We need to know the propagators of the complex system, $\mathbf{D}_{nn'}^0$, giving the amplitude of the wave incident on the sphere at $\mathbf{R}_n$ due to an outgoing wave from the sphere at $\mathbf{R}_{n'}$. They are the sum of all possible ways (paths) for a wave to travel from n' to n, as shown in Fig. 3.13d: it can propagate directly without any scattering, or after being scattered once, by the site n'', or after being scattered twice, by sites n'' and n''', and so on. Thus we can write $\mathbf{D}_{nn'}^0 = \mathbf{\Omega}_{nn'} + \sum_{n''} \mathbf{\Omega}_{nn''}\mathbf{T}_{n''}^0\mathbf{\Omega}_{n''n'} + \sum_{n'',n'''} \mathbf{\Omega}_{nn''}\mathbf{T}_{n''}^0\mathbf{\Omega}_{n''n'''}\mathbf{T}_{n'''}^0\mathbf{\Omega}_{n'''n'} + \cdots$ which after an easy matrix manipulation gives [44]

$$\mathbf{D}_{nn'}^0(\omega) = \mathbf{\Omega}_{nn'}(\omega) + \sum_{n''} \mathbf{\Omega}_{nn''}(\omega)\mathbf{T}_{n''}^0(\omega)\mathbf{D}_{n''n'}^0(\omega) \ . \tag{3.2}$$

We note here that the superscript "0" in $\mathbf{D}$ and $\mathbf{T}$ refers to unperturbed system to be distinguished from the perturbed one (see below). Equation (3.2) describes in its present form any disorder and/or defect case: we can assume any kind (material and/or size) and any arrangement of spherical scatterers in space. When applied to a periodic system (as we need here), every site n lies on a lattice point and $\mathbf{T}_n^0 = \mathbf{T}^0$ is the same for every point.

We next introduce to the periodic (unperturbed) system some defects (i.e., spheres of different size and/or material) which will be described by a new T-matrix, denoted by $\mathbf{T}_n$. By analogy to Eq. (3.2) we can then find [44] the corresponding propagator functions for the "perturbed" system, $\mathbf{D}_{nn'}$, to be

$$\mathbf{D}_{nn'}(\omega) = \mathbf{D}_{nn'}^0(\omega) + \sum_{n''} \mathbf{D}_{nn''}^0(\omega)\Delta\mathbf{T}_{n''}(\omega)\mathbf{D}_{n''n'}(\omega) \ , \tag{3.3}$$

with $\Delta\mathbf{T}_n = \mathbf{T}_n - \mathbf{T}_n^0$, where $\mathbf{T}_n$ is the T-matrix at the site $\mathbf{R}_n$ of the perturbed system and $\mathbf{T}_n^0$ is that of the original (reference) system. Obviously, the lattice sum of Eq. (3.3) is now restricted only to the sites where the spheres of the reference system have been substituted. The importance of Eq. (3.3) will become more evident in the next section.

We note that all matrices appearing here are defined in $\{\mathrm{P}\ell m\}$ space, i.e. including the polarization type P of the waves (one longitudinal and two transverse) and the angular momentum quantum numbers (ℓ, m) used in the spherical-wave expansions of the elastic waves.

3.3.2.2 Waveguiding

Waveguiding in 3D phononic crystals has been very little studied up to now [4, 49]. Waveguides in phononic-crystal structures are usually created by removing a row (line) of cylinders (spheres), so that the elastic waves of angular frequency ω can be guided through those channels, being reflected on their walls (in analogy to classical waveguides) in condition that ω falls within the frequency region of the absolute gap of the perfect phononic crystal. This type of waveguiding has been widely studied both theoretically and experimentally in 2D phononic crystals, but we know of only one work [4] that transferred this idea in 3D phononic crystals using the FDTD method in a solid/solid composite (lead spheres in epoxy). An alternative mechanism for waveguiding relying on weak coupling through defects along a chain, known as coupled-cavity waveguide (CCW), will be presented here. First proposed in photonic crystals [55], it has been transferred in the acoustic case for a bubbly liquid crystal [49] which supports only longitudinal acoustic waves.

For an efficient waveguiding the operating frequencies must lie within an absolute frequency gap of the phononic crystal. A simple cubic crystal, of lattice constant a, consisting of spherical air bubbles of radius $S_0 = 0.20a$ in water exhibits a wide absolute gap extending from $\omega a/c = 0.087$ to $\omega a/c = 2.228$ (see left panel of Fig. 3.14a), c being the sound velocity in water. This hybridization-type gap originates from the monopole Minnaert resonances of the individual scatterers. The monopole approximation (angular momentum $\ell = 0$ in the spherical-wave expansions of the longitudinal field) compares very well to the exact results obtained by truncating at $\ell_{\max} = 3$ in the said frequency region, as shown in the complex band structure of the above crystal along the [001] direction (see right panel of Fig. 3.14a). Therefore, in what follows, we assume only s-wave scattering (monopole approximation, $\ell_{\max} = 0$).

For the above reasons all matrices related to the multiple-scattering of the previous section will be shrunk to scalar quantities. This system, for the purpose we need it here, allows an exact, almost analytic solution with the consequent physical insight. We calculate from Eq. (3.2) the propagator functions $D^0_{nn'}$ for s waves for the phononic crystal under consideration, where now T^0 is the element of the T-matrix of a single sphere associated with s scattering, having a very simple form [49]; the propagator functions for s waves in the host medium are

$$\Omega_{nn'} = \frac{\exp(\mathrm{i}q|\mathbf{R}_n - \mathbf{R}_{n'}|)}{\mathrm{i}q|\mathbf{R}_n - \mathbf{R}_{n'}|}\ , \qquad n \neq n' \tag{3.4}$$

with $q = \omega/c$, while by definition, $\Omega_{nn} = 0$. The calculation of $D^0_{nn'}$ on the basis of Eq. (3.2) involves, in general, a summation over a large number of lattice sites because the $\Omega_{nn'}$ decay slowly with the distance $|\mathbf{R}_n - \mathbf{R}_{n'}|$, according to Eq. (3.4). However, the lattice sum in Eq. (3.2) is rapidly convergent at frequencies within an absolute gap because, in this case, the propagator functions $D^0_{nn'}$ decay exponentially with distance, as shown in Fig. 3.14b for a specific frequency at about the middle of the absolute gap of the system under consideration. Moreover, $D^0_{nn'}$ varies practically in isotropic manner and can be well fitted by the function $C\exp(-\beta|\mathbf{R}_n - \mathbf{R}_{n'}|)/(\beta|\mathbf{R}_n - \mathbf{R}_{n'}|)$ for $n \neq n'$, where C is a complex constant [compare here the analogy between $D^0_{nn'}$ and $\Omega_{nn'}$ of Eq. (3.4), when $q = \mathrm{i}\beta$]. The value of β obtained from this fit agrees within less than 2 % with the value of $\mathrm{Im}k_z$

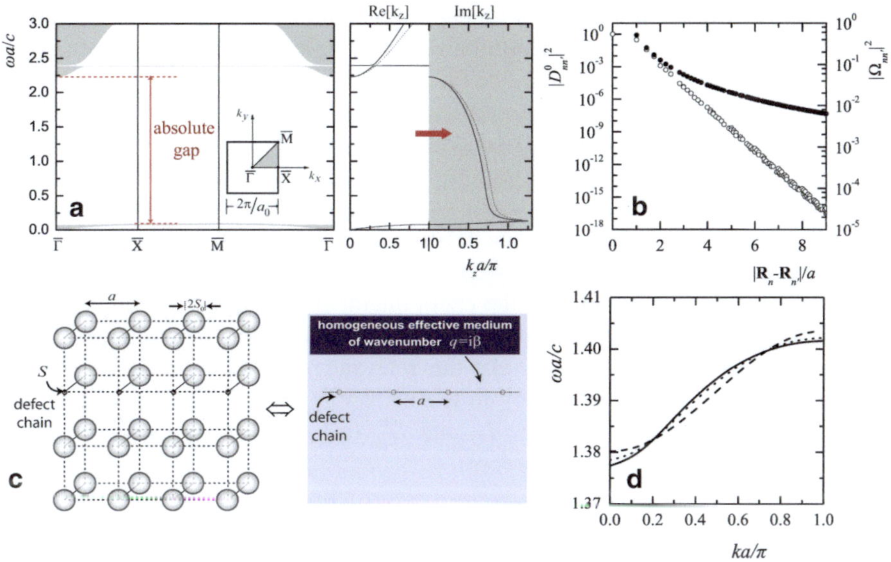

Fig. 3.14 After [49] (**a**) *Left*: Projection of the phononic band structure of a simple cubic crystal of spherical air bubbles ($S_0 = 0.20a$) in water on the high symmetry lines of the SBZ of its (001) surface (shown in the *inset*). *Right*: The corresponding complex band structure along the [001] direction. In the gap region the $k_z(\omega)$ with the smallest imaginary part is plotted (*shaded area*). *Dotted lines*: monopole approximation. The blank areas correspond to frequency gaps [all $k_z(\omega)$ are complex]. (**b**) Propagator functions versus distance (*empty circles*), for the crystal in (**a**) and free-host propagator functions (*filled circles*), both obtained for $\omega a/c = 1.39$ [at about the middle of the absolute band gap shown in (**a**) by the *red arrow*]. (**c**) The real waveguide under study (*left*) and its equivalent tight-binding-based description (*right*). (**d**) Defect band generated by an infinitely long linear chain of defect spheres ($S = 0.01a$) along the [001] direction of the phononic crystal shown in (**b**). *Solid line*: exact, *dotted line*: nearest-neighbor and frequency-independent propagator approximation. *Dashed line*: best fit to Eq. (3.7). The negative-k section of the band is symmetric to the one shown

at the given frequency (see red arrow in the right panel of Fig. 3.14a). In this respect we can say that *the phononic crystal under consideration behaves, for frequencies within the gap, as a homogeneous and isotropic effective medium characterized by an imaginary wavenumber* $i\beta \cong i\mathrm{Im}[k_z(\omega)]$.

We now introduce the perturbed system by replacing a number of spheres along a line (say along the [001] direction) with bubbles of a different radius $S = 0.01a \neq S_0$ (see Fig. 3.14c). The propagators $D_{nn'}$ for the perturbed system are obtained by solving Eq. (3.3), with $\Delta T_n = T - T^0$, where T and T^0 are the T-matrices for s waves of the perturbed and the original (reference) system, respectively, restricted only to the sites $\mathbf{R}_n = (0, 0, na)$ where defects exist. Note here that $T_n = T$ is independent of n (all defects are taken the same). Taking advantage of the 1D periodicity of this defect chain, Eq. (3.3) is transformed to $D(\omega, k) = [1 - D^0(\omega, k)\Delta T(\omega)]^{-1} D^0(\omega, k)$, where $D^0(\omega, k)$ and $D(\omega, k)$ are the Fourier transforms of $D^0_{nn'}(\omega)$ and $D_{nn'}(\omega)$, respectively. The poles of $D(\omega, k)$ determine the eigenfrequencies of the acoustic field associated with the chain as a function of k. And since $D^0(\omega, k)$ has no poles over the frequency gap of the unperturbed crystal, these eigenfrequencies are given by the roots of

$$1 - D^0(\omega, k)\Delta T(\omega) = 0 \tag{3.5}$$

for every k in the 1D BZ: $-1 < ka/\pi \leq 1$. The frequency (defect) band of the linear infinite defect chain is calculated from Eq. (3.5) (see solid line of Fig. 3.14d) appearing within the gap. This nondegenerate narrow band results from s modes localized on the defect spheres of the chain interacting weakly between them. We can therefore assume that only nearest-neighbor interaction between the defect spheres matters in Eq. (3.3) and take the relevant propagator functions $D^0_{nn'}$ as constants independent of frequency in the narrow region of the defect band, i.e. put $D^0_{nn}(\omega) \simeq d_0$, $D^0_{n;n\pm1}(\omega) \simeq d_1$, and $D^0_{nn'}(\omega) \simeq 0$ otherwise. With these approximations, Eq. (3.5) takes the form

$$\frac{1}{\Delta T(\omega)} = d_0 + 2d_1 \cos(ka) , \tag{3.6}$$

which gives indeed very good results (dotted line in Fig. 3.14d).

Here, it is worth comparing the dispersion curve obtained from Eq. (3.6), shown in Fig. 3.14d, with that obtained using a phenomenological tight-binding approach. As already pointed in relation to Fig. 3.14b, for a narrow frequency range about the middle of the gap the crystal can be replaced by an effective homogeneous medium characterized by an imaginary wavenumber $\mathrm{i}\beta$ (see Fig. 3.14c), independent of direction and frequency. When a single defect sphere is embedded in it, a localized (about the defect) state of the wavefield appears, decaying as $\exp(-\beta r)/r$ outside the defect sphere, with r the radial distance from its center. The eigenfrequency ω_0 of such a state is given by a pole of the T-matrix for an air bubble in the homogeneous effective medium, and, near the pole we can write $T(\omega) \simeq A/(\omega - \omega_0)$, A being a real quantity. In the case of a chain of spheres in the above effective medium, the wave outgoing from the nth sphere, which behaves like $a_n \exp(-\beta|\mathbf{r}-\mathbf{R}_n|)/|\mathbf{r}-\mathbf{R}_n|$, is generated by the waves incident on it outgoing from the $(n-1)$th and the $(n+1)$th spheres (nearest-neighbor approximation). The latter can be expanded about the nth sphere using the $\Omega_{nn'}$ for the effective medium [Eq. (3.4) for $q = \mathrm{i}\beta$]. We obtain: $a_n = T(\Omega_{n;n-1}a_{n-1} + \Omega_{n;n+1}a_{n+1})$. For an infinite periodic chain the Bloch theorem requires $a_{n\pm1} = a_n \exp(\pm \mathrm{i}ka)$, which leads to the dispersion relation

$$\omega = \omega_0 + 2W \cos(ka) , \tag{3.7}$$

where $W = -A\exp(-\beta a)/(\beta a)$. The values of ω_0 and W in Eq. (3.7) are in general treated as adjustable parameters (leading to the dashed line in Fig. 3.14d) or obtained following the effective-medium approximation as described above [49].

It is worth noting that the same results [Eq. (3.6)] describe the case of a finite chain consisting of N defect spheres, which can also bend any number of times [49].

The above could be generalized to more complicated systems including solid components and whose individual scatterers involve spherical waves of higher angular momentum.

3.4 Multi-component 3D Phononic Crystals (Locally Resonant and Acoustic Metamaterials)

In this section, we first present an overview on 3D locally resonant phononic crystals concerning both 3D bulk systems and finite plates (considered as 2.5D). We next conclude with a short outline on acoustic metamaterials. We will define and describe those types of periodic acoustic materials, and discuss some of their unusual properties, allowing the appearance of new physics phenomena and new appealing applications.

3.4.1 Locally Resonant Phononic Crystals

3.4.1.1 Bulk Structures

During the last decade, the term *Locally Resonant phononic crystals* (LRPC), or sometimes *Locally Resonant sonic materials* (LRSM), established in the pioneer work of P. Sheng and co-workers [25] is tightly associated with systems exhibiting narrow hybridization gaps (HG) at very low frequencies compared to those at which the corresponding Bragg gaps (BG) of the structure occur. We remember here that BG appear at about an angular frequency ω of the order of $\pi c_{l(t)}/a$, a being the lattice period of the structure, and at higher frequencies, which tells us how to choose a if we want a gap in a certain frequency region. On the other hand, the HG originating from localized resonances in the structural units (scatterers) of the system (see relevant discussion in Sect. 3.2.1) open up at frequencies very close to the corresponding resonant eigenfrequencies. The latter (the lower-order mode in an infinite series) occur when the wavelength inside the scatterer is of the order of its characteristic length, i.e. $\lambda_s \sim \Gamma$, Γ being the circumference for the case of spheres. Thus, it is evident that the position of the HG depends strongly on the material(s) and size of the scatterer and is quite robust against structural changes [57], leading to absolute gaps. A proper choice of the constituent materials can therefore tune these HG in a desired frequency region without the need to change the dimensions of the unit cell of the crystal.

P. Sheng and co-workers proposed the use of multilayered spheres within a phononic crystal [24, 25], in particular, scatterers made of a relatively hard elastic core and a soft shell, embedded in an elastically hard host medium, exhibiting localized (in core or shell region) resonant modes which lead to HG in phononic crystals realized from such building components [24, 25, 56]. More precisely for a

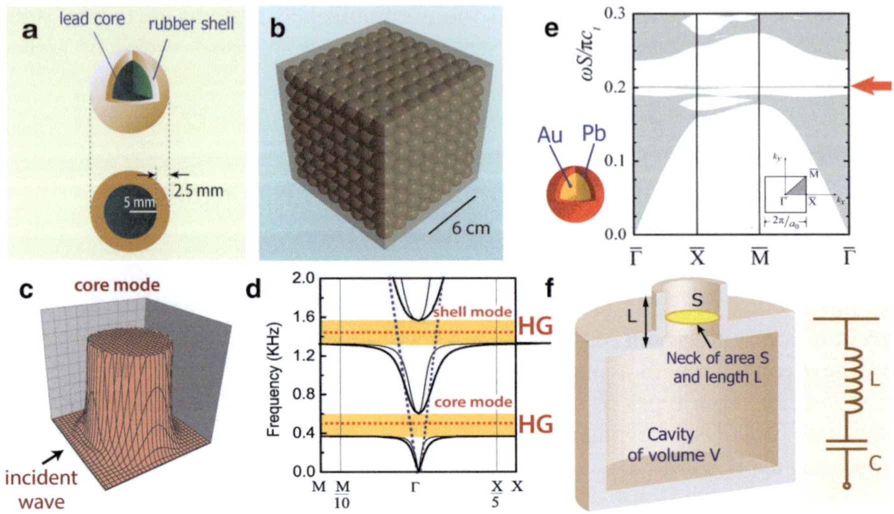

Fig. 3.15 Cross section of a rubber-coated lead sphere (**a**) that forms the basic structure unit for an 8 × 8 × 8 sonic crystal (**b**). The core-shell sphere exhibits localized resonances, an example is given in (**c**), responsible for the appearance of hybridization gaps at very low frequencies, as shown in the calculated band structure (**d**). *Dotted lines* denote schematically the unhybridized bands, with the *red lines* originating from core- or shell-localized resonant states. All results in (**c**) and (**d**) were calculated by using the computer code of [46]. (**e**) An example of LR crystal whose resonant state results in a flat band (*red arrow*) within an absolute gap. (**f**) Cross-sectional view of a Helmholtz resonator. A cavity is carved out of a rigid material (*gray*) and connected to the outside through a neck. The inset illustrates its LC circuit analog. (After [7, 25, 56])

cubic lattice of lead spheres coated by soft silicone rubber and embedded in epoxy matrix (see Fig. 3.15a, b) they managed to obtain narrow HG at wavelengths in the host matrix, λ, two orders of magnitude larger than the lattice constant a [25], as explained in Fig. 3.15c, d. Note here that these wavelengths correspond to the linear dispersion region (long wavelength limit) where the whole crystal can be described, away from the resonances, by an effective homogeneous medium (blue lines in Fig. 3.15d). More importantly, at frequencies close to these HG this linear description is not valid any more and a negative effective behavior for some elastic constants [25] is observed. These frequency ranges can be tuned at will by varying the size and geometry of the structural unit.

Following the same scheme for the multilayered sphere, i.e. hard-soft-hard, but with relatively harder materials than those described above, one can obtain, with an appropriately choice of constituents and shell size, a higher in frequency HG together with a narrow flat band within it, as shown in Fig. 3.15e for the specific case of a phononic crystal of gold-core and lead-shell spherical inclusions in a silicon matrix [24, 56]. This could be very useful in applications such as filtering, waveguiding, etc. More or less the same band structure image (i.e., a flat band originating from localized inside the complex structured scatterer mode) has been also observed in systems consisting of hollow polymer spheres and cylinders, these resonance modes being tunable up to very low frequencies by varying the shell size [48]. As a remark, we can say that, in general, inclusions of two or more components with an alternation of hard and soft materials act as mass-spring-damper oscillators in each unit cell [7, 20, 24, 25, 48].

We note in passing that, besides multilayered scatterers, a spring-mass behavior can be also achieved by using Helmholtz resonators [7]. They consist of a rigid-wall cavity connected to a fluid matrix through a much narrower neck (see Fig. 3.15f). The fluid in the neck acts approximately as a mass, whereas the compressible fluid in the cavity performs the function of a spring. With the appropriate choice of neck and cavity dimensions, Helmholtz resonators can be made deep-subwavelength at resonance [8]. This type of resonators have not been used up to now in 2D or 3D arrays, as far as we know.

An alternative method to create LRSM without having multicomponent units relies on the use of inclusions with phase speed much lower than that of the matrix [8]. And again, at resonance, the lowest order eigenmodes of an inclusion nominally occur when $\lambda_s \sim \Gamma$, and since $\Gamma \sim a$, we conclude again that in order to have $\lambda \gg a$, a material with very low phase velocity is needed. The most common choice is a soft silicone rubber, which has phase speeds that are two orders of magnitude lower than those of typical solids. Very few studies have considered inclusions consisting entirely of soft rubber [21, 63].

3.4.1.2 Finite Plates

We know of only one work dealing with thin finite slabs of 3D phononic crystals, precisely a square monolayer of polymer spheres placed on a thin glass substrate and immersed in water [41]. In such a system, the resonance modes of the spheres interact with the folded (because of the 2D periodicity) Lamb modes of the plate giving rise to very narrow hybridization gaps (HG) close to the corresponding Bragg gaps (BG). A theoretical study of the several guided and quasi-guided elastic waves together with a group-theory symmetry analysis of the corresponding modes is given in [41], however this structure cannot be typically classified as an LRSM, since the HG do not occur at very low frequencies compared to the BG.

These systems, i.e. thin finite plates containing a 2D array of scatterers along their characteristic surface (instead of spheres we can imagine any type of finite-size scatterers such as cylinders of finite length, disks, ellipsoids, etc.), are usually considered as 2.5D systems, but they remain in fact 3D ones. They attract a lot of interest, since they offer a natural acoustic energy confinement and a simple fabrication [13, 29, 64]. The difference with the corresponding thick finite PC slabs is that now we have to deal with Lamb waves instead of bulk waves.

In order to create locally resonant (thin) phononic slabs we need to choose appropriately their scatterers as described in Sect. 3.4.1.1; we can mainly distinguish two types for these systems: embedded resonators in plates and stubbed plates. An example of the former was reported by T-T. Wu and co-workers who investigated

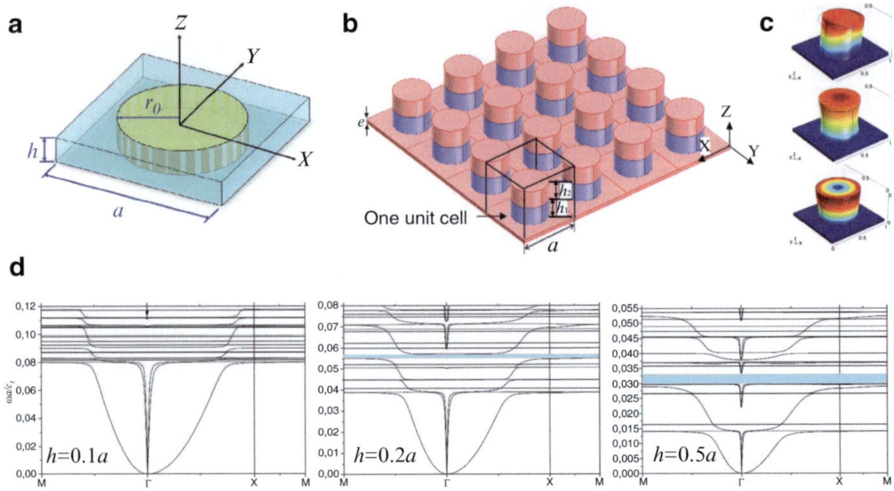

Fig. 3.16 In (**a**) the unit cell of a PC plate, shown schematically in (**b**), consisting of cylinder stubs arranged in square lattice. The stubs of radius r and of height h, deposited on a thin epoxy plate, are one-layer (rubber) or two-layer (rubber layer with a Pb cap, with thicknesses h_1 and h_2, respectively, and $h = h_1 + h_2$). The displacement distribution of some localized modes of different polarization (shear, elongated and breathing modes from *top* to *bottom*) is shown in (**c**). (**d**) Band structure for the PC of simple stub, for $r = 0.48a$ and $h = h_1$ varying. (After [13, 29])

a binary LR plate composed of soft rubber inclusions in epoxy matrix (see Fig. 3.16a) [13]. They reported on tuning the low frequency band gap by varying the plate thickness and radius of inclusions. Regarding the LR stubbed plates, binary- and ternary-component LR systems have been recently studied [29, 64]. They are composed, for instance, of simple stubs (silicon rubber) or composed stubs (lead on rubber) (see Fig. 3.16b). The stubs act as spring-mass resonators on a homogenous elastic finite plate and exhibit localized eigenmodes of different polarizations (Fig. 3.16c). These eigenmodes interact with the Lamb modes of the plate resulting in very narrow HG opening up around the corresponding eigenfrequencies, which can be pushed down to extremely low values compared to the frequencies of the BG. Figure 3.16d displays a series of computed band structures showing the opening of HG as a function of the height of stubs. The evolution of the position and width of these HG can be easily explained using a spring-mass model [29]. The LR stubbed plates can be easily fabricated and their physics can be understood more straightforwardly, since we can separately change the effective "spring" constant and the total mass.

The major limitation of locally resonant phononic crystals is their operation only in a narrow frequency regime. To overcome this limitation, we can, as a first step, imagine a 3D system composed of a stack of stubbed plates. The main advantage of this structure besides its potentiality to give a wide LR band gap remains its air/solid composition. A low weight and sturdy 3D structure for sonic regime can then be targeted. The first set of proposed 3D structures (see Fig. 3.17a) are based on rubber

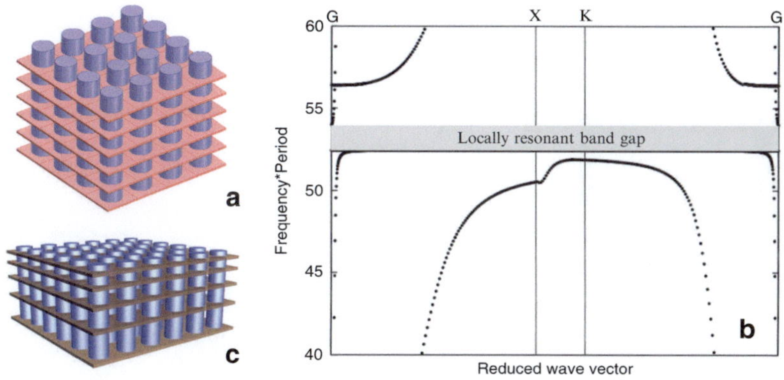

Fig. 3.17 (**a**) Schematic representation of a 3D LRPC composed of silicone rubber stubs hosted in aluminium frame, whose computed band structure is shown in (**b**). A complete band gap is observed and *gray shaded*. In (**c**) an example of a "Gradient" 3D LRPC structure made of silicone rubber stubs on metallic frame

(soft, medium, or hard) stub resonators hosted in a metallic frame (aluminium, steel, brass). The layer-by-layer form of these 3D structures makes their fabrication over a large volume possible. More importantly, they can exhibit omnidirectional (complete) band gaps together with slow sound lines (flat bands). An example is illustrated in Fig. 3.17b for a structure composed of rubber resonators and aluminum plates with a lattice constant of $a = 1$ cm for in-plane and out-of-plane directions; the diameter of rubber resonators is 0.6 cm. The band gap occurs at a normalized frequency of about $fa = 52\,\mathrm{ms}^{-1}$ ($f = 5.2$ kHz for $a = 1$ cm) [3]. The flexibility of this structure enables its engineering by modulating the geometry of phononic crystal in order to achieve more complex structures with desired functionalities. For example, a "gradient structure" can be formed by stacking multiple layers of LR stubbed plates with different thicknesses or periods, as shown in Fig. 3.17c. The principle is the same with the one presented in Sect. 3.3.1.2 for a heterostructure: each layer operating at different frequency range, the whole LR phononic crystal can be, consequently, effective in a large bandwidth. In that manner, wider 3D complete band gaps can be obtained, which can potentially cover the whole desired frequency range for applications such as acoustic isolation or shielding.

Other novel structures are investigated to allow fabrication of wide 3D LR band gap materials, as well. Z. Yang et al. [66] proposed recently a new 3D system composed of stack of different LR membranes called membrane-type acoustic metamaterials. They are composed of an elastic thin membrane extended and fixed by a relatively rigid plastic grid, with a small weight attached at the center of each cell. Figure 3.18a shows a front view of a planar array of the structure. In Fig. 3.18b we show the sound transmission losses (STL) spectra for different samples: two different single membranes, a stack of two membranes and a stack of four membranes. All stacks are composed of different membranes, each one having a different band gap. The transmission of the stacked panel (two membranes)

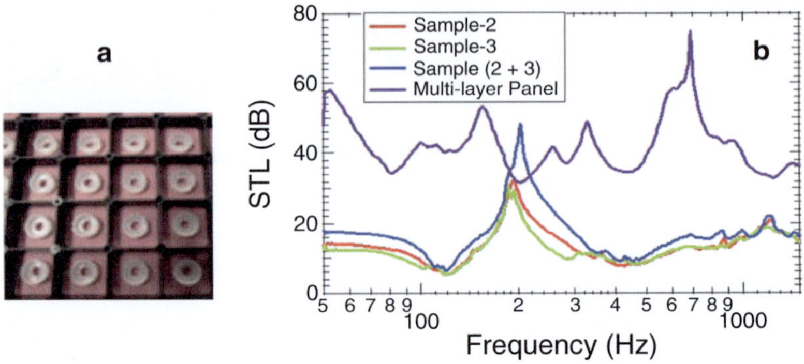

Fig. 3.18 After [66]: (**a**) Front view of a planar array of square unit cells composing the elastic thin membrane. In (**b**) some examples of the sound transmission loss of single membranes and of stack of different membranes, each one of them having different band gaps

shows a simple superposition in the individual spectra without any interaction between individual layers, since the membranes spacing (15 mm) is much larger than the evanescent wave decay length (0.8 mm). Broadband shields can therefore be constructed by simply stacking several single-layer panels with strategically selected dip frequencies. In addition, samples with multiweights per cell exhibit multimode vibration patterns and their transmission spectra contain multiple dips, making them advantageous for broadband applications.

3.4.2 Acoustic Metamaterials

Advances in electromagnetic metamaterials [54] offer excellent opportunities to realize new and exotic phenomena, such as negative refraction [51] and artificial magnetism [23]. The acoustic analog of these materials, namely *acoustic metamaterials*, usually includes all those complex media exhibiting unusual properties and thus producing novel phenomena, the most famous, probably, example being that of materials with negative elastic properties [7, 21, 25]. Acoustic metamaterials with LR elements, appropriately designed, is such a case: they can display anomalous response at selected frequencies, leading to negative effective behavior for their elastic parameters [21], which are frequency-dependent and highly dispersive [1]. Therefore, sound waves can be manipulated in surprising ways through such materials, including negative refraction, collimation, focusing, cloaking, sonic screening, and extraordinary transmission [22]. Additionally, these materials can provide a major step toward an effective-medium description, since the spatial periodic modulation of the elastic properties, a, is much smaller than the wavelength in the matrix, λ.

We will briefly describe here a few examples related to some main applications of acoustic metamaterials: negative refraction, focusing, and subwavelength imaging. The first example is related to the negative refraction and focusing in a 3D phononic crystal structure. S. Yang et al. [65] have shown both theoretically and experimentally how a dramatic variation in wave propagation with both frequency and propagation direction leads to novel focusing phenomena associated with large negative refraction. They evidenced that this variation is due to the propagation anisotropy by which the periodic structure "bends" ultrasound beams in unusual ways. The details of the experiment used are illustrated in Fig. 3.19a. A focusing of sound by a 3D periodic acoustic metamaterial was then demonstrated (see Fig. 3.19b, c). On the other hand, experimental evidence of deep-subwavelength imaging using a 2D holey-structured plate was recently reported by Z. Zhu et al. [68]. The basic structure, which consists of a rigid block of brass (impenetrable for sound waves) of thickness h, perforated with deep-subwavelength square holes of side a forming a periodic array with lattice parameter Λ (see Fig. 3.20a), is surrounded by air. They achieved acoustic imaging resolution down to a feature size of $\lambda/50$ (see Fig. 3.20b).

The emergence of LR phononic crystals and acoustic metamaterials, and especially their 3D configurations can offer some new approaches to consider and to deal

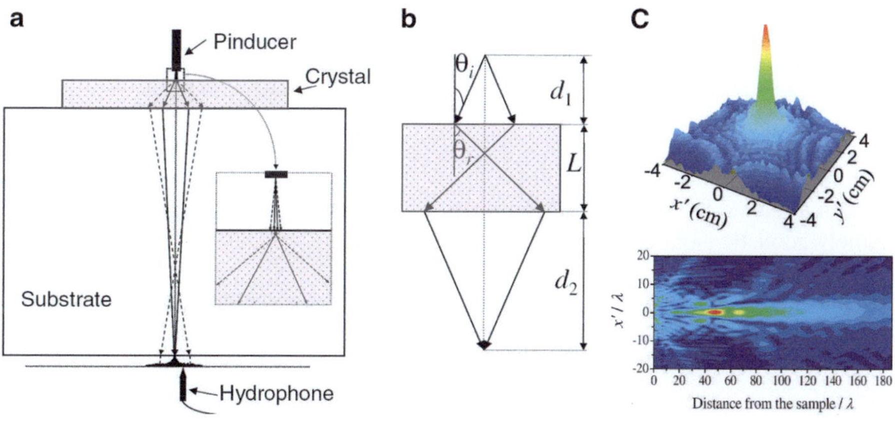

Fig. 3.19 After [65]: (**a**) Experimental setup, with rays indicating the predicted directions of the group velocity for specific frequency and angles of incidence. (**b**) Diagram showing the focusing condition in a medium with negative refraction effect. (**c**) Calculated field plots demonstrating the focusing effect at 1.57 MHz

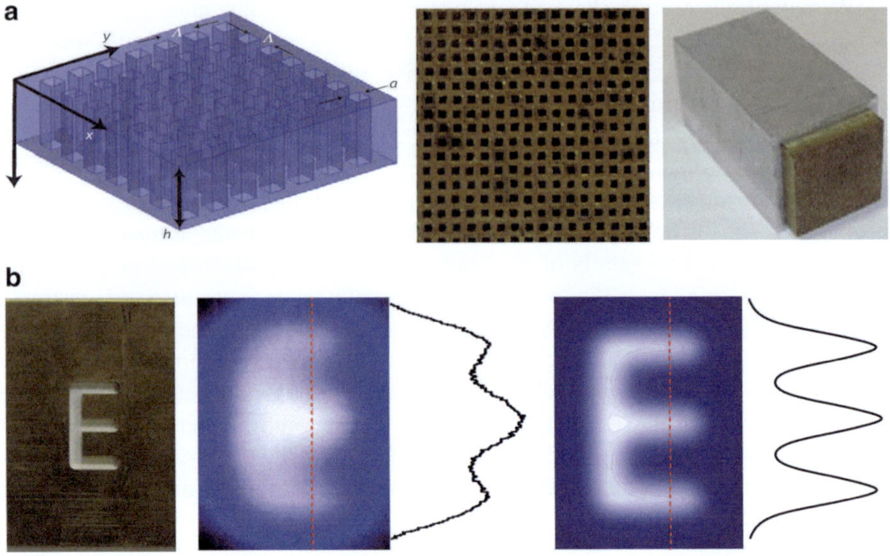

Fig. 3.20 After [68]: (**a**) Schematic representation (*left*) and top view image (*middle*) of a square array of holes (*square tubes*) drilled in a brass alloy. The holey block is fitted in parallel into 4-inch-wide square aluminium tube and clamped firmly together (*right*). (**b**) An example of imaging for a deep-subwavelength sized letter "E." *Left*: the imaging object (letter "E") of linewidth 3.18 mm, perforated in an ultrathin brass plate. Measured (*middle*) and simulated (*right*) image of letter "E," obtained at a distance of 1.58 mm from the output plane, together with the acoustic field distribution along the cross-section indicated by the *red dashed line*. The operating frequency is 2.18 kHz ($\lambda = 158$ mm). A $\lambda/50$ linewidth of the object can still be observed in the experiment

with new and appealing applications in low frequency regime. Those materials can present a real added value for a significant breakthrough in some application fields, namely acoustic isolation, seismic shielding, etc.

Acknowledgements R. Sainidou thanks Professors A. Modinos and N. Stefanou for the valuable, enriching, and fruitful collaboration on this topic, from 2000 to 2006.

References

1. M. Ambati, N. Fang, C. Sun, X. Zhang, Surface resonant states and superlensing in acoustic metamaterials. Phys. Rev. B **75**, 195447 (2007)
2. P.W. Anderson, Absence of diffusion in certain random lattices. Phys. Rev. **109**, 1492 (1958)
3. M.B. Assouar, A. Khelif, A.A. Eftekhar, A. Adibi, Unpublished data
4. H. Chandra, P.A. Deymier, J.O. Vasseur, Elastic wave propagation along waveguides in three-dimensional phononic crystals. Phys. Rev. B **70**, 054302 (2004)
5. W. Cheng, J. Wang, U. Jonas, G. Fytas, N. Stefanou, Observation and tuning of hypersonic bandgaps in colloidal crystals. Nat. Mater. **5**, 830–836 (2006)

6. M. Eichenfield, J. Chan, R.M. Camacho, K.J. Kerry Vahala, O. Painter, Optomechanical crystals. Nature **462**, 78–82 (2009)
7. N. Fang, D. Xi, J. Xu, M. Ambati, W. Srituravanich, C. Sun, X. Zhang, Ultrasonic metamaterials with negative modulus. Nat. Mater. **5**, 452 (2006)
8. L. Fok, M. Ambati, X. Zhang, Acoustic metamaterials. MRS Bull. **33**, 931 (2008)
9. G. Gantzounis, N. Papanikolaou, N. Stefanou, Multiple-scattering calculations for layered phononic structures of nonspherical particles. Phys. Rev. B **83**, 214301 (2011)
10. D. García-Pablos, M. Sigalas, F.R. Montero de Espinoza, M. Torres, M. Kafesaki, N. García, Theory and experiments on elastic band gaps. Phys. Rev. Lett. **82**, 4349–4352 (2000)
11. H. Hefei Hu, A. Strybulevych, J.H. Page, S.E. Skipetrov, B.A. van Tiggelen, Localization of ultrasound in a three-dimensional elastic network. Nat. Phys. **4**, 945–948 (2008)
12. B.K. Henderson, K. Maslov, V.K. Kinra, Experimental investigation of acoustic band structures in tetragonal periodic particulate composite structures. J. Mech. Phys. Solids **49**, 2369–2383 (2001)
13. J.-S. Hsu, T.-T. Wu, Lamb waves in binary locally resonant phononic plates with two-dimensional lattices. Appl. Phys. Lett. **90**, 201904 (2007)
14. M.I. Hussein, M.J. Frazier, Band structure of phononic crystals with general damping. J. Appl. Phys. **108**, 093506 (2010)
15. S.M. Ivansson, Numerical design of Alberich anechoic coatings with superellipsoidal cavities of mixed sizes. J. Acoust. Soc. Am. **124**, 1974–1984 (2008)
16. M. Kafesaki, E.N. Economou, Multiple-scattering theory for three-dimensional periodic acoustic composites. Phys. Rev. B **60**, 11993 (1999)
17. M. Kafesaki, M. Sigalas, E. Enonomou, Elastic wave band gaps in 3-D periodic polymer matrix composites. Solid State Commun. **96**, 285–289 (1995)
18. V.K. Kinra, E.L. Ker, An experimental investigation of pass bands and stop bands in two periodic particulate composites. Int. J. Solids Struct. **19**, 393–410 (1983)
19. M.S. Kushwaha, P. Halevi, L. Dobrzynski, B. Djafari-Rouhani, Acoustic band structure of periodic elastic composites. Phys. Rev. Lett. **71**, 2022–2025 (1993)
20. H. Larabi, Y. Pennec, B. Djafari-Rouhani, J.O. Vasseur, Multicoaxial cylindrical inclusions in locally resonant phononic crystals. Phys. Rev. E **75**, 066601 (2007)
21. J. Li, C.T. Chan, Double-negative acoustic metamaterial. Phys. Rev. E **70**, 055602 (2004)
22. J. Li, L. Fok, X. Yin, G. Bartal, X. Zhang, Experimental demonstration of an acoustic magnifying hyperlens. Nat. Mater. **8**, 931 (2009)
23. S. Linden, C. Enkrich, M. Wegener, J. Zhou, T. Koschny, C.M. Soukoulis, Magnetic response of metamaterials at 100 terahertz. Science **306**, 1351 (2004)
24. Z. Liu, C.T. Chan, P. Sheng, Three-component elastic wave band-gap material. Phys. Rev. B **65**, 165116 (2002)
25. Z. Liu, X. Zhang, Y. Mao, Y.Y. Zhu, Z. Yang, C.T. Chan, P. Sheng, Locally resonant sonic materials. Science **289**, 1734–1736 (2000)
26. K. Maslov, V.K. Kinra, B.K. Henderson, Elastodynamic response of a coplanar periodic layer of elastic spherical inclusions. Mech. Mater. **32**, 785–795 (2000)
27. A. Modinos, *Field, Thermionic and Secondary Electron Emission Spectroscopy* (Plenum Press, New York, 1984)
28. A. Modinos, N. Stefanou, I.E. Psarobas, V. Yannopapas, On wave propagation in inhomogeneous systems. Physica B **296**, 167–173 (2001)
29. M. Oudich, Y. Li, M.B. Assouar, Z. Hou, A sonic band gap based on the locallly resonant phononic plates with stubs. New J. Phys. **12**, 083049 (2010)
30. J.H. Page, S.X. Yang, Z.Y. Liu, M.L. Cowan, C.T. Chan, P. Sheng, Tunneling and dispersion in 3D phononic crystals. Z. Kristallogr. **220**(9–10), 859–870 (2005)
31. N. Papanikolaou, I.E. Psarobas, N. Stefanou, Absolute spectral gaps for infrared light and hypersound in three-dimensional metallodielectric phoxonic crystals. Appl. Phys. Lett. **96**, 231917 (2010)
32. N. Papanikolaou, I.E. Psarobas, B. Djafari-Rouhani, B. Bonello, V. Laude, Light modulation in phoxonic nanocavities. Microelectron. Eng. **90**, 155–158 (2011)

33. J.B. Pendry, *Low Energy Electron Diffraction* (Academic Press, London, 1974)
34. I.E. Psarobas, Viscoelastic response of sonic band-gap materials. Phys. Rev. B **64**, 012303 (2001)
35. I.E. Psarobas, Phononic crystals - sonic band-gap materials (editorial). Z. Kristallogr. **220**(9–10), IV (R4) (2005)
36. I.E. Psarobas, M.M. Sigalas, Elastic band gaps in a fcc lattice of mercury spheres in aluminum. Phys. Rev. B **66**, 052302 (2002)
37. I.E. Psarobas, N. Stefanou, A. Modinos, Scattering of elastic waves by periodic arrays of spherical bodies. Phys. Rev. B **62**, 278–291 (2000)
38. I.E. Psarobas, N. Stefanou, A. Modinos, Phononic crystals with planar defects. Phys. Rev. B **62**, 5536–5540 (2000)
39. I.E. Psarobas, R. Sainidou, N. Stefanou, A. Modinos, Acoustic properties of colloidal crystals. Phys. Rev. B **65**, 064307 (2002)
40. I.E. Psarobas, N. Papanikolaou, N. Stefanou, B. Djafari-Rouhani, B. Bonello, V. Laude, Enhanced acousto-optic interactions in a one-dimensional phoxonic cavity. Phys. Rev. B **82**, 174303 (2010)
41. R. Sainidou, N. Stefanou, Guided and quasiguided elastic waves in phononic crystal slabs. Phys. Rev. B **73**, 184301 (2006)
42. R. Sainidou, N. Stefanou, I.E. Psarobas, A. Modinos, Scattering of elastic waves by a periodic monolayer of spheres. Phys. Rev. B **66**, 024303 (2002)
43. R. Sainidou, N. Stefanou, A. Modinos, Formation of absolute frequency gaps in three-dimensional solid phononic crystals. Phys. Rev. B **66**, 212301 (2002)
44. R. Sainidou, N. Stefanou, A. Modinos, Green's function formalism for phononic crystals. Phys. Rev. B **69**, 064301 (2004)
45. R. Sainidou, N. Stefanou, I.E. Psarobas, A. Modinos, The layer multiple-scattering method applied to phononic crystals. Z. Kristallogr. **220**, 848–858 (2005)
46. R. Sainidou, N. Stefanou, I.E. Psarobas, A. Modinos, A layer-multiple-scattering method for phononic crystals and heterostructures of such. Comput. Phys. Commun. **166**, 197–240 (2005)
47. R. Sainidou, N. Stefanou, A. Modinos, Widening of phononic transmission gaps via Anderson localization. Phys. Rev. Lett. **92**, 205503 (2005)
48. R. Sainidou, B. Djafari-Rouhani, Y. Pennec, J.O. Vasseur, Locally resonant phononic crystals made of hollow spheres or cylinders. Phys. Rev. B. **73**, 024302 (2006)
49. R. Sainidou, N. Stefanou, A. Modinos, Linear chain of weakly coupled defects in a three-dimensional phononic crystal: A model acoustic waveguide. Phys. Rev. B **74**, 172302 (2006)
50. R. Sainidou, B. Djafari-Rouhani, J.O. Vasseur, Surface acoustic waves in finite slabs of three-dimensional phononic crystals. Phys. Rev. B **77**, 094304 (2008)
51. R.A. Shelby, D.R. Smith, S. Schultz, Experimental verification of a negative index of refraction. Science **292**, 77 (2001)
52. M. Sigalas, E.N. Economou, Elastic and acoustic wave band structure. J. Sound Vib. **158**, 377 (1992)
53. M. Sigalas, M. Kushwaha, E.N. Economou, M. Kafesaki, I.E. Psarobas, W. Steurer, Classical vibrational modes in phononic lattices: theory and experiment. Z. Kristallogr. **220**, 765–809 (2005)
54. D.R. Smith, J.B. Pendry, M.C.K. Wiltshire, Metamaterials and negative refractive index. Science **305**, 788 (2004)
55. N. Stefanou, A. Modinos, Impurity bands in photonic insulators. Phys. Rev. B **57**, 12127 (1998)
56. N. Stefanou, R. Sainidou, A. Modinos, Low-frequency absolute gaps in the phonon spectrum of macrostructured elastic media. Rev. Adv. Mater. Sci. **12**, 46–50 (2006)
57. T. Still, W. Cheng, M. Retsch, R. Sainidou, J. Wang, U. Jonas, N. Stefanou, G. Fytas, Simultaneous occurrence of structure-directed and particle resonance-induced phononic gaps in colloidal films. Phys. Rev. Lett. **100**, 194301 (2008)
58. T. Still, R. Sainidou, M. Retsch, U. Jonas, P. Spahn, G.P. Hellmann, G. Fytas, The "Music" of core-shell spheres and hollow capsules: influence of the architecture on the mechanical properties at the nanoscale. Nano Lett. **8**, 3194–3199 (2008)

59. T. Still, M. Retsch, U. Jonas, R. Sainidou, P. Rembert, K. Mpoukouvalas, G. Fytas, Vibrational eigenfrequencies and mechanical properties of mesoscopic copolymer latex particles. Macromolecules **43**, 3422–3428 (2010)
60. T. Still, G. Gantzounis, D. Kiefer, G. Hellmann, R. Sainidou, G. Fytas, N. Stefanou, Collective hypersonic excitations in strongly multiple scattering colloids. Phys. Rev. Lett. **106**, 175505 (2011)
61. D. Sutter-Widmer, P. Neves, P. Itten, R. Sainidou, W. Steurer, Distinct band gaps and isotropy combined in icosahedral band gap materials. Appl. Phys. Lett. **92**, 073308 (2008)
62. Y. Tanaka, Y. Tomoyasu, S. Tamura, Band structure of acoustic waves in phononic lattices: two-dimensional composites with large acoustic mismatch. Phys. Rev. B **62**, 7378–7393 (2000)
63. G. Wang, X.S. Wen, J.H. Wen, L.H. Shao, Y.Z. Liu, Two-dimensional locally resonant phononic crystals with binary structures. Phys. Rev. Lett. **93**, 154302 (2004)
64. W. Xiao, G.W. Zeng, Y.S. Cheng, Flexural vibration band gaps in a thin plate containing a periodic array of hemmed discs. Appl. Acoust. **69**, 255 (2008)
65. S. Yang, J.H. Page, Z. Liu, M.L. Cowan, C.T. Chan, P. Sheng, Focusing of sound in a 3D phononic crystal. Phys. Rev. Lett. **93**, 024301 (2004)
66. Z. Yang, H.M. Dai, N.H. Chan, G.C. Ma, P. Sheng, Acoustic metamaterial panels for sound attenuation in the 50–1000 Hz regime. Appl. Phys. Lett. **96**, 041906 (2010)
67. Z. Ye, H. Hsu, E. Hoskinson, Phase order and energy localization in acoustic propagation in random bubbly liquids. Phys. Lett. A **275**, 452–458 (2000)
68. J. Zhu, J. Christensen, J. Jung, L. Martin-Moreno, X. Yin, L. Fok, X. Zhang, A holey-structured metamaterial for acoustic deep-subwavelenght imaging. Nat. Phys. **52**, 7 (2011)

Chapter 4
Computational Problems and Numerical Techniques for the Analysis of Phononic Crystals

Vincent Laude and Abdelkrim Khelif

4.1 Basic Equations for Wave Propagation

As a general rule, computational problems for phononic crystals are the combination of equations in a domain of a given geometry, supplemented with some given boundary conditions. This introductory section focuses on the basic equations that govern wave propagation inside a domain composed of either, or both, fluids and solids. It also briefly introduces how material loss can be represented, and exposes the Bloch-Floquet theorem and its consequences.

4.1.1 Equations for Solids

In the absence of internal sources, the basic equations for elastic waves propagating in solid media are written as

$$T_{ij}(\mathbf{r}, t) = c_{ijkl}(\mathbf{r})\ u_{k,l}(\mathbf{r}, t), \tag{4.1}$$

$$T_{ij,j}(\mathbf{r}, t) = \rho(\mathbf{r})\frac{\partial^2 u_i(\mathbf{r}, t)}{\partial t^2}, \tag{4.2}$$

V. Laude
Institut FEMTO-ST, Université de Franche-Comté, Centre national de la Recherche Scientifique, 32 avenue de l'Observatoire, 25044 Besançon cedex, France
e-mail: vincent.laude@femto-st.fr

A. Khelif (✉)
Institut FEMTO-ST, Centre National de la Recherche Scientifique, Besançon Cedex, France
e-mail: abdelkrim.khelif@femto-st.fr

A. Khelif, A. Adibi (eds.), *Phononic Crystals*, DOI 10.1007/978-1-4614-9393-8_4

with u_i the three displacements in space and T_{ij} the stress tensor. In these equations, we have explicitly written the dependence on spatial coordinates and time. Note that we assume that the solid material with density ρ (in kg m^{-3} units) and elastic constants c_{ijkl} (in Pa units) is static but has a spatially varying composition. As a general rule, indices like i, j, k, or l run between 1 and 3 for the three directions of space, a comma before an index indicates derivation (e.g., $u_{k,l} = \partial u_k/\partial x_l$), and summation over repeated indices is implied (e.g., we have $T_{ij,j} = \sum_{j=1}^{3} \partial T_{ij}/\partial x_j$). Such tensorial notations are useful to keep expressions in tractable length and are easily transposed as loops in computer programs.

The rank-4 elastic tensor c_{ijkl} has symmetries that are directly dictated by the lattice symmetry of the solid material. It is not the purpose of this chapter to discuss these symmetries and the corresponding number of independent elastic constants, as such a discussion and practical values for usual materials can be found in classical textbooks [1, 2]. In many papers on wave propagation in solid phononic crystals, the equations are specified for specific cases such as isotropic or cubic media; we note however that as long as the tensor formulation is understood and correctly applied, such simplifications do not generate any practical computational gains.

Stresses can be eliminated from Eqs. (4.1)–(4.2) to obtain a wave equation for the displacements only

$$(c_{ijkl}(\mathbf{r})\, u_{k,l}(\mathbf{r},t))_{,j} = \rho(\mathbf{r}) \frac{\partial^2 u_i(\mathbf{r},t)}{\partial t^2}. \tag{4.3}$$

This last equation is often taken as the basis for formulating models for phononic crystals, because it only involves one unknown field, but we stress that it is not directly suited to surface and evanescent wave problems, for which mixed displacement/stress formulations are both elegant and efficient.

In case the propagation medium (i.e., the phononic crystal) is perfectly periodic and the fields are monochromatic, we can further make use of the Bloch-Floquet theorem to obtain equations limited to the unit-cell (please see Chap. 2 for a definition of the unit-cell). Specifically, if $\mathbf{a}$ is any vector giving the repetition of the structure in space, we have $\rho(\mathbf{r}) = \rho(\mathbf{r}+\mathbf{a})$ and $c_{ijkl}(\mathbf{r}) = c_{ijkl}(\mathbf{r}+\mathbf{a})$ and all elastic quantities describing an eigenmode can be written in the form

$$u_i(\mathbf{r},t) = \bar{u}_i(\mathbf{r}) \exp(\imath(\omega t - \mathbf{k}\cdot\mathbf{r})), \tag{4.4}$$

with ω the angular frequency, $\mathbf{k}$ the Bloch-Floquet wave vector, and $\bar{u}_i(\mathbf{r})$ a periodic function defined in the unit-cell ($\bar{u}_i(\mathbf{r}) = \bar{u}_i(\mathbf{r}+\mathbf{a})$). A similar expression holds for the stresses. We can write for the periodic part of the solution

$$\bar{T}_{ij}(\mathbf{r}) = c_{ijkl}(\mathbf{r})(\bar{u}_{k,l}(\mathbf{r}) - \imath k_l \bar{u}_k(\mathbf{r})), \tag{4.5}$$

$$\bar{T}_{ij,j}(\mathbf{r}) - \imath k_j \bar{T}_{ij}(\mathbf{r}) = -\rho(\mathbf{r})\omega^2 \bar{u}_i(\mathbf{r}). \tag{4.6}$$

The case of piezoelectric solids is a bit more involved but actually follows exactly the same line. In piezoelectric materials, an electromagnetic wave travels along with the elastic wave and provides a very efficient way to generate and detect elastic waves with electrical signals. In the quasi-static approximation of Maxwell equations, it suffices to consider the electric field vector to derive from a scalar potential ϕ, and the constitutive laws are taken in the linear limit as [3]

$$T_{ij}(\mathbf{r},t) = c_{ijkl}(\mathbf{r})\, u_{k,l}(\mathbf{r},t) + e_{kij}\phi_{,k}(\mathbf{r},t), \tag{4.7}$$

$$D_i(\mathbf{r},t) = e_{ikl}u_{k,l}(\mathbf{r},t) - \epsilon_{ij}\phi_{,j}(\mathbf{r},t). \tag{4.8}$$

Here $\mathbf{D}$ is the electric induction vector, e_{ikl} is the piezoelectric tensor, and ϵ_{ij} is the dielectric tensor considered at low frequency (i.e., as opposed to optical frequencies). In addition to the above two equations, Eq. (4.2) still holds, and the auxiliary Maxwell equation $D_{i,i} = 0$ must be added. As explained, for instance, in [4], the equations for piezoelectric solids can be cast in a form similar to those of elastic solids by defining a general displacement vector with $u_4 = \phi$ and a generalized stress tensor with $T_{4i} = D_i$.

To summarize this subsection, elastic wave propagation in solids can be described by either a constitutive relation relating stresses and displacements and the fundamental relation of dynamics, such as the set of equations (4.1–4.2), or by a wave equation for the displacement only, such as Eq. (4.3). As long as only linear phenomena are considered, this model can be enriched in a straightforward manner to include effects such as piezoelectricity, piezomagneticity, chemical potentials, and so on.

4.1.2 Equations for Fluids

It is well known that wave propagation in fluids (gases, liquids) only involves longitudinal displacements as long as viscoelasticity can be neglected. As a consequence, it might be thought that propagation in fluids can be treated as a limiting case of elastic wave propagation in isotropic solids with only the longitudinal displacements considered; but it turns out that this simplification is not correct. In the absence of applied external forces, the basic equations for propagation of acoustic waves in a still fluid can instead be taken as

$$-\nabla p(\mathbf{r},t) = \rho(\mathbf{r})\frac{\partial \mathbf{v}(\mathbf{r},t)}{\partial t}, \tag{4.9}$$

$$\frac{\partial p(\mathbf{r},t)}{\partial t} = -B(\mathbf{r})\nabla\cdot\mathbf{v}(\mathbf{r},t), \tag{4.10}$$

where variable dependence has been stressed and will not be repeated in the following. ρ is again the mass density and B is a bulk modulus expressed with the same units as the elastic constants. We have purposely employed vector notations instead of tensorial notations because it is the customary way to present these equations, but of course both are equivalent. The velocity can be eliminated from Eq. (4.10) to get a scalar wave equation for the pressure

$$\frac{1}{B}\frac{\partial^2 p}{\partial t^2} = \nabla \cdot \left(\frac{1}{\rho}\nabla p\right). \tag{4.11}$$

If we had eliminated the pressure instead, we would have obtained a vector wave equation for the velocity (or the displacement) similar (but not equivalent) to Eq. (4.3). Obviously, solving the pressure wave equation is the best choice in practice, in particular for band structure computations.

Nevertheless, when surface and evanescent wave problems are considered, the set of first-order differential equations (4.9)–(4.10) relating pressure and velocity is the logical choice. Further assuming monochromaticity and making use of the Bloch-Floquet theorem as in the previous subsection, we have for the periodic part of the solution

$$-\frac{1}{\rho}(\bar{p}_{,i} - \imath k_i \bar{p}) = \imath\omega\bar{v}_i, \tag{4.12}$$

$$\imath\omega\frac{1}{B}\bar{p} = -(\bar{v}_{i,i} - \imath k_i \bar{v}_i). \tag{4.13}$$

The case of phononic crystals containing both solid and fluid materials has been considered quite often in the literature, especially in relation to experiments, for instance for the steel rods in water or air cases. The reason is that macroscopic realizations of phononic crystals are rather easy with these material systems. As is apparent from the equations given above for the solid and the fluid cases, it is not possible to use just one set of equations with coefficients assuming different values in different regions of space. The rigorous way to consider the problem is to solve the problem known as fluid–structure interaction. For instance, if the displacement field has been computed inside the solid, Eq. (4.10) and use of the divergence theorem from vector calculus lead to a boundary condition relating the pressure created in the fluid with the normal acceleration of the solid boundary. Such boundary conditions can be employed naturally with FDTD and FEM methods described later in this chapter. The reverse reasoning shows that the pressure exerted by the fluid on the solid boundary leads to a boundary condition for the acceleration inside the solid. Simpler but only approximate methods can be followed if the solid is considered very rigid: either the solid boundary can be considered to have no motion (hence it is assumed there are no elastic waves in the solid), or the solid material can be regarded as an equivalent fluid supporting only longitudinal waves (and thus having an equivalent bulk modulus). The latter approximate solution gives results less distant from the exact solution than the former. Such approximate solutions have been employed mostly with PWE methods, but are not restricted to them.

4.1.3 Consideration of Material Loss

We end this section on basic equations with a short discussion of the representation of material losses. There are several physical origins for loss occurring during the propagation of elastic waves and thus not one unique solution to represent them. Losses occurring because of microscopic defects in the materials or because of scattering at random rough surfaces can be described in statistical terms and are thus in some limit amenable to macroscopic ad hoc loss terms. Elastic waves in solids naturally interact with thermal acoustic phonons and this loss mechanism can be viewed as dominating in perfectly ordered crystalline solids such as silicon, quartz, or lithium niobate, to cite a few materials that are used for the realization of hypersonic phononic crystals by microfabrication techniques. In general, it is found that losses in the best elastic materials increase with the square of the frequency. A simple approach to viscous losses that results in this behavior is to consider that at a fixed temperature we can add a restoring term depending on a viscosity tensor η_{ijkl} and the time derivative of the strain tensor so that the constitutive relation is modified to [2]

$$T_{ij} = c_{ijkl}\, u_{k,l} + \eta_{ijkl}\frac{\partial u_{k,l}}{\partial t}. \tag{4.14}$$

The viscosity tensor has the same symmetries as the elastic tensor. Such an expression is well suited to time-domain methods such as FDTD. For monochromatic waves, the elastic constants are then seen to be modified from real to complex valued as

$$c'_{ijkl} = c_{ijkl} + \imath\omega\eta_{ijkl}. \tag{4.15}$$

This modification offers a simple means to include the effect of losses at a fixed frequency, e.g. with the extended PWE method [5].

4.2 Computational Problems of Phononic Crystals

Before turning to the different type of methods that are used to describe wave propagation in phononic crystals, we outline in this section the different kinds of problems that one encounters and their characteristics in terms of numerical computation. There is a first obvious separation of problems depending on whether they are considered in the spectral or the time domain. However, it is in principle always possible to go from one type to the other with the help of Fourier transforms. A deeper separation of problems, in our opinion, stems from the boundary conditions that are considered, as these lead to different types of waves (i.e., bulk, surface, and plate waves). This is exposed in Sect. 4.2.1. Less fundamental in character, but very important for the understanding of the physics behind phononic crystals are the different problem types that can be considered: band structures, waveguides, cavities, scattering problems. These are considered in Sect. 4.2.2.

4.2.1 Classification by Geometry

In case the phononic crystal is unbounded, there are no boundary conditions required to terminate the infinite domain of definition. We speak of the propagation of bulk elastic or acoustic waves. In a physical three-dimensional space, we can consider from 1 to 3 periodicities and it is customary to refer to 1D, 2D, and 3D phononic crystals. When the phononic crystal fills a semi-infinite domain terminated by a given surface, there are additional waves that can be guided along the surface that we term surface waves. In case there can be radiation to the remaining half space, we speak of interface waves. Plate (or slab) phononic crystals are limited by two parallel planes and thus two surface or interface boundary conditions. Rather than attempting a lengthy description, we depict in Fig. 4.1 the different possibilities that arise from this discussion. Note that there are even more complicated cases that can be considered (e.g., finite phononic crystals with arbitrary enclosing surfaces), so that our tentative classification is clearly not exhaustive.

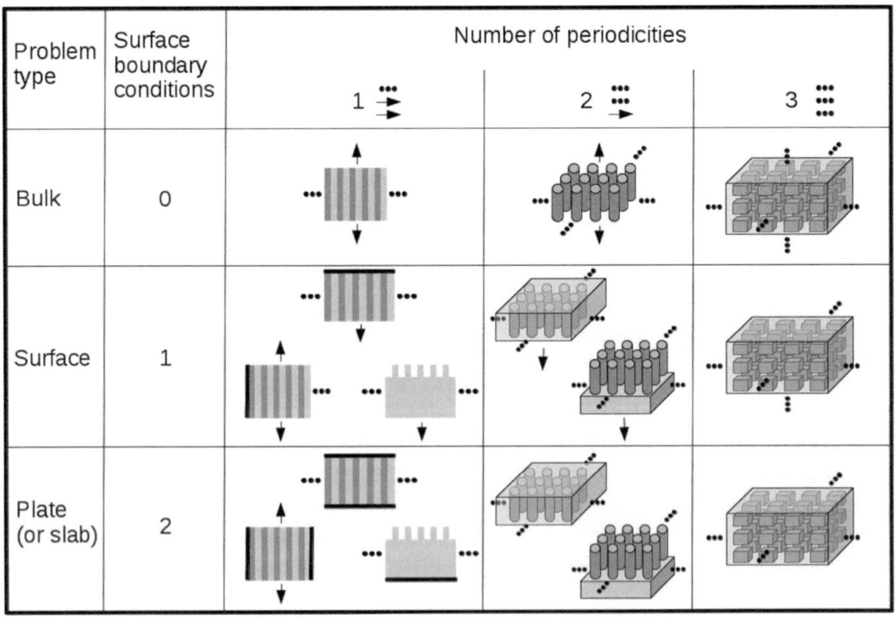

Fig. 4.1 Classification of phononic crystals according to their geometry. *Three consecutive dots* indicate a direction of periodic repetition of the unit-cell, while an *arrow* means the structure extends homogeneously to infinity

Table 4.1 Basic computational problems for perfectly periodic and defect-based phononic crystals

Structure		Dispersion (infinite structure)	Scattering (finite structure)
Phononic crystal	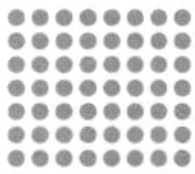	Band structure $\omega(k)$ Complex BS $k(\omega)$	Transmission, reflection, and diffraction of an incident plane wave
Waveguide	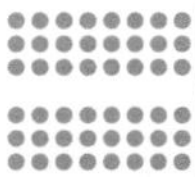	Guided modes	Waveguide transmission
Cavity	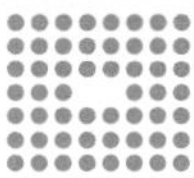	Confined modes	Cavity transmission

4.2.2 *Classification by Problem Type*

The previous classification by geometry insists upon the physical properties of phononic crystals, and especially upon the types of waves that they can support. A second classification can be performed based on the phononic crystal structure, as attempted in Table 4.1. Perfectly periodic phononic crystal can be thought ideally as extending to infinity, in which case any solution of the wave equation can be written as a superposition of Bloch waves. The dispersion relation of all Bloch waves forms the band structure which can in general be obtained by solving a generalized eigenvalue problem, possibly supplemented with surface or interface boundary conditions. Note that Bloch waves constitute a complete basis (in the monochromatic sense) only if all complex $k(\omega)$ solutions are considered [6]. If a perfectly periodic phononic crystal of finite size is considered, for instance for comparison with experiments, then the scattering of an incident plane wave to all orders of diffraction is an important basic problem. Phononic crystal cavities and waveguides can be formed by breaking locally the periodicity of the phononic crystal structure. Defect-based phononic structures are generally mostly interesting for frequencies that fall inside a complete phononic band gap, in which case perfect guidance or confinement can be achieved. Otherwise, phononic crystal waveguides and cavities are leaky. By defining a supercell that artificially restores periodicity,[1] waveguide and cavity modes can be estimated, and their band structures

[1] A supercell extends a few periods away from the defect it encloses. Modal computations then give physically meaningful results when only evanescent Bloch waves of the elementary phononic crystal exist, i.e., inside a complete band gap. Furthermore, the number of phononic crystal rows must be sufficient so that the Bloch wave with the least imaginary part of the wave vector can be considered negligible on the boundary of the supercell.

can be obtained. Finite-size defect-based phononic crystal structures are in general considered for comparison with experiments, with the accent on obtaining the transmission of the waveguide or the resonance properties of the cavity.

4.3 Multiple Scattering Theory and Layer Multiple Scattering Methods

Multiple scattering type computation methods (MST) have a long history in condensed matter and particle physics. The methods used for phononic crystals have a close relation with those used for the treatment of electron scattering in solids and computing the band structure of photonic crystals and transmission through them. The basic idea is to make use of the superposition principle for non-overlapping scatterers. Considering a single scatterer with a given shape, one first obtains the field scattered from an incident plane wave with given angular frequency and wave vector. Then the scattered field from an assembly of scatterers is constructed by summing properly all individual contributions. The method thus works very efficiently if the single scatterer problem can be solved analytically, as is the case of isotropic spherical scatterers in a homogeneous isotropic medium (for which spheroidal expansions are known) or of infinite isotropic cylinders in a homogeneous isotropic medium. Multiple scattering methods have been employed with great success for the computation of band structures, density of states, and transmission through finite phononic crystals, though almost always for isotropic spheres or cylinders embedded in an isotropic background, as is the case for sonic crystals. Generalization of multiple scattering methods to arbitrary anisotropic periodic materials still stands as a highly difficult problem.

Arguably the most developed and versatile MST method is the layer-multiple-scattering (LMS) method [7–9]. We refer the reader to these papers for a full description of the method and only outline here the method. The basic principle of the method is that the wave incident on a given scatterer is the sum of the waves outgoing from all the other scatterers and the externally incident wave. A scattering matrix relating the amplitudes (which are expansion coefficients in a given basis, e.g., spheroidal harmonics) of the scattered wave with those of the incident wave is obtained for the composite system from the corresponding matrices of the individual scatterers and proper propagator functions in the host medium. The LMS method considers a phononic crystal to be constituted by a stack of phononic crystal layers. It proceeds layer by layer by evaluating the scattering properties of the phononic crystal from those of the constituent phononic crystal layers, for given frequency and wave vector. The layers can be either planes of 3D scatterers with the same 2D periodicity, or homogeneous layers. For each plane of scatterers, the method calculates the full multipole expansion of the total multiply scattered wave field and deduces the corresponding transmission and reflection matrices in the plane-wave basis for the surrounding media. For homogeneous layers, the transmission and reflection matrices are directly obtained in the plane-wave basis of the homogenous material.

In any multiple scattering calculation, the angular frequency is a fixed quantity. The wave vector only appears as a eigenvalue in band structure computations or as a particular component of the wave vector of the wave incident on a finite phononic crystal. As such, MST and LMS methods give access to complex band structures and to the consideration of propagation loss in the form of frequency-dependent viscous elastic constants [10].

4.4 Plane Wave Expansion Method

The discussion in this section relies on the presentation of the plane wave expansion (PWE) method proposed in [4, 11]. However, whereas these papers contained the full piezoelectric formulation, we restrict it here to the (anisotropic) elastic case for simplicity of the presentation. The PWE method relies on a literate application of the Bloch-Floquet theorem. As discussed in Sect. 4.1.1, any eigenmode of a perfectly periodic medium (termed a Bloch wave) is the product of a plane-wave like term and a function periodic in the unit cell. In the PWE method, the periodic function is considered explicitly through its Fourier series expansion. Specifically, we write for any displacement component

$$u(\mathbf{r},t) = \sum_{\mathbf{G}} u_{\mathbf{G}}(\omega,\mathbf{k}) \exp(\imath(\omega t - \mathbf{k}\cdot\mathbf{r} - \mathbf{G}\cdot\mathbf{r})), \tag{4.16}$$

where the vectors of the reciprocal lattice are (for the square lattice case) $\mathbf{G} = (2\pi m_1/a, 2\pi m_2/a, 2\pi m_3/a)^T$. One should note that the Fourier coefficients $u_{\mathbf{G}}(\omega,\mathbf{k})$ are those of the periodic part of the Bloch wave solution, $\bar{u}$.

The periodicity of the structure is also used to expand the material constants as Fourier series

$$\alpha(\mathbf{r}) = \sum_{\mathbf{G}} \alpha_{\mathbf{G}} \exp(-\imath\mathbf{G}\cdot\mathbf{r}), \tag{4.17}$$

where α is either ρ or c_{ijkl}. The Fourier harmonics $\alpha_{\mathbf{G}}$ are easily calculated for various scatterers and lattice geometries [12, 13].

4.4.1 Band Structures with PWE

Assuming a truncation to a total of N harmonics in the Fourier expansions, the following vector notations are considered for the harmonics of the generalized stress and displacement fields ($3N$ components each)

$$\mathbf{T}_i = \left(\mathbf{t}_{i\mathbf{G}_1} \ldots \mathbf{t}_{i\mathbf{G}_N}\right)^T, \tag{4.18}$$

$$\mathbf{U} = \left(\mathbf{u}_{\mathbf{G}_1} \ldots \mathbf{u}_{\mathbf{G}_N}\right)^T, \tag{4.19}$$

where the vectors of the reciprocal lattice, $\mathbf{G}_m$, are labeled using a single index m. Bulk waves are then obtained as the eigensolutions of the secular equation

$$\omega^2 R\mathbf{U} = \sum_{i,l=1,3} \Gamma_i A_{il} \Gamma_l \mathbf{U}, \tag{4.20}$$

with the $3N \times 3N$ matrices Γ_i, A_{il} and R defined by the $N \times N$ blocks with 3×3 elements

$$(\Gamma_i)_{mn} = \delta_{mn}(k_i + G_{im})I_3, \tag{4.21}$$

$$(A_{il})_{mn} = A_{il\mathbf{G}_m - \mathbf{G}_n}, \tag{4.22}$$

$$(R)_{mn} = \rho_{\mathbf{G}_m - \mathbf{G}_n} I_3, \tag{4.23}$$

with I_3 the 3×3 identity matrix and

$$A_{il\mathbf{G}}(j,k) = c_{ijkl\mathbf{G}}, \tag{4.24}$$

with $i, j, k, l = 1, 2, 3$ and $m, n = 1 \ldots N$. Equation (4.20) defines a generalized eigenvalue problem which can be solved for ω^2 as a function of $\mathbf{k}$ to obtain the band structure of bulk waves. This formulation encompasses 1D, 2D, and 3D bulk wave problems, plus the out-of-plane 2D bulk wave propagation problem [14].

It is also instructive to consider the problem of surface waves propagating on the surface of a 2D phononic crystal. The phononic crystal is assumed to be periodic in the directions x_1 and x_2 and invariant along direction x_3, except for the presence of a surface at $x_3 = 0$. Surface waves are defined by the angular frequency ω and a wave vector $\mathbf{k} = (k_1, k_2, 0)^T$ defined in the plane of the surface. The two-dimensional phononic crystal is not periodic along the x_3 axis and k_3 can be obtained as a function of the other parameters of the model—i.e., k_1, k_2, and ω. We group the displacements and the stresses normal to the surface in the $6N$-components state vector $\mathbf{H} = (\mathbf{U}, \mathbf{T}_3)^T$ and obtain k_3 as the eigenvalue of the equation

$$\begin{bmatrix} \omega^2 \tilde{R} - B & 0 \\ -C_2 & I_d \end{bmatrix} \mathbf{H} = k_3 \begin{bmatrix} C_1 & I_d \\ D & 0 \end{bmatrix} \mathbf{H}, \tag{4.25}$$

where

$$B = \sum_{i,j=1,2} \Gamma_i \tilde{A}_{ij} \Gamma_j, \; C_1 = \sum_{i=1,2} \Gamma_i \tilde{A}_{i3},$$

$$C_2 = \sum_{j=1,2} \tilde{A}_{3j} \Gamma_j, \qquad D = \tilde{A}_{33}. \tag{4.26}$$

Solving this system yields $6N$ complex-valued eigenvalues k_{3q} and eigenvectors $\mathbf{H}_q$. By grouping in the eigenvectors the six components corresponding to the mth harmonic, we introduce the notation

$$\mathbf{h}_{mq} = \begin{pmatrix} (u_i)_{\mathbf{G}_{mq}} \\ (T_{3j})_{\mathbf{G}_{mq}} \end{pmatrix}, \tag{4.27}$$

with $i, j = 1, 2, 3$, $m = 1, \ldots, N$, and $q = 1, \ldots, 6N$. The generalized displacement and normal stress fields are obtained from the superposition with relative amplitudes A_q

$$\mathbf{h}(\mathbf{r}, t) = \sum_{m=1}^{N} \sum_{q=1}^{6N} A_q \tilde{\mathbf{h}}_{mq} \exp(\jmath(\omega t - (\mathbf{G}_m + \mathbf{k}_q) \cdot \mathbf{r})). \tag{4.28}$$

This superposition is a finite approximation to the infinite series (4.16). From this partial wave expansion, boundary conditions can be constructed to solve surface and plate problems [4, 11].

The PWE method has one drawback that is apparent in Eq. (4.16). The Fourier series expansion of the displacement and the stress fields implicitly makes the solution continuous everywhere inside the unit-cell. Whereas the displacements and the normal stresses can be considered continuous at an interface between different solid materials, such is not the case for shear stresses applying along the interface. Also, the boundary condition at the interface of a solid and a fluid cannot in general be satisfied, which makes the PWE method rather unsuitable for solid–fluid problems (its application can result in the appearance of spurious modes in the fluid part).

4.4.2 Evanescent Waves in Phononic Crystals

A limitation of the PWE method for band structure computations is that the eigenvalue problem of Eq. (4.20) gives eigenfrequencies at fixed values of the Bloch wave vector. Thus, it is not easy to consider materials with frequency-dependent elastic constants describing viscous losses, contrary to the LMS case. The extended PWE method relieves this particular limitation and enables the computation of the full complex band structure [5, 6], as we outline here.

It is well-known in the theory of diffraction gratings and near-field optics or acoustics that evanescent waves play an essential role in all wave propagation problems. The set of all propagative and evanescent Bloch waves constitutes a complete basis for phononic crystal problems [6]. Evanescent Bloch waves can be characterized as eigensolutions of the periodic Helmholtz equation describing monochromatic wave propagation, for which the wave vector can be complex-valued. Usual band structures are obtained assuming the wave vector to be a real-valued quantity (through the $\omega(\mathbf{k})$ eigenvalue problem); complex band structures are obtained by assuming real-valued frequencies and solving for the (generally complex-valued) wave vector (the $\mathbf{k}(\omega)$ problem).

As an example, let us outline how the complex band structure for solid phononic crystals can be obtained from the equations we have written in Sect. 4.1. Multiplying Eq. (4.3) with α_j we get

$$\bar{T}'_i = \alpha_j \bar{T}_{ij} = c_{ijkl}\alpha_j \bar{u}_{k,l} - \imath k c_{ijkl}\alpha_j \alpha_l \bar{u}_k. \tag{4.29}$$

Inserting Eq. (4.3) into Eq. (4.4) we also get

$$\rho\omega^2 \bar{u}_i = -c_{ijkl}\bar{u}_{k,jl} + \imath k c_{ijkl}\alpha_l \bar{u}_{k,j} + \imath k \bar{T}'_i. \tag{4.30}$$

$\bar{T}'_i$ is the component along x_i of the stress tensor projected along the propagation direction. In the two last equations, the modulus of the wave vector, k, appears in factor of terms that are linear in the unknown periodic fields $\bar{u}$ and $\bar{\phi}$. Thus these equations can be used to formulate a generalized eigenvalue problem for the eigenvalue k. Written in terms of the PWE notations given in Sect. 4.3, the eigenvalue problem is as follows [6]:

$$\begin{pmatrix} -C_2 & I_d \\ \omega^2 R - B & 0 \end{pmatrix} \begin{pmatrix} U \\ \imath T' \end{pmatrix} = k \begin{pmatrix} D & 0 \\ C_1 & I_d \end{pmatrix} \begin{pmatrix} U \\ \imath T' \end{pmatrix}, \tag{4.31}$$

with $B = \Gamma_i A_{ij} \Gamma_j$, $C_1 = \Gamma_i A_{ij} \alpha_j$, $C_2 = \alpha_i A_{ij} \Gamma_j$, $D = \alpha_i A_{ij} \alpha_j$, and $(\Gamma_i)_{mn} = (k_{0i} + G_i^m)\delta_{mn}$. It can be checked easily that Eq. (4.25) appears as a special case of the above generalized eigenvalue problem. Equations (4.29)–(4.30) can also be used to formulate the finite element version of the generalized eigenvalue problem, by using a mixed displacement-stress formulation.

4.5 Finite Element Method

The finite element method (FEM) is a numerical technique to solve partial differential equations (PDE) and integral equations in the time domain as well as in the spectral domain. The primary challenge is to create an equation that approximates the equation to be studied and that is numerically convergent. FEM is a powerful method suitable for solving PDE over complicated domains, when the domain changes—e.g., for moving boundaries—and when the desired precision varies over the entire domain. Several authors of the phononic community have recently proposed FEM to study different geometries and compositions, through the calculation of band structures [15, 16] or the design of cavities and waveguides—defect modes—in perfect structures [17]. The commercial software COMSOL, for instance, is now widely used to compute band structures of phononic crystals. In the following, we describe an example of such a calculation in the case of the phononic crystal slab.

The geometry of a square lattice phononic crystal slab is depicted in Fig. 4.2. The phononic crystal is assumed to be infinite and arranged periodically in the x and y directions. a_1 and a_2 are the pitches of the array, with $a_1 = a_2 = a$ in

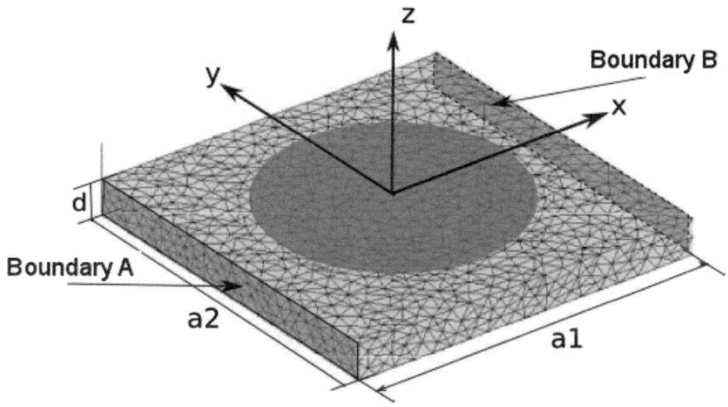

Fig. 4.2 Unit cell of phononic crystal slab

the square lattice case. The slab has a finite thickness d in the z direction. Each unit cell is indexed by (m, p). The whole domain is split into successive unit cells, consisting of a single cylinder surrounded by a matrix. The inclusions are assumed to have a circular cross section so that the filling fraction is $F = \pi r^2/a^2 = 0.5$, where r is the radius of the inclusion. The unit cell is meshed and divided into finite elements connected by nodes as shown in Fig. 4.2. According to the Bloch-Floquet theorem, all fields obey a periodicity law, yielding, for instance, the following relation between the mechanical displacements u_i for nodes lying on the boundary of the unit cell

$$u_i(x + ma_1, y + pa_2, z) = u_i(x, y, z) \exp\left(-\imath\left(k_x ma_1 + k_y pa_2\right)\right), \tag{4.32}$$

where k_x and k_y are the components of the Bloch wave vectors in the x and y directions, respectively. Considering the periodic boundary conditions above allows us to reduce the model to a single unit cell which can be meshed using finite elements. A mechanical displacement (for elastic solids) and electrical potential (for piezoelectric solids) FE scheme is used. Considering a monochromatic variation of mechanical and electrical fields with a time dependence $\exp(\imath\omega t)$, where ω is the angular frequency, the general piezoelectric problem with no applied external force can be written

$$\begin{pmatrix} K_{uu} - \omega^2 M_{uu} & K_{u\phi} \\ K_{\phi u} & K_{\phi\phi} \end{pmatrix} \begin{pmatrix} u \\ \phi \end{pmatrix} = \begin{pmatrix} 0 \\ 0 \end{pmatrix}, \tag{4.33}$$

where K_{uu} and M_{uu} are the stiffness and mass matrices of the purely elastic part of the problem, $K_{u\phi}$ and $K_{\phi u}$ are piezoelectric-coupling matrices, and $K_{\phi\phi}$ accounts for the purely dielectric problem. u and ϕ represent, respectively, all displacements and electrical potential at the nodes of the mesh, gathered together in vector form. As the angular frequency ω is a periodical function of the wave vector, the problem can be reduced to the first Brillouin zone. Practically, we relate all the degrees of freedom on boundary A to those on boundary B along the x direction, according to

$$\begin{pmatrix} u_{i(B)} \\ \phi_{(B)} \end{pmatrix} = \begin{pmatrix} u_{i(A)} \\ \phi_{(A)} \end{pmatrix} \exp\left(-\imath\left(k_x a\right)\right), \tag{4.34}$$

in which k_x varies in the interval $(0, \pi/a)$ along the x direction. Similar periodic boundary conditions are applied to the boundaries orthogonal to the y axis. The dispersion curves are then obtained by varying the wave vector in the first Brillouin zone for a given propagation direction and by solving the eigenvalue problem to obtain the eigenfrequency solutions, or angular pulsations, ω. The full band structure is then deduced using symmetries. An example of a dispersion calculation is presented in Fig. 4.3.

In transmission calculations involving a phononic crystal of finite size, an acoustic wave with specific polarization (u_x, u_z, u_y) can be generated by applying a line source vibration on the upper surface. We apply in addition in the y direction periodic boundary conditions to account for the lateral size of the structure. The line source generates monochromatic waves propagating along the x direction. To prevent reflections of the scattering wave from the edge, a perfectly matched layer (PML) [18] is applied. The PMLs have the property that the mechanical disturbances are gradually absorbed in the layers before they reach the outer boundaries. In this way, there will be no reflections that can disturb the propagation of the source wave (Fig. 4.4). Indeed, we can write the dynamical equation as

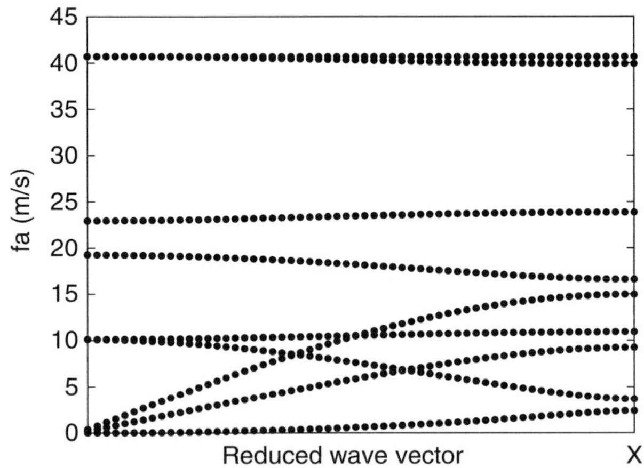

Fig. 4.3 Band structure of a periodic array of tungsten rods in epoxy. The lattice is based on square symmetry. The filling fraction is about 0.35. A large band gap extending from $fa = 23$ to 42 m/s is obtained

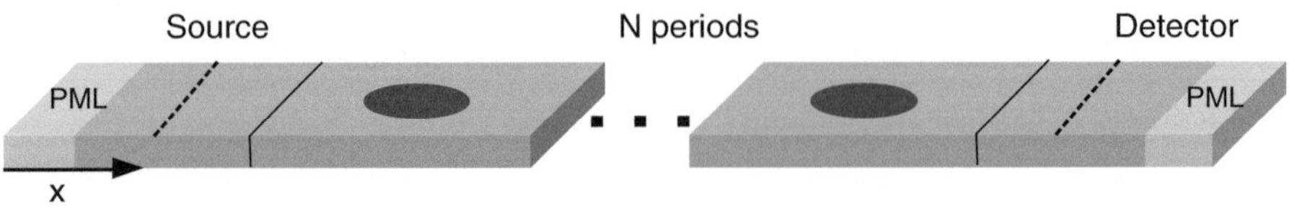

Fig. 4.4 Phononic crystal slab geometry for transmission calculation

$$\frac{1}{\gamma_j}\frac{\partial T_{ij}}{\partial x_j} = -\rho\omega^2 u_i, \tag{4.35}$$

where ρ is the density of the material. The parameter γ_j is an artificial damping factor at position x_j in the PML. As the PML is added to attenuate the acoustic wave propagating in the plane of the structure, for propagation in direction x only γ_x is different from 1 and is given by

$$\gamma_x(x) = 1 - \imath\sigma_x(x - x_l)^2, \tag{4.36}$$

where x_l is the coordinate at the interface between the regular domain and the PML, and σ_x is a suitable constant. There is no damping outside the PMLs and $\gamma_x = 1$ there. A suitable thickness of the PML as well as the value of σ_x must be found by calculations such that the mechanical disturbances are absorbed before reaching the outer boundaries. Absorption, however, must also be sufficiently low as reflections will occur at the interface between the regular domain and the PML if their material properties are not comparable. Mechanical stresses depend on the strains as usual

$$T_{jk} = c_{jklm}S_{lm}, \tag{4.37}$$

where c_{jklm} are the elastic stiffness constants, but strains are now defined from the displacements as

$$S_{ij} = \frac{1}{2}\left(\frac{1}{\gamma_j}\frac{\partial u_i}{\partial x_j} + \frac{1}{\gamma_i}\frac{\partial u_j}{\partial x_i}\right). \tag{4.38}$$

4.6 Finite-Difference Time-Domain Method

Finite-difference time-domain (FDTD) is a well-known computational electro- and elastodynamics modeling technique. It is usually considered easy to understand and easy to implement. Since it is a time-domain method, with the advantage that the solutions can cover a wide frequency range with a single simulation run. The FDTD method belongs in the general class of grid-based differential time-domain numerical modeling methods. Generally, the time-dependent wave equations—in partial differential form—are discretized using central-difference approximations to the space and time partial derivatives. The resulting finite-difference equations are solved in a leapfrog manner: in the case of elastic wave, the displacement field vector components in a volume of space are solved at a given instant in time; then, the stress field tensor components in the same spatial volume are solved at the next instant in time; and the process is repeated until the desired transient or steady-state acoustic or elastic displacement field behavior has fully evolved. Since the rise of phononic crystal field, FDTD has emerged as an alternative technique to PWE when dealing with finite size structure—to evaluate the transmission through a finite number of period—and more important with acoustic/elastic wave interactions in solid/fluid mixed phononic structures. In the latter case, PWE fails in getting stable solutions due to the non-consideration of shear waves in the equations describing the fluid.

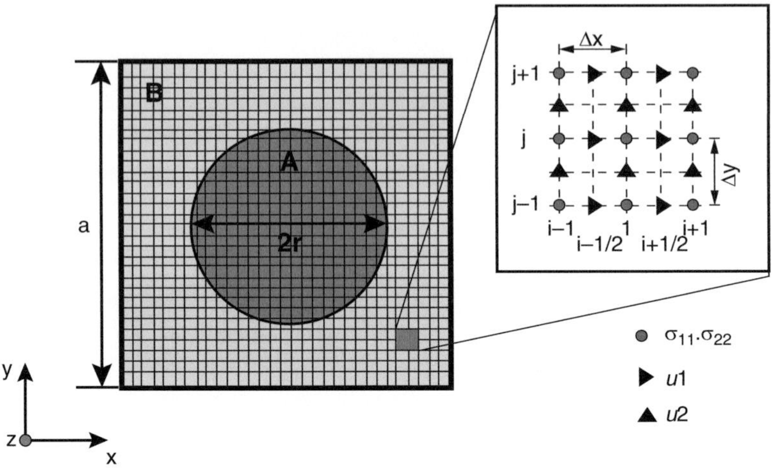

Fig. 4.5 Gridded unit cell of a phononic crystal composed of a cylinder (*A*) embedded in a background (*B*) and arranged according to a square lattice. A closer view of spacial discretization of elastic fields is shown

In the following, we describe how FDTD can obtain the band structure of infinite periodic 2D phononic crystals composed and the spectral transmission through a finite size structure.

We first consider a perfect 2D elastic or acoustic system composed of a periodic array of cylinders of a material A embedded in a background material B according to a square lattice array (see Fig. 4.5). Both materials are assumed to be elastically isotropic. We choose the z $(= x_3)$ direction parallel to the cylinder axis of the 2D phononic crystal. The mass density ρ and the elastic tensor c_{ijmn} of the system depend only upon coordinates $(x, y) = (x_1, x_2)$ in the plane perpendicular to the cylinder axis. The equations of motion for the lattice displacement are those of Eqs. (4.1)–(4.2). In the phononic crystals composed of isotropic materials and for acoustic waves propagating in the (x, y) plane, the equations of motion can be decomposed into two independent sets of equations. One is in explicit form

$$\rho(x,y)\frac{\partial^2 u_1(x,y,t)}{\partial t^2} = \frac{\partial T_{11}(x,y,t)}{\partial x} + \frac{\partial T_{12}(x,y,t)}{\partial y}, \tag{4.39}$$

$$\rho(x,y)\frac{\partial^2 u_2(x,y,t)}{\partial t^2} = \frac{\partial T_{21}(x,y,t)}{\partial x} + \frac{\partial T_{22}(x,y,t)}{\partial y}, \tag{4.40}$$

$$T_{11}(x,y,t) = c_{11}(x)\frac{\partial u_1(x,y,t)}{\partial x} + c_{12}(x,y)\frac{\partial u_2(x,y,t)}{\partial y}, \tag{4.41}$$

$$T_{12}(x,y,t) = T_{21}(x,y,t) = c_{44}(x,y)\left(\frac{\partial u_2(x,y,t)}{\partial x} + \frac{\partial u_1(x,y,t)}{\partial y}\right), \tag{4.42}$$

$$T_{22}(x,y,t) = c_{12}(x,y)\frac{\partial u_1(x,y,t)}{\partial x} + c_{11}(x,y)\frac{\partial u_2(x,y,t)}{\partial y}. \tag{4.43}$$

In these equations, the c_{IJ} are the contracted elastic constants with two indices and the relations $c_{21} = c_{21}$, $c_{22} = c_{11}$, and $c_{66} = c_{44}$ have been used. The equations are restricted to mixed modes consisting of both longitudinal and transverse vibrations, which means that shear modes with a pure transverse vibration along z is not included in these equations.

The basic FDTD space grid and time-stepping algorithm traces back to a paper written by K. Yee, which was originally applied to electromagnetic wave propagation in dielectric media [19]. To solve Eqs. (4.39)–(4.43), we extend Yee's discretization scheme in the framework of the FDTD method. The variables involved in those equations are now defined on rectangular grids with side lengths Δx and Δy. Displacement fields are separated in time by a unit time step Δt and the displacement and stress fields are spatially interleaved by half a grid cell. Based on this scheme, center differences in both space and time are used to approximate the equations of motion for mixed modes. The explicit expressions for this discretization version are given as

$$\begin{aligned} u_1^{i+\frac{1}{2},j;n+1} &= 2u_1^{i+\frac{1}{2},j;n} - u_1^{i+\frac{1}{2},j;n-1} \\ &\quad + \frac{(\Delta t)^2}{\rho^{i+\frac{1}{2},j}}\left[\frac{\sigma_{11}^{i+1,j;n} - \sigma_{11}^{i,j;n}}{\Delta x} + \frac{\sigma_{12}^{i+\frac{1}{2},j+\frac{1}{2};n} - \sigma_{12}^{i+\frac{1}{2},j-\frac{1}{2};n}}{\Delta y}\right], \end{aligned} \tag{4.44}$$

$$\begin{aligned} u_2^{i,j+\frac{1}{2};n+1} &= 2u_2^{i,j+\frac{1}{2};n} - u_2^{i,j+\frac{1}{2};n-1} \\ &\quad + \frac{(\Delta t)^2}{\rho^{i,j+\frac{1}{2}}}\left[\frac{\sigma_{21}^{i+\frac{1}{2},j+\frac{1}{2};n} - \sigma_{21}^{i-\frac{1}{2},j+\frac{1}{2};n}}{\Delta x} + \frac{\sigma_{22}^{i,j+1;n} - \sigma_{12}^{i,j;n}}{\Delta y}\right], \end{aligned} \tag{4.45}$$

$$\sigma_{11}^{i,j;n} = c_{11}^{i,j}\frac{u_1^{i+\frac{1}{2},j;n} - u_1^{i-\frac{1}{2},j;n}}{\Delta x} + c_{12}^{i,j}\frac{u_2^{i,j+\frac{1}{2};n} - u_2^{i,j-\frac{1}{2};n}}{\Delta y}, \tag{4.46}$$

$$\sigma_{22}^{i,j;n} = c_{12}^{i,j}\frac{u_1^{i+\frac{1}{2},j;n} - u_1^{i-\frac{1}{2},j;n}}{\Delta x} + c_{11}^{i,j}\frac{u_2^{i,j+\frac{1}{2};n} - u_2^{i,j-\frac{1}{2};n}}{\Delta y}, \tag{4.47}$$

$$\begin{aligned} \sigma_{12}^{i+\frac{1}{2},j+\frac{1}{2};n} &= \sigma_{21}^{i+\frac{1}{2},j+\frac{1}{2};n} \\ &= C_{44}^{i+\frac{1}{2},j+\frac{1}{2}}\left[\frac{u_1^{i+\frac{1}{2},j+1;n} - u_1^{i-\frac{1}{2},j;n}}{\Delta y} + \frac{u_2^{i+1,j+\frac{1}{2};n} - u_2^{i,j+\frac{1}{2};n}}{\Delta x}\right], \end{aligned} \tag{4.48}$$

where (i, j) defines a 2D grid point (grid spacings are Δx and Δy), n specifies the time step with an interval Δt. In the fluid region, $c_{44} = 0$ and $c_{11} = c_{12}$, which simplifies the equations without disturbing the stability of the solution process as no matrix inversion is required.

4.6.1 Boundary Conditions

Two major boundary conditions can be distinguished depending on the purpose of calculations: (1) the periodic boundary condition simulating an infinite structure and (2) the absorbing boundary condition or absorbing region applied when a truncation of an open region is needed.

1. When we calculate the dispersion relation of bulk waves with a 2D wave vector in the (x, y) plane, we may consider a 2D unit cell of a square shape involving just one cylinder and may impose periodic boundary conditions in both x and y directions. Here we note that the periodic boundary conditions in the (x, y) plane are expressed as

$$u_i(x + L_x, y + L_y, t) = \exp(-\imath(k_x L_x + k_y L_y))u_i(x, y, t)), i = 1, 2 \tag{4.49}$$

   where (L_x, L_y) are the periods in the x and y directions, respectively, and in the present case $L_x = L_y = a$. These equations define the two-dimensional wave vector $k = (k_x, k_y)$ in the first Brillouin zone $-\pi/L_x < k_x < \pi/L_x$ and $-\pi/L_y < k_y < \pi/L_y$.
2. When we compute the transmission of acoustic or elastic waves through a finite phononic crystal structure with, say, five periods, homogeneous regions consisting of the background material are added to the lattice in the regions $x < 0$ and $x > 5a$. At the far ends of these regions in the longitudinal (x) direction, absorbing boundary conditions are imposed and periodic boundary conditions are applied at boundaries orthogonal to the lateral y direction. Basically, the most commonly used grid truncation techniques for open-region FDTD modeling problems are the Mur absorbing boundary condition (ABC), the Liao ABC and various perfectly matched layer (PML) formulations. The Mur and Liao techniques are simpler than the PML. However, PML (which is technically an absorbing region rather than a boundary condition) can provide orders-of-magnitude lower reflections. In this work, we use Mur absorbing boundary conditions (ABC) where we impose that the elastic or acoustic wave is propagating—at the ends of the medium in the x direction for instance—in the forward direction and is leaving the medium without reflexion. The condition is, therefore,

$$V_L \frac{\partial u_i(x, y, t)}{\partial x} + \frac{\partial u_i(x, y, t)}{\partial t} = 0, \tag{4.50}$$

where the longitudinal speed of sound in the radiation medium V_L is used for both the x and y components of the displacement field. A finite-difference approximation of that condition may be used.

4.6.2 Calculation of Dispersion Relations

In the example of Fig. 4.6, we consider a perfect 2D periodic crystal where elastic circular steel cylinders of radius r are embedded periodically in a water medium, forming a square lattice with lattice spacing a. For the calculation of dispersion relations, a small disturbance at a randomly chosen position of the structure is set as the initial condition. Thus all possible wave modes are excited inside the considered 2D unit cell, and the displacement can be recorded and subsequently expanded into Fourier series. Then the eigenfrequencies for a given wave vector k_x, k_y (imposed via the periodic boundary conditions) are obtained by selecting the resonance peaks of the spectrum. This procedure allows us to find all possible types of bulk waves. In the numerical calculations, the spacings of the grid points are chosen as $\Delta x = \Delta y = a/60$ and the unit time step Δt is selected as $\Delta t = 0.95\Delta t_{\max}$, where the stability criterion for the FDTD method requires

$$\Delta t < \Delta t_{\max} = \frac{1}{v_{\max}\sqrt{(\frac{1}{\Delta x})^2 + (\frac{1}{\Delta y})^2}}, \tag{4.51}$$

with $v_{\max} = v_{\text{steel}}$. In practical cases, after $t = 2^{17}\Delta t = 131,100\Delta t$ time steps, the vibration of the system becomes stationary and is composed of a superposition of eigenmodes.

Figure 4.6 shows the low frequency part of the dispersion curves of bulk waves along the boundary of the irreducible Brillouin zone. The filling fraction is $\mathrm{f} = 0.41$. We can notice the existence of a large complete gap that prohibits bulk wave propagation in the (x, y) plane.

4.6.3 Calculation of Transmission Spectra

To gain a better insight into the effects of band gaps, and also to investigate the qualitative behavior of transmission inside pass bands, the FDTD method is a powerful tool to calculate the transmission through finite size samples. To do so, the samples are composed of three adjacent regions (see the inset in Fig. 4.7). The probing signal, a longitudinal wave that propagates along the x direction, is launched from the first region and detected in the third one. The signal is the superposition of a few sinusoidal waves with frequencies centered around the central frequency of interest and is usually weighed by a Gaussian profile. Transmission of this signal

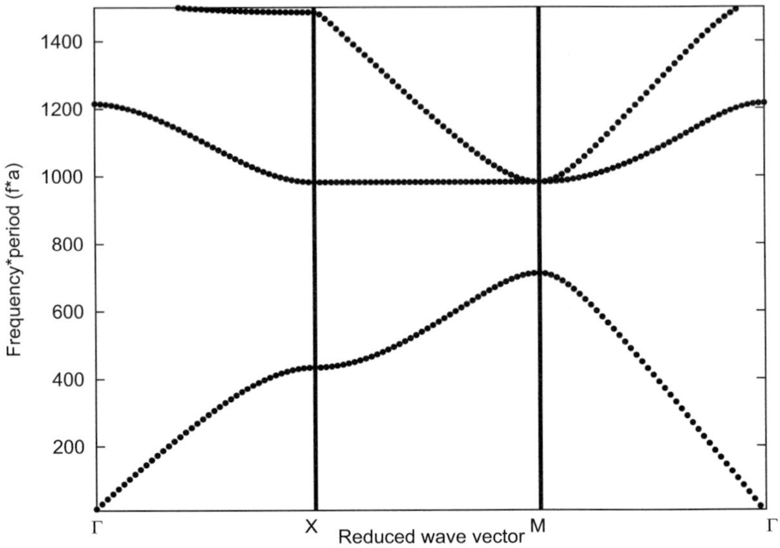

Fig. 4.6 Band structure of a square lattice array of steel rods in water computed with FDTD. The filling fraction is 0.41 approximately. A full band gap extending from 700 to 1,000 m/s is obtained

through a homogeneous water medium produces a broad band spectrum that covers the frequency range of interest. The central region contains the phononic crystal. To probe the transmission, several output displacements (longitudinal vibrations) are detected at different locations in the third region. Transmission spectra are calculated as an average of their Fourier transforms. As an example, Fig. 4.7a shows the time evolution of bulk waves received at a spatial position located in the third region. The Fast Fourier transform of the time signal gives the transmission spectrum of wave propagation through the finite size phononic crystal structure.

In addition, the FDTD method allows to visualize the spatial profiles of the displacement amplitudes in the structure at a precise frequency. Indeed, displacement amplitudes averaged over one period of oscillation give the displacement fields of the mode. Figure 4.8 shows a bended waveguide in steel/water structure for a frequency inside the complete band gap. We can notice the high confinement of the acoustic field inside the defect waveguide.

4.7 Conclusion

In this chapter, we have presented an overview of the different computational problems that are encountered when studying wave propagation in phononic crystals. We have outlined the characteristics, and pros and cons of four dominant methods: the MST method (and its improvement, LMS), the PWE method (and the

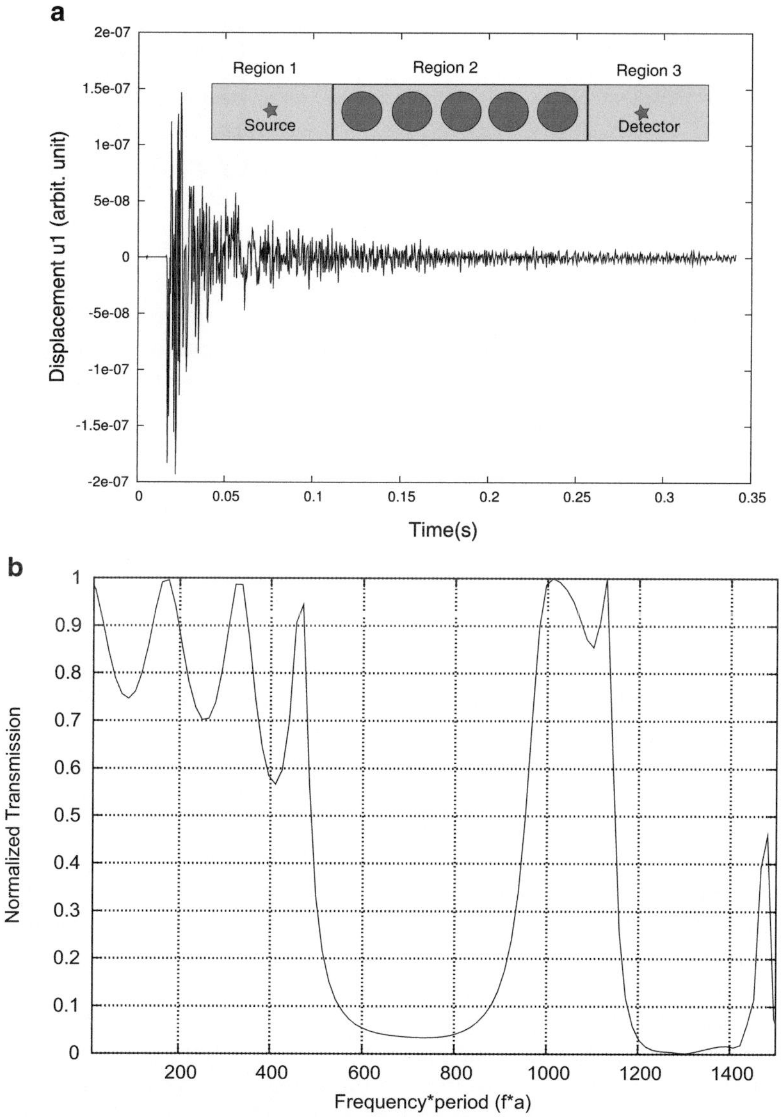

Fig. 4.7 (**a**) Time response of displacement detected at some position after the phononic crystal. (**b**) Calculated transmission power spectrum along the ΓX direction of the irreducible Brillouin zone of the square lattice of steel cylinders in water. A strong attenuation in transmission extending from 700 to 1,000 m/s in reduced frequency is obtained

extended PWE method for complex band structures), the FDTD method, and the FEM. As a final useful comment for the reader, we may summarize the usage of these methods as follows.

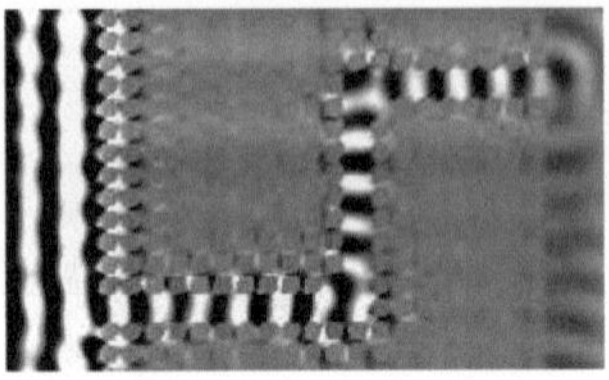

Fig. 4.8 Calculated longitudinal displacement amplitude averaged over one period of oscillation in bended waveguide

- The layer multiple scattering (LMS) method is the method of choice for both infinite periodic and finite phononic crystals composed of non-overlapping spheres or cylinders in a uniform background, provided all materials are isotropic. Its usage for surface or plate problems is possible but much more involved, and its usage for arbitrary anisotropic scatterers is presently out of reach. Sainidou et al. propose a free LMS code [9].
- The plane wave expansion (PWE) method is very popular for band structure computations because it is general purpose (arbitrary scatterers and background, anisotropy, piezoelectricity). However, it has slow convergence properties (the number of Fourier harmonics that have to be included can make the computation rather slow). To the authors' opinion, the finite element method (FEM) is to be preferred over classical PWE in most cases. The extended PWE was introduced recently and enables the computation of complex band structures and the consideration of losses without sacrificing the general purpose character of the original PWE method. No free or commercial phononic PWE codes are available to the authors' knowledge.
- The finite-difference time-domain (FDTD) method is very popular because it allows to solve the wave equation directly in space and time, thus mimicking real experiments. It is a general purpose method, though the consideration of anisotropy and piezoelectricity is not trivial in the general case. Convergence can be very slow, especially for resonant structures with high quality factors (Q-values). No free or commercial phononic FDTD codes are available to the authors' knowledge.
- The finite element method (FEM) is a general purpose method that solves all drawbacks of the PWE method. It allows the consideration of various boundary conditions (e.g., solid–fluid phononic crystals), the computation of band structures (it is especially efficient for thin plate problems because of the reduced size of the 3D mesh in this case), and the study of finite phononic crystals. FEM is more often used for time-harmonic or band structure problems. However, it can also be employed in the time-domain, much as the FDTD method, though this property has not been really exploited so far in the literature. Comsol Multiphysics®, a commercial FEM software, has recently become increasingly popular because of its ease of use (the obvious drawback being its "black-box" character). Many free or even open-source softwares can be employed as well, and one of the authors (VL) has, for instance, been using FreeFem++ (http://www.freefem.org/ff++/) with success.

References

1. B.A. Auld, *Acoustic Fields and Waves in Solids* (Wiley, New York, 1973)
2. D. Royer, E. Dieulesaint, *Elastic Waves in Solids* (Wiley, New York, 1999)
3. IEEE standard on piezoelectricity 176-1987. IEEE Trans. Ultrason. Ferroelectr. Freq. Control **43**(5), 717 (1996)
4. M. Wilm, S. Ballandras, V. Laude, Th. Pastureaud, A full 3-D plane-wave-expansion model for piezocomposite structures. J. Acoust. Soc. Am. **112**, 943–952 (2002)
5. R.P. Moiseyenko, V. Laude, Material loss influence on the complex band structure and group velocity in phononic crystals. Phys. Rev. B **83**(6), 064301 (2011)
6. V. Laude, Y. Achaoui, S. Benchabane, A. Khelif, Evanescent Bloch waves and the complex band structure of phononic crystals. Phys. Rev. B **80**(9), 092301 (2009)
7. I.E. Psarobas, N. Stefanou, A. Modinos, Phononic crystals with planar defects. Phys. Rev. B **62**(9), 5536–5540 (2000)
8. R. Sainidou, N. Stefanou, A. Modinos, Formation of absolute frequency gaps in three-dimensional solid phononic crystals. Phys. Rev. B **66**(21), 212301 (2002)
9. R. Sainidou, N. Stefanou, I.E. Psarobas, A. Modinos, A layer-multiple-scattering method for phononic crystals and heterostructures of such. Comput. Phys. Commun. **166**, 197–240 (2005)
10. I.E. Psarobas, Viscoelastic response of sonic band-gap materials. Phys. Rev. B **64**(1), 012303 (2001)
11. V. Laude, M. Wilm, S. Benchabane, A. Khelif, Full band gap for surface acoustic waves in a piezoelectric phononic crystal. Phys. Rev. E **71**, 036607 (2005)
12. J.O. Vasseur, B. Djafari-Rouhani, L. Dobrzynski, M.S. Kushwaha, P. Halevi, Complete acoustic band gaps in periodic fibre reinforced composite materials: the carbon/epoxy composite and some metallic systems. J. Phys. Condens. Matter **6**(42), 8759–8770 (1994)
13. T.-T. Wu, Z.-G. Huang, S. Lin, Surface and bulk acoustic waves in two-dimensional phononic crystal consisting of materials with general anisotropy. Phys. Rev. B **69**(9), 094301 (2004)
14. M. Wilm, A. Khelif, S. Ballandras, V. Laude, B. Djafari-Rouhani, Out-of-plane propagation of elastic waves in two-dimensional phononic band-gap materials. Phys. Rev. E **67**, 065602 (2003)
15. A. Khelif, B. Aoubiza, S. Mohammadi, A. Adibi, V. Laude, Complete band gaps in two-dimensional phononic crystal slabs. Phys. Rev. E **74**, 046610 (2006)
16. M.I. Hussein, Reduced bloch mode expansion for periodic media band structure calculations, Proc. R. Soc. A Math. Phys. Eng. Sci. **465**(2109), 2825–2848 (2009)
17. V. Laude, J.C. Beugnot, S. Benchabane, Y. Pennec, B. Djafari-Rouhani, N. Papanikolaou, J.M. Escalante, A. Martinez, Simultaneous guidance of slow photons and slow acoustic phonons in silicon phoxonic crystal slabs. Opt. Express **19**(10), 9690–9698 (2011)
18. J.-P. Berenger, A perfectly matched layer for the absorption of electromagnetic waves. J. Comput. Phys. **114**, 185 (1994)
19. K. Yee, Numerical solution of initial boundary value problems involving maxwell's equations in isotropic media. IEEE Trans. Antennas Propag. **14**(3), 302–330 (1966)

Chapter 5
Phononic Crystal Membranes (Slabs or Plates)

Saeed Mohammadi and Ali Adibi

5.1 Introduction

In the previous chapters, an overview of different types of phononic crystal (PnC) structures and their properties was given. Methods for the analysis and simulation of such structures were also discussed. Among different types of PnC structures, planar PnCs are of particular interest due to their flexibility in fabrication and the possibility of mass production using lithography-based fabrication techniques. Two of the main categories of planar acoustic wave devices are surface acoustic wave (SAW) devices and the devices with a finite thickness (called plates, slabs, or membranes). SAWs form on a free surface of a thick (compared to the wavelength of interest) substrate while most of the energy is near the surface of the substrate. SAWs are of great interest in design of wireless communications devices. In the previous chapter, some of the properties, methods of fabrication, and characterization of PnC structures for handling SAWs (for abbreviation we will use "SAW PnCs" for referring to them from now on) were discussed. In this chapter, we discuss an important drawback that SAW PnCs are prone to and discuss PnC slab structures which are periodic structures with a finite thickness of the order of the wavelengths of pertinence. The available methods of simulation, fabrication, and characterization of PnC slab devices and their advantages over SAW PnC devices will be addressed. Then an example of a practical PnC slab structure with the stages of design simulation,

S. Mohammadi
Multimodality Molecular Imaging Lab, The James H Clark Center, 318 Campus Drive, East Wing, 1st Floor, Stanford, CA 94305-5427, USA
e-mail: saeed.mohammadi@gmail.com

A. Adibi (✉)
School of Electrical Engineering, Georgia Institute of Technology, Ferst Dr NW 266, Atlanta, GA 30332, USA
e-mail: adibi@ee.gatech.edu

A. Khelif, A. Adibi (eds.), *Phononic Crystals*, DOI 10.1007/978-1-4614-9393-8_5

fabrication, bandgap engineering is given followed by a brief description of functional devices developed based on these structures for wireless communications and sensing applications at high frequencies.

5.1.1 A Brief History of the PnC Slab Structures

Although the first evidences of the existence of phononic band gaps (PnBGs) for SAW PnCs were given much before than those for PnC slabs, SAW PnCs have so far been unable to show excellent properties expected from PnC structures and subsequently, devices based on SAW PnBGs have proven to suffer from excessive losses and showed below-expectation performance. As will be subsequently discussed, the problem of loss of acoustic waves into the substrate causes this poor performance of SAW PnC devices, however, this problem is avoided in PnC slabs, and therefore, the demonstration of PnBGs in PnC slabs leads to PnC slab-based devices with exquisite performance such as high quality resonators, waveguides, filters, and (de)multiplexers. While periodic plates with 1D periodicity may have their own limited applications [1], here we focus on two-dimensionally periodic slabs that have their inclusions in the plane of propagation as they provide higher flexibility in the design of complex devices.

Originally, periodic or non-periodic inclusions were inserted in a membrane made of a softer material such as epoxy for purposes such as impedance matching (e.g., with water) for creating better transducers [2]. The main purpose of such analyses was to avoid problems caused by the stopbands and vibration resonances in these transducers. However, no effort was reported to analyze these structures for creating complete (for all-directions) stopbands nor for utilizing these stopbands or strong dispersions caused by the periodicity of these structures for other applications.

There are two different known mechanisms that can lead to PnBGs or strong dispersions in PnC structures. One is the mechanism of Bragg scattering in which the periodicity of the lattice is of the order of wavelength of the propagating wave and the PnBG is governed mainly by the periodicity of the crystal [3]. The other mechanism is the local resonance of scatterers that couple through the host material and where the frequency of the PnBG does not have a strong dependence on the period of the crystal lattice [4].

For PnC slabs and on the scattering side, firstly, Sigalas and Economou theoretically studied the possibilities of obtaining partial and complete PnBGs in very thick or very thin plates with solid/solid structure (solid inclusions in solid hosts) [3]. They offered possibilities of obtaining complete PnBGs in such structures. Long after, a seminal work by Khelif et al. theoretically demonstrated the possibility of the existence of complete PnBGs (for all polarizations and for all propagation directions) for piezoelectric square lattice of quartz inclusions in an epoxy slab with a finite thickness [5]; PnBG maps for different geometrical parameters were also generated. This was followed by a work by Bonello et al. demonstrating propagation of Lamb waves in a thin plate of Si coated with a thin periodic metallic film has some evidence of stop bands in one direction at relatively low frequencies [6]. An important turning point in the development of high frequency PnC slabs was the theoretical [7] and experimental [8] demonstration of the possibility of obtaining large high-frequency PnBGs in micromachined structures made of an array of holes embedded in a Si slab by Mohammadi et al. These observations lead to the demonstration of a variety of devices based on PnC slabs including waveguides and resonators [9] as well as the demonstration of simultaneous photonic and phononic band gaps for optomechanical applications [10–12]. We will briefly cover some of these findings in this chapter. Meanwhile, demonstration of high frequency PnC slabs with solid/solid structure also emerged [13, 14]. Today, the studies and demonstrations of PnC slabs based on Bragg scattering is expanding at an even faster pace some of which will be explored in what follows.

On the locally resonant PnC plates front, Fung et al. showed that transmission gaps can exist in a plate made of silicone-rubber coated lead spheres embedded in an epoxy matrix [15], while Yu et al. showed the possibility of complete flexural band gaps in thin plates of cylindrical rubber inclusions using FEM [16]. A similar publication by Hsu and Wu confirmed the existence of such PnBGs using PWE [17]. Ever since, the utilization of locally resonant PnCs to obtain PnBGs and other dispersion engineering applications has been on the rise [18–20].

5.1.2 Problem of Leaky SAWs in PnC Structures

In PnC structures with one free surface, SAW branches usually cross the branches of the in-plane bulk acoustic waves in the 2D structure and can therefore be coupled to a spectrum of bulk modes propagating far from the free surface. Also in the transmission profile of the Rayleigh SAWs, it is observed that the transmission of SAWs drops at the PnBG, but does not fully recover at frequencies higher than the PnBG region [21, 22]. This means SAWs are lossy at such structures. Although a few efforts have been made to resolve this issue [23, 24], the problem of excessive loss of SAWs in PnC half spaces has prevented them from being used in high performance PnC devices. Recently, there have been evidences reported that under certain conditions, SAWs can keep most of their energy near the surface using some band gap engineering techniques in the bulk sound cone [25]. However, practical realization of devices based on such techniques is still under scrutiny.

The mechanism of coupling of SAWs to bulk modes because of the periodic structure of PnCs is demonstrated in Fig. 5.1.

It can be seen from Fig. 5.1 that SAWs may couple to the bulk waves in the presence of a periodic structure. The reason of this phenomenon can be explained using the properties of wave vectors and the coupling requirements as follows.

SAWs can propagate on the free surface of a substrate when their wave vector is larger (phase velocity is smaller) than any bulk wave vector in the substrate. Therefore, when the SAW wave vector is larger, the phase-matching condition

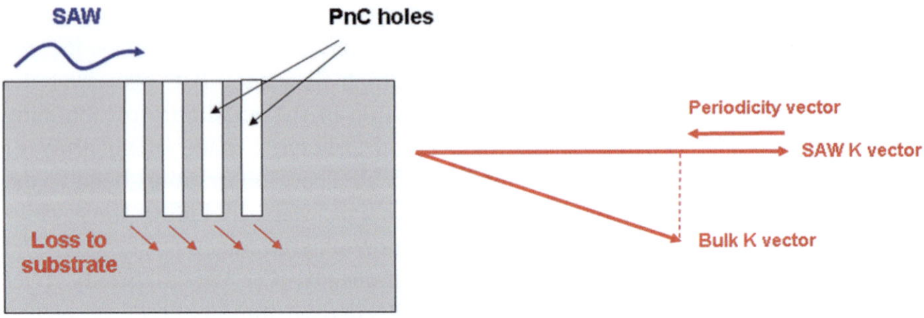

Fig. 5.1 (*Left*) A schematic of SAWs propagating through a PnC SAW structure. (*Right*) A wave vector representation of how the periodic structure of the PnC can result in the coupling of SAWs to the bulk modes

Fig. 5.2 Schematic of the side view of a PnC slab in which elastic modes are decoupled from the substrate modes

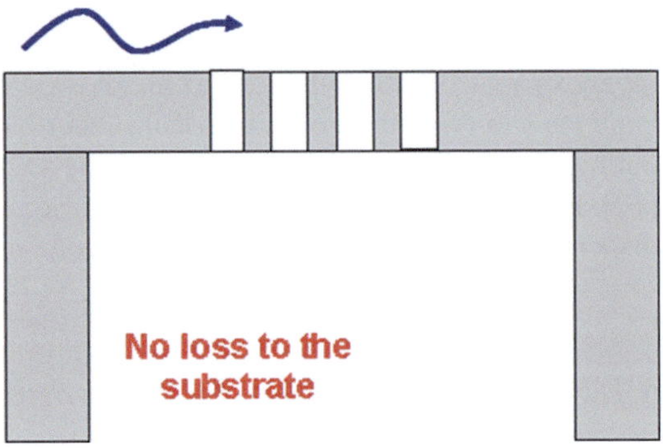

between SAW wave vectors and bulk wave vectors cannot be satisfied and no coupling to the bulk modes can occur in the absence of a PnC structure (of course, it should be noted that in certain directions of propagation in anisotropic crystals, SAWs may inherently be lossy even without the presence of PnCs). However, as can be seen in Fig. 5.1, in the presence of a PnC structure, the PnC structure may cause SAWs to be phase matched with bulk waves propagating toward the depth of the substrate. This folding effect can cause SAWs to couple to the continuum of bulk mode branches that have a wave vector component toward the depth of the substrate.

This type of loss for SAWs makes the use of PnC structures for controlling SAWs very limited. As can be seen what follows, the use of PnC slabs and slab modes mitigates this problem and efficient PnC structures with very low loss can therefore be developed. A schematic of such a PnC slab is shown in Fig. 5.2.

5.2 PnC Slab (Membrane) Structures

As discussed above, PnC SAW structures may not be the most efficient structures due to excessive loss that may occur for SAWs propagating in PnC SAWs. In this section, PnC slab structures, which can bear merits such as simplicity, compatibility of fabrication with CMOS and standard MEMS technology, versatility, and low loss are discussed. This type of PnC structure is made by applying 2D in-plane periodicity to a solid slab rather than a solid half space. Although the fabrication of such a structure is more complicated than that of SAWs, its merits and the provision of better control over the propagation of elastic vibrations make PnC slabs advantageous over PnC SAW configurations. In addition, as will be discussed, these structures can be combined with photonic crystals to obtain acousto-optic and optomechanical functionalities.

In what follows, different PnC slab structure types for high-frequency applications that can be engineered to provide the desired properties such as PnBGs in their band structures will be discussed and their advantages and properties will be described. Further, the fabrication of some types of these structures with emphasis on high-frequency applications will be explained. Finally, the performance of such devices in controlling the flow of elastic waves and vibrations to form PnC-based devices will be mentioned.

5.2.1 Types of PnC Slab Structures Based on Geometry

There are different methods for realizing a membrane or slab acoustically periodic. Based on the location of the periodic inclusions we can categorize PnC slabs to the following categories:

1. Inclusions are embedded inside the PnC slab (inclusions can be made out of another material or can be void)
2. Inclusions are decorated on the slab

Types 1 and 2 are demonstrated in Fig. 5.3a, b, respectively. Each of first and second categories has its own advantages.

5.2.2 Inclusion-Based PnC Slabs

In inclusion-based PnC slabs, the periodicity of the PnC is made by either fully or partially embedding inclusions inside the thickness of the slab. These inclusions can be either filled with another material (e.g., a solid) [5], be void [7], or partially filled and partially void [26]. The physics of modifying the dispersion properties of elastic

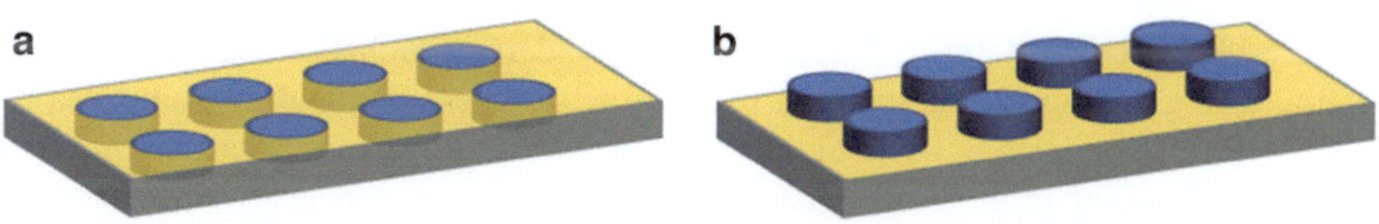

Fig. 5.3 Schematic of a PnC membrane with (**a**) embedded inclusions and (**b**) decorations (stubs)

waves in these structures is usually based on scattering and a better contrast between the properties of the materials can provide better control over the properties of the band structure. For example to obtain PnBGs when using solid inclusions inside the slab, usually heavier and more stiff inclusions compared to the background material is required and there is an optimum size for the inclusions inside the slab for achieving maximum PnBG widths [27], however, for the void inclusions, usually a larger inclusion, provides a larger PnBG in the crystal, although an optimum thickness for the slab exists [7].

Depending on the particular application, any of the above-mentioned PnC slabs might be the best choice. In particular PnC slabs with void inclusions have unique advantages such as the possibility of fabricating single crystalline PnC structures for low loss at high-frequency applications, microelectronic compatibility, simpler fabrication procedure [8], and the possibility of combining with photonic crystals [10], while the ones with solid inclusions might be of better use in cases where limitations on size of the inclusions and minimum obtainable feature size in fabrication is an issue.

5.2.3 Decorated Stubs PnC Membranes

Another type of PnC slab structure can be made by fabricating arrays of stubs on top of solid slabs. In this case the mechanism of scattering and modification of the dispersion properties of the slab is mostly based on local resonance. In this case the interaction of locally resonant stubs (or dots) with the modes of the slab leads to strong modification of the band structure. Therefore, as the local resonance of the stub can be significantly lower than the band gaps obtained based on scattering, much lower PnBGs compared to that of the Bragg scattering can be achieved [28]. In this case, there is another degree of freedom added to the engineering parameters of the band structure, which is the thickness of the stubs. Further, it is also possible to make composite stubs to have additional engineering parameters in the slab [19]. Formation of waveguides and energy confinement in this type of PnC slab would also be possible. For an overview of the types of modes and different forms of waveguides based on PnC slabs with stubs, the readers can refer to [29].

5.2.4 Examples of PnC Slab Structures

The two above-mentioned categories of PnC slabs encompass almost all of the reported PnC slab structures. The choice of cylinders as inclusions or stubs makes defining (e.g., using lithography) of these structures less demanding compared to other types of inclusions by avoiding sharper edges. However structures that use more than one, or more fancier inclusion or stub shapes can also be developed. Windowed and connected structures that are usually used at low frequencies and for applications such as vibration isolation are examples of such PnC slabs. These

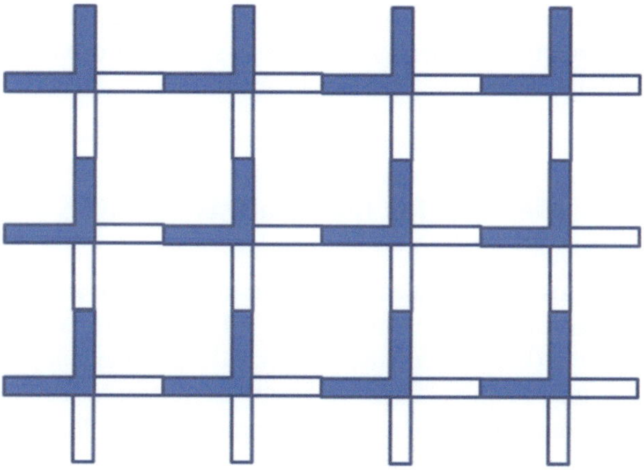

Fig. 5.4 Example of a windowed PnC structure discussed in [16, 30]

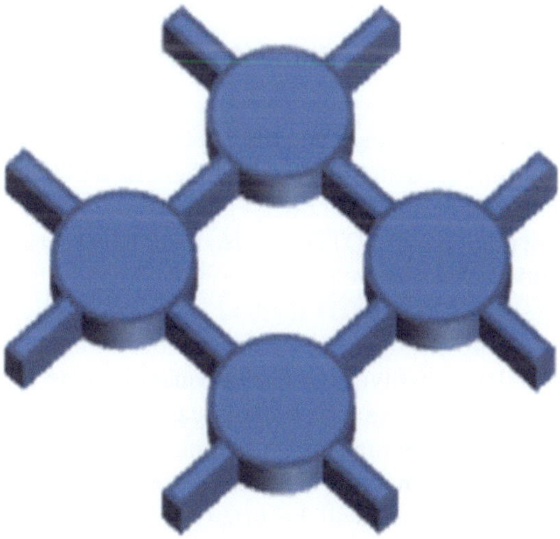

Fig. 5.5 A PnC slab in which objects are connected through tethers

structures are made by interconnecting rods of materials together to form topologies such as triangles or squares and can constitute of several materials [16, 30]. Figure 5.4 shows a schematic example of a windowed-type structure.

Another type of PnC slab structure can be made by placing spherical inclusions (e.g. made of steel) inside a host material (such as solid epoxy). It has been shown that complete PnBGs and waveguiding in defects can be obtained at frequencies around 300 kHz [31]. These structures are challenging for implementation at high frequencies due to incompatibility with lithography-based fabrication techniques, however may be well suited for applications such as noise control at low frequencies.

Recently, PnC slab structures that are made by an array of cylinders connected through supporting tethers as shown in Fig. 5.5 are dubbed "inverse acoustic bandgap structures." It is shown that such structures can also show complete PnBGs with appropriate design [32]. These structures can also be categorized as PnC slabs with non-cylindrical void inclusions.

5.3 Methods for the Analysis of PnC Slabs

The most important analysis techniques for the analysis of PnC slab structures are the finite difference time domain (FDTD), plane wave expansion (PWE), and finite element (FEM) methods. The basics of these techniques have been discussed in the previous chapters. Here we briefly explain each method and their modifications to suit simulation of acoustic wave propagation in PnC slabs.

5.3.1 FDTD for the Analysis of PnC Slabs

The FDTD method has long been used to solve wave propagation problems in the fields of acoustics and electromagnetism [33]. With the advances in the computation technology, this method has gained more attention for solving complex problems, on the other hand, FDTD has also been widely used for the analysis of electromagnetic and photonic wave propagation in complex structures. This is because of the unique merits of the FDTD method such as simplicity, the ability to simulate complex structures, straightforward parallelization, possibility of simulating propagation problems involving a wide frequency range in a single run, and the possibility of implementing various boundary conditions. In addition, one of the main advantages of FDTD over many other computation techniques is the fact that it does not involve any matrix inversion. In this section, the FDTD scheme for the analysis of PnC plate structures will be briefly discussed. It should be noted that the following FDTD scheme is designed to handle materials with a minimum of orthorhombic symmetry (which includes isotropic and cubic-symmetry materials), and without piezoelectricity.

Figure 5.6a shows the designed distribution of elastic vibration components (stress components T and velocity components v) in the FDTD grid. As can be seen in this figure, different components are staggered in space to satisfy the second-order approximation. In this scheme, if the stress components are updated at time step n, the velocity components are updated at time step $n + 1/2$, where n is an integer. The stress components are updated using the previously computed (or initial) stress values at time step $n - 1$ and the computed (or initial) values of velocity components at time step $n - 1/2$. Similarly, the velocity components are updated using the velocity values and the values of the stress fields at time steps $n - 1/2$ and n, respectively. This method of updating field values is called *leapfrog* scheme because of the hopping mechanism from one set of components to the other as time advances. The leapfrog scheme in this FDTD algorithm is demonstrated in a flow

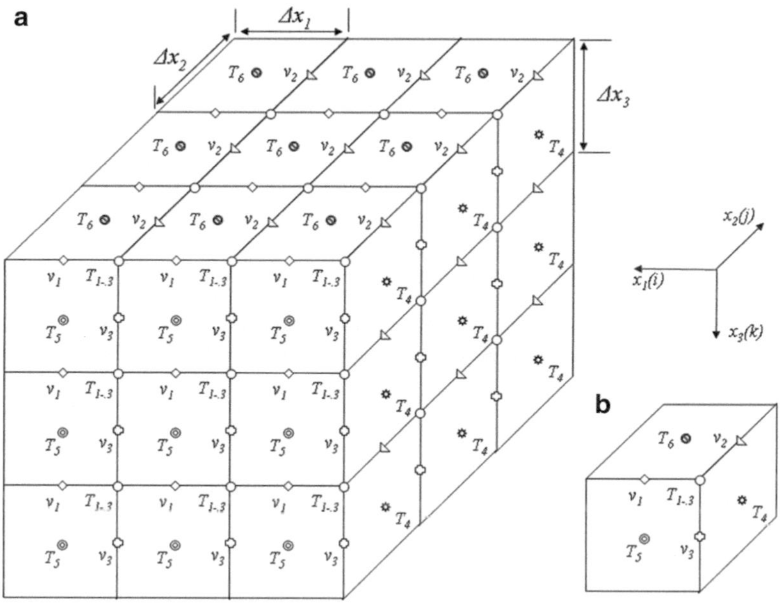

Fig. 5.6 (**a**) Distribution of the elastic wave components for orthorhombic solids in the developed FDTD scheme is demonstrated. As can be seen from this figure, the distribution of the field components supports a discretization scheme accurate to the second order. (**b**) A unit cell of our FDTD grid is shown. As can be seen in this figure, all field components lie on the three adjacent faces of the cube

chart in Fig. 5.7. For more details on the basics of the FDTD method for solving acoustic wave propagation problems, the reader is encouraged to refer to [34–36].

5.3.1.1 Free Surface Boundary Condition

In problems involving wave propagation in PnC slabs, it is often possible that the solid material is interfaced with materials with much lower elastic impedance such as a gas, air, or vacuum. It is also possible that the intended simulation domain ends on such a boundary. The propagation of elastic waves in such materials is much slower than the propagation of waves in solids, and therefore, the minimum wavelength in a given frequency range reduces significantly. This requires the grid cell to be chosen much smaller than the system consisting of only the solid material according, and increase the number of points to be simulated significantly. For example, since the phase velocity of sound waves in air is approximately 340 m/s, which is about an order of magnitude smaller than that in average solids, the number of grids cells for a certain 3D structure increases by a factor of 1,000. Further, due to the Courant stability condition requirements, the required time step also decreases by an order of magnitude, increasing the computation burden by another factor

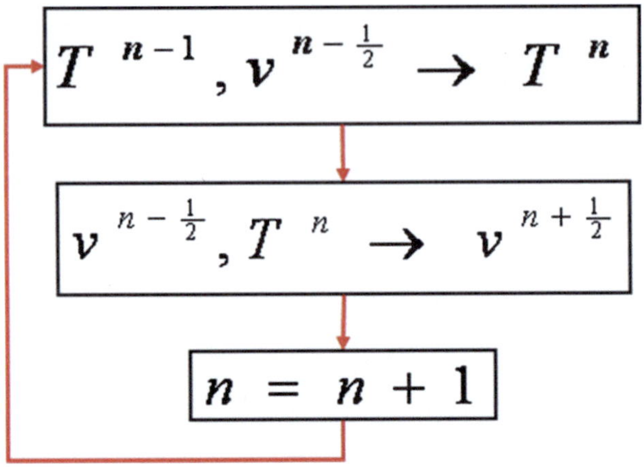

Fig. 5.7 Flow chart of updating elastic wave components as the time advances in the FDTD algorithm. At each time step, the stress components are all updated using their values at the previous time step and the velocity components at half a time step before. Similarly, the velocity components are updated using their value and the value of the stress fields at one and half a time step earlier, respectively. This scheme of updating different field components in alternating time steps is called leapfrog scheme

of 10. Because of such inefficiency, it is better to treat the interface with such low-impedance materials as a boundary condition rather than an internal boundary. In such a scheme, we can ignore the simulation of air/vacuum and limit the simulation domain to the solid structure. The air/vacuum structure interface can be replaced with a free-surface boundary condition to emulate the effect of air/vacuum.

Exemplarily, a free slab with one unit cell thickness is assumed for simulation in Fig. 5.8. Similar to the internal material interfaces, the free interface faces are assumed to occur at the middle of the grid cells. For more details on the implementation of the FDTD method for analyzing PnC slabs, readers can refer to [36].

5.3.2 *PWE for the Analysis of PnC Slabs*

While FDTD and FE methods are versatile methods that can be used to solve a large class of general problems, they are not specially optimized for calculation of the modes and the band structure of the periodic structures. Therefore, when it comes to intense calculation of the modes of different periodic structures and their band structure, it is computationally advantageous to implement a method that fits such a need more appropriately. PWE has been proven to be the reference method for the calculation of band structures in PtCs [37] and is widely used in PnCs as well.

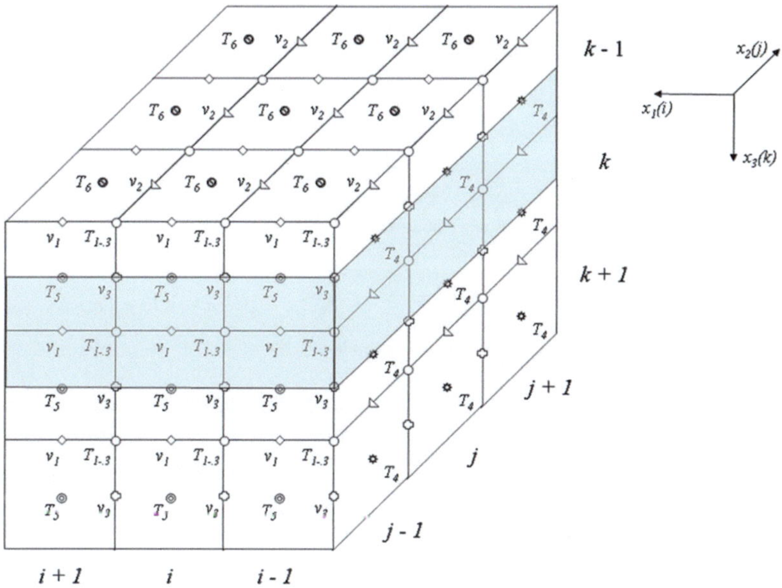

Fig. 5.8 Demonstration of the FDTD grid for a slab with a one-cell thickness in the x_3 direction with surrounding air or vacuum

To calculate the band structure of the modes propagating in a periodic structure using the PWE method, the Bloch theorem discussed is used with a slightly different notation. According to the Bloch's theorem, any field (such as the displacement $u(r, t)$ in a periodic medium can be expanded as an infinite series:

$$u\left(r,t\right)=\sum_{G}u_{K+G}e^{j(\omega t-k.r-G.r)}, \tag{5.1}$$

where $r=(x_1, x_2, x_3)$ is the position vector and $G=(2\pi m/p_1, 2\pi n/p_2, 2\pi n/p_3)$ are the vectors of the reciprocal lattice and k is the wave vector. In addition, any physical parameter $\alpha(r)$ in the periodic structure can be expanded as:

$$\alpha(r)=\sum_{G}\alpha_{G}e^{-jG.r} \tag{5.2}$$

If the wave vector of the propagation is fixed, we can then find the values of eigen-frequencies (i.e., the frequencies of the modes in the structure). Having the harmonic wave components in the form of phasors and eliminating the time from the equations, we can infer:

$$u(r)=\sum_{G}u_{K+G}e^{-j(k.r+G.r)} \tag{5.3}$$

By inserting these equations into the elastic wave equation (immediately following) and using the index-based notation we will have:

$$\nabla_{iJ} c_{JI} \nabla_{Ik} u_k = \rho \omega^2 u_i \tag{5.4}$$

$$\sum_{M-m,N-n,P-p=-\infty}^{\infty} \sum_{m,n,p=-\infty}^{\infty} \sum_{J=1}^{6} \sum_{k=1}^{3} \nabla_{iJ} {c_{JI}}^{M-m,N-n,P-p} \nabla_{Jk} {u_k}^{m,n,p} = \omega^2 \sum_{M-m,N-n,P-p=-\infty}^{\infty} \sum_{m,n,p=-\infty}^{\infty} \rho^{M-m,N-n,P-p} {u_i}^{m,n,p} \tag{5.5}$$

In the above equations, ω and $u_i^{m,n,p}$ are the unknowns that need to be found, and the rest are known parameters. To solve these equations, they can be combined and form an eigenvalue problem with $u_i^{m,n,p}$ as the vector elements and ω^2 as the eigenvalues. The components of the wave vector (i.e., k_x, k_y, and k_z) are fixed for solving the each eigenvalue problem, which can be solved by re-arranging the summations in the form of matrices and the calculation of eigenvalues can be done numerically using software like MATLAB as a standard eigenvalue problem. The PWE technique is a very effective technique for calculation of the modes of a periodic structure. The numerical precision of the PWE is limited by the number of terms kept in the Fourier expansions. For more detailed explanation of this method, please refer to [38].

In order to apply the PWE technique for extracting the modes of the PnC slabs, the periodic unit cell in the vertical direction needs to include two regions of vacuum above and underneath the slab. This technique is demonstrated in Fig. 5.9.

A more efficient technique developed by Hsu and Wu is to expand the modes propagating in the plane of the slab based on the modes of the slab. This method is based on the Mindlin's theory which can save a substantial amount of computational power in calculating the band structure of PnC slabs [39].

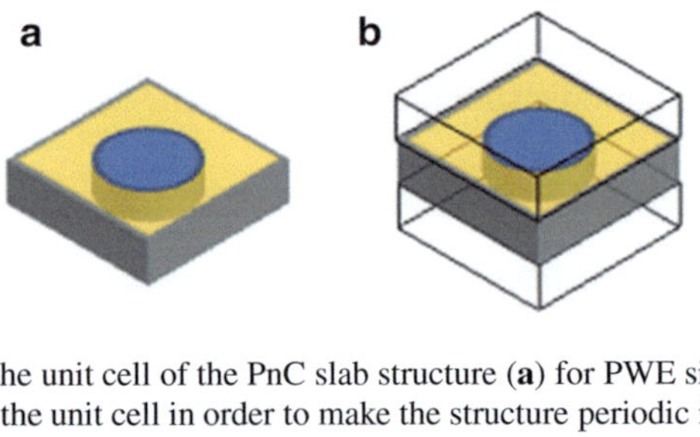

Fig. 5.9 Modification the unit cell of the PnC slab structure (**a**) for PWE simulation. Two regions of vacuum are added to the unit cell in order to make the structure periodic in the vertical direction so that it can be fit into the PWE formulations

5.3.3 FE Method for the Analysis of PnC Slabs

FE method is one of the most effective and diverse methods used to solve integral form of differential equations in real-life problems in various areas of science and technology such as aeronautics, biomechanics, acoustics, fluid dynamics, and photonics. The FE method is especially a good choice for solving differential equations over complicated domains, where the required precision varies over the solution domain. FE method, in general, can be more accurate and can be applied to a larger class of solid materials compared to FDTD for solving elastic wave propagation equations. However, it cannot be easily applied to very large domains of simulations as its parallelization can be challenging due to the need for matrix inversions.

The time harmonic form of the wave equation can be utilized in the commercially available software such as COMSOL for solving some of the problems pertaining to PnC slab structures. The equation that may be solved using FEM is of the form [40]:

$$e_a \frac{\partial^2 u}{\partial t^2} + d_a \frac{\partial u}{\partial t} + \nabla \cdot (-c\nabla u - \alpha u + \gamma) + \beta \cdot \nabla u + au = f \quad \text{in } \Omega \tag{5.6}$$

The boundary conditions that can be applied to the boundaries of the simulation domain are:

$$n \cdot (-c\nabla u - \alpha u + \gamma) + qu = g - h^T \mu \quad \text{in} \partial\Omega \tag{5.7}$$

and

$$hu = r \quad \text{in} \partial\Omega, \tag{5.8}$$

where Ω is the computational domain, $\partial\Omega$ is the domain boundary, and n is the outward normal vector on $\partial\Omega$. Equation (5.6) is the main PDE to be solved while Eqs. (5.7) and (5.8) are the boundary conditions. By comparing Eq. (5.6) and the main acoustic equation, we can find the similarities between the two. By replacing c and e_a in Eq. (5.6) with c^E and ρ in the wave equation, and setting all the other constants in Eq. (5.6) to zero, Eq. (5.6) and the wave equation will be virtually equal. However, it should be noted that u in this case is a vector of three components u_1, u_2, and u_3, which are the displacement components in x_1, x_2, and x_3, respectively. It is worth noting that Eq. (5.6) is very general, and it can be applied to even more general elastic equations that can involve damping or piezoelectricity. The boundary equations of (5.7) and (5.8) can also be modified to fit the problem of interest. As these boundary conditions can be easily adapted to satisfy the free surface or clamped boundary conditions. This is an advantage of FE method in satisfying boundary conditions compared to other simulation techniques.

5.3.4 Other Significant Techniques for the Analysis of PnC Slab Structures

Here we briefly survey over other important numerical methods that can be used for the analysis of PnC slab structures. Although these methods have not yet been used or studied extensively for the analysis of PnC slab structures, detailed study and utilization of their capabilities adds to the available analysis tools for the analysis of PnC slab structures.

5.3.4.1 Transfer, Reflection, and Transmission Matrix Methods

Transfer, reflection and transmission matrix methods have also been used extensively in studying periodic structures with finite sizes, especially to analyze PtC structures. Besides the fact that these methods are extremely reliable, they have also been demonstrated to be efficient in terms of computation time [37]. In addition, such methods have the advantage of having the capability of predicting the attenuation of the field propagating in the crystal as a function of the number of layers.

5.3.4.2 Scattering Matrix (Modal) Method

Also known as the multi-pole method, or the modal method, scattering matrix method has proved to be particularly efficient for the resolution of scattering problems involving finite number of objects for the analysis of PtCs. Fundamentally, there are two approaches to analyze PnC structures. One approach is fundamentally analytical and is carried out with the aid of calculators (e.g., like the PWE method) and the other approach is numerical in nature and the equations are solved by discretizing the computation domains (e.g., like the FDTD method). The scattering matrix method belongs to the first category. This method has also been applied to the analysis of PnCs [41, 42]. The matrix method is especially effective in the analysis of 2D PnCs with cylindrical scatterers, or in the case of 3D PnCs with spherical scatterers. However, this method has some fundamental limitations [43]: (1) the objects in this method must be isolated from each other, form a convex shape on the exterior, and have to be uniform on the interior. Most of the calculations are restricted to the cylindrical or spherical objects as the calculation task is much simpler. (2) Only static solutions of the fields can be obtained, the time response cannot be obtained and the nonlinearity calculations are impossible. Major modifications to this method need to be performed to make it suitable for PnC slab applications.

5.4 Fabrication and Characterization of PnC Slab Structures

Real-life realization of PnC slab structures mainly depends on the feature size requirements of the developed PnCs. PnC slabs have been demonstrated with feature sizes of the order of a few centimeters to a few nanometers. Of course, there is a strong correlation between the feature sizes and the range of frequencies, and subsequently, the application of such PnC slab structures. Various methods can be used to fabricate a variety of different types of PnC slabs.

These methods include bonding by melting an adhesive to decorate blocks of Lucite on a steel plate [4], drilling holes in a substrate using drilling machines [44], decorating the surface of a slab with composite metals [6], interference lithography [45], or manual placement of the structures [31].

However, the most advantageous and versatile methods of fabricating PnC slabs is the lithography-based methods [8, 26], where patterns and locations of PnC inclusions can be realized accurately in the surface of the slab. The lithography-based methods allow for free introduction of defects and variations of the inclusions in the structure and allow for mass production of the structures. Comparing the solid/solid PnC slabs and the void/solid PnC slabs, the void/solid one has shown the advantages of simpler fabrication, more flexibility, possibility of obtaining single-crystal PnCs for higher quality factors, and the possibility of combining with mainstream photonic crystal structures.

Various methods for characterization of PnC slabs are also available can be used. These methods include piezoelectric transduction [8, 26], laser excitation and interferometry [46], light interferometry combination with other transduction mechanisms [31], electrostatic excitation, and Brillouin scattering [45]. While laser excitation and detection is probably the most versatile method with non-invasive characterization, piezoelectric excitation is considered the most practical method of realizing PnC slab actual devices for real-life applications.

An effective method of characterizing PnC structures that can be used to more accurately understand the propagation of elastic waves in PnC slabs is described in Chap. 8 of this book. In the following subsection, using a thin layer of piezoelectric transducers, engineering functional PnC slab devices based on void/silicon PnC slab structures are discussed.

5.5 Case Study: Engineering in PnC Slabs with Void Inclusions

As a practical example, in this section we will address the design, fabrication, and characterization of a PnC slab structure based on void inclusions in a silicon membrane. The void/Si PnC structures have many merits. The good quality of Si as a mechanical material [47], the low-cost fabrication of this structure, and the possibility of integration of these devices with electronic circuitry and Si photonic devices are to name a few. The first goal in void/Si structure is to obtain large complete PnBGs at frequencies appropriate for wireless communications and sensing applications. Based on this developed PnC structure, the possibility of the development of novel devices will be investigated and showed.

5.5.1 *Design of Void/Solid Si PnC Slab Structures with High-Frequency Complete PnBGs*

To realize efficient PnC-based MM resonators and confinement-based devices operating at high frequencies, a fundamental requirement is to obtain a large complete PnBG in micro/nano-fabricated structures. While partial PnBGs (PnBGs in some directions or for some types of waves) have some special applications, complete PnBGs are of a greater interest as they provide full control over the propagation of all types of waves and in all propagation directions. Accurate low-cost fabrication of these structures using a CMOS-compatible fabrication technology would be desirable. The possibility of extending the frequency range of operation of such devices to the frequencies appropriate for wireless communication and sensing applications is also an important demand.

Probably the most basic 2D arrangement of holes in a PnC is the square-lattice arrangement; therefore, we first examine the PnC slab structure with a square-lattice arrangement of holes in Si. A schematic of a 2D PnC slab with square-lattice arrangement of cylindrical holes in a Si slab is shown in Fig. 5.10a, in which the main crystalline axes of Si are aligned to x, y, and z directions. The structure

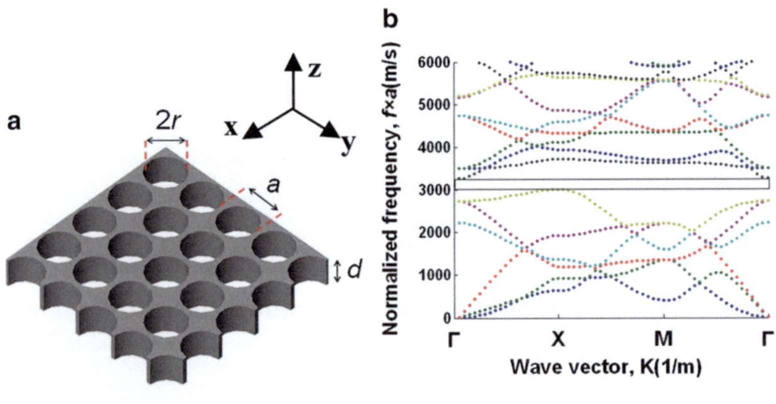

Fig. 5.10 (**a**) Schematic of the PnC slab structure with square-lattice of air holes in a solid slab (e.g., Si). a stands for the lattice constant, r is the radius of the holes, and d is the slab thickness. (**b**) Band structure of elastic waves associated with the structure shown in (**a**) with $r/a = 0.45$ and $d/a = 0.5$. A complete PnBG with 8.1 % band gap to midgap ratio is indicated in this figure. The band structure is calculated using the PWE method

can be fabricated on a Si-on-insulator (SOI) wafer by etching holes through the slab using standard dry etching techniques, and then removing the underlying insulator by wet or dry etching. The actual fabrication procedure is discussed in the following sections. As shown in Fig. 5.10a, a is the spacing between the adjacent holes (or the lattice constant), d is the thickness of the slab, and r is the radius of the holes.

To calculate the 2D band structure of the PnCs, the best way is to use the PWE technique discussed in the previous sections, as it clearly suits the band structure calculation. The FEM and FDTD may be used for verification purposes and subsequent uses. A complete PnBG can be found in the square-lattice PnC slab by carefully choosing the thickness of the slab and the radius of the holes. As can be seen from Fig. 5.10b, a complete PnBG for the structure with $r/a = 0.45$ and $d/a = 0.5$ is obtained and the complete PnBG covers the normalized frequency ($f \times a$) range of 3,000 m/s $< f \times a <$ 3,261 m/s, which corresponds to a gap to midgap ratio of 8.3 %.

Finding this promising complete PnBG, the dependence of the complete PnBG width and position on the thickness of the slab and radius of the holes around this working point would be worth studying. A band gap map is a useful tool for investigating the conditions under which a PnBG is obtained and for designing the structure based on the PnBG width required. The dependence of the complete PnBG on the normalized PnC hole radius (r/a), for a constant slab thickness ($d/a = 0.5$) is shown in Fig. 5.11a. It is seen from Fig. 5.11a that the complete PnBG opens up at around $r/a = 0.43$, and its width constantly increases by increasing r/a. In Fig. 5.11b, the extent of the complete PnBG of the square-lattice PnC as a function of the normalized slab thickness (d/a) for a constant normalized radius ($r/a = 0.45$) is shown. It can be seen in the figure that the complete PnBG opens up at $d/a \approx 0.4$, reaches its maximum width at $d/a \approx 0.55$, and closes again at $d/a \approx 0.7$.

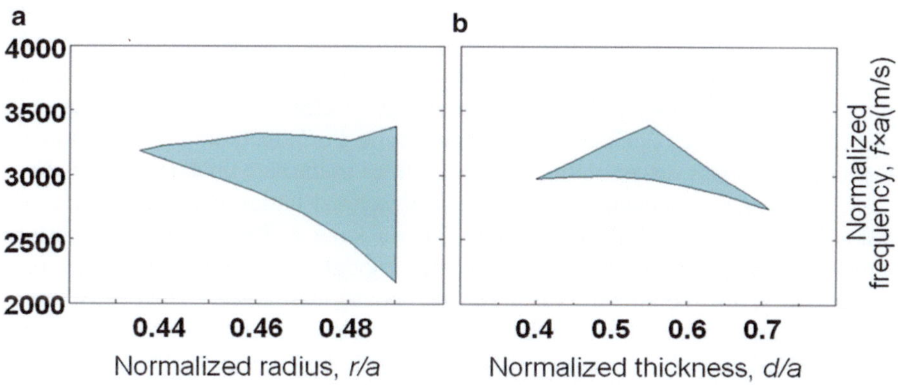

Fig. 5.11 Complete PnBG extent for square-lattice PnC structure shown in Fig. 5.10a as a function of (**a**) normalized holes radius (r/a) for a constant normalized thickness of $d/a = 0.5$ and (**b**) normalized thickness (d/a) for a constant normalized radius of $r/a = 0.45$

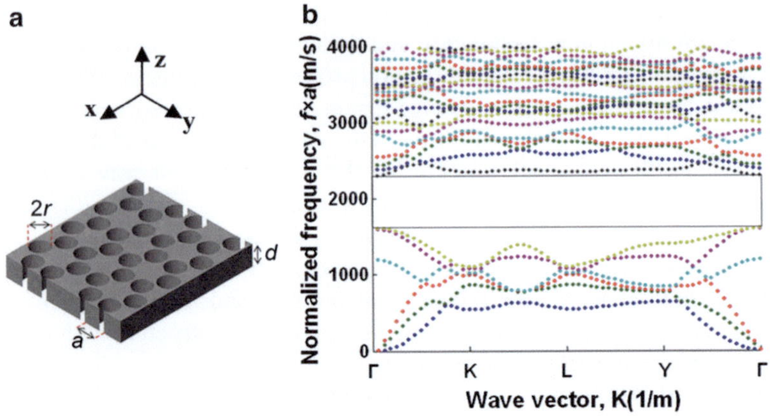

Fig. 5.12 (**a**) Schematic of a hexagonal (honeycomb) PnC made by embedding a hexagonal array of holes in a solid slab (e.g., Si). In this figure, *a* is the spacing between the closest holes, *r* is the radius of the holes, and *d* is the thickness of the slab. *x*, *y*, and *z* denote the main crystalline directions of the crystal. (**b**) Normalized band structure of elastic waves propagating through the PnC shown in (**a**) for $r/a = 0.45$ and $d/a = 1$. As can be seen in this figure, a complete PnBG exists for this structure, in which no mechanical energy is allowed to propagate. The band gap to midgap ratio of this complete PnBG is 35 %

The results shown in Fig. 5.11 clearly illustrate the importance of the PnC geometry on the extent of the complete PnBG. It is also seen that a square-lattice PnC has a limited complete PnBG unless r/a reaches close to its maximum (i.e., $r/a = 0.5$), for which severe fabrication limitations and mechanical-stability issues for the PnC slab exist.

As an alternative PnC structure, we also investigated PnCs with a hexagonal lattice shown in Fig. 5.12a. Again the main crystalline axes of Si are aligned with *x*, *y*, and *z* directions. The band structure of a hexagonal-lattice PnC with $r/a = 0.45$ and $d/a = 1$ is shown in Fig. 5.12b, and the associated complete PnBG maps of the PnC structure are shown in Fig. 5.13. These figures clearly show the advantages of the hexagonal-lattice PnCs over the square-lattice PnCs including larger complete PnBG size and less susceptibility to the variations in the slab thickness.

For example, as shown in Fig. 5.13, a hexagonal lattice with $r/a = 0.45$, and $d/a = 1$ has a complete PnBG extent of $1{,}608 < f \times a < 2{,}298$, which corresponds to a gap to midgap ratio of 35 %, which is large enough for all practical applications. Furthermore, the complete PnBG for the hexagonal-lattice PnC opens up at $r/a \approx 0.37$ and a reasonable complete PnBG with gap to midgap ratio of 18 % can be achieved with $r/a = 0.4$, and $d/a = 1$. This example shows that such PnC parameters impose considerably less fabrication limitations and mechanical-stability issues than those for a square-lattice PnC. Finally, another interesting feature of the hexagonal-lattice PnC is the existence of more than one complete PnBG for a wide range of PnC geometries (i.e., d/a, and r/a) as can be found from Fig. 5.13b.

Unfortunately, the triangular arrangement of holes does not provide a reasonable PnBG in PnC slabs per our extensive investigations.

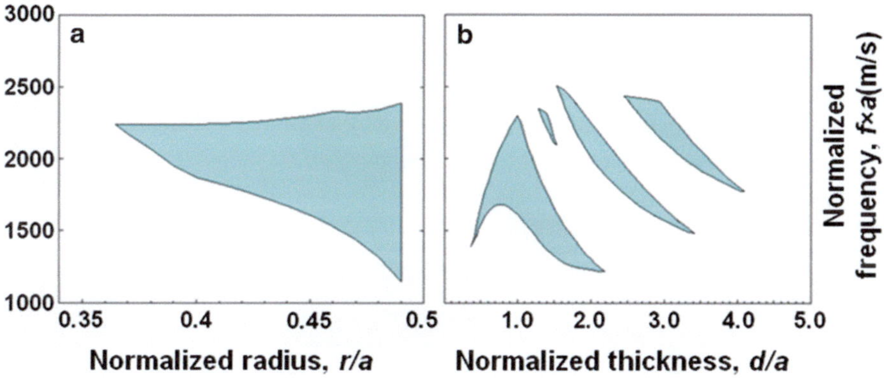

Fig. 5.13 Complete PnBG maps of the PnC structure shown in Fig. 5.12a as a function of (**a**) normalized radius for a constant normalized thickness of $d/a = 1$, and (**b**) normalized thickness with a constant normalized radius of $r/a = 0.45$

5.5.2 *Fabrication and Characterization of Si PnC Slab Structures*

In this section, a fabrication procedure for developing functional PnC slab devices made of arrays of holes in a Si membrane for high-frequency applications [8] will be explained. The fabrication parameters of PnC slabs discussed in the previous section are chosen for fabrication of the device. The distance between the centers of the nearest holes and the hole radius are $a = 15$ μm and $r = 6.4$ μm, respectively. The thickness of the Si layer is $d = 15$ μm. These values are easily obtainable using optical lithography and the conventional micromachining techniques. However, for higher frequencies the device can be scaled and other advanced lithographic techniques such as electron-beam lithography and imprinting can also be used. The simulations show that the resulted complete PnGB extends in the range of 117 MHz $< f <$ 151 MHz. This corresponds to a gap to midgap frequency ratio of about 25 %, which is large enough for all practical applications envisioned for PnCs. The fabrication steps that can be used to realize the device are shown and explained in Fig. 5.14a–f. SEM images of the top view and the cross sectional view of a typical fabricated device are shown in Fig. 5.15a, b, respectively.

To characterize the band gap properties of the fabricated PnCs, a network analyzer is used to find the transmission in the ΓK direction through a structure with eight PnC periods.

The normalized average transmission response of different modes is shown in Fig. 5.16(left) along with the theoretical calculation of the band structure using PWE in Fig. 5.16(right), both for the ΓK direction. Figure 5.16(left) shows a window of low transmission (with more than 30 dB reduction in transmission) in

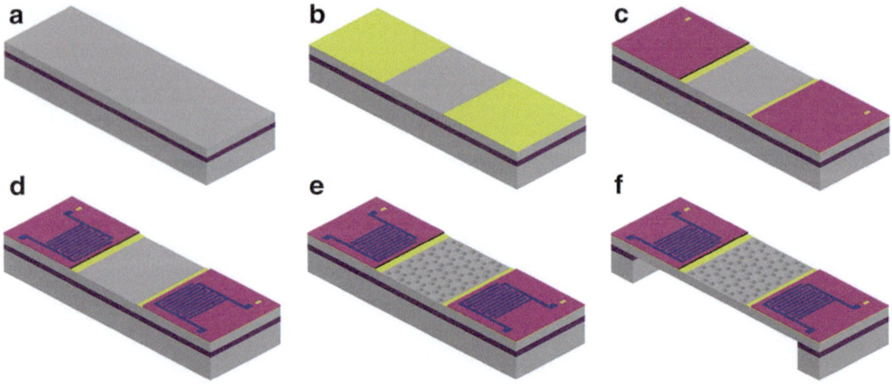

Fig. 5.14 Fabrication steps for realizing the Si PnC slab structures: (**a**) the original SOI substrate. (**b**) Lower electrode is deposited and patterned. (**c**) ZnO layer is sputtered and patterned. (**d**) Top metal electrode is deposited and patterned. (**e**) PnC holes are etched through the device layer, and (**f**) the handle Si substrate and the insulator layers are etched away using plasma etching to form the final structure

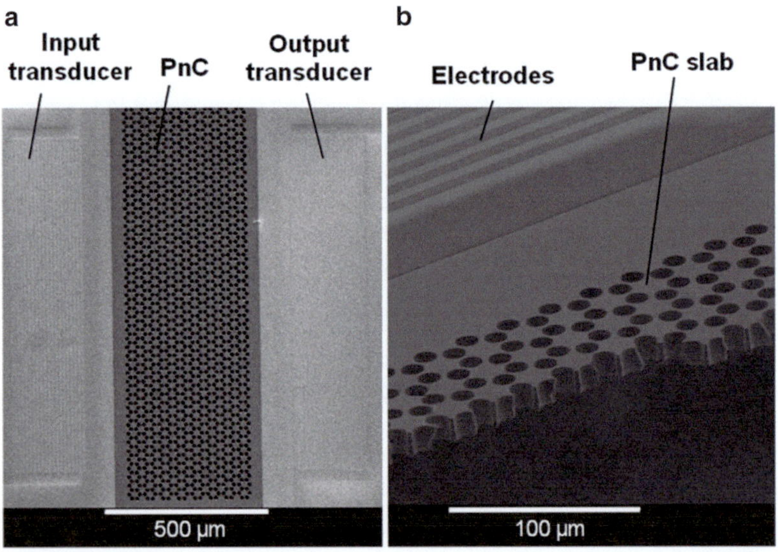

Fig. 5.15 (**a**) Top view of one of the fabricated devices with the hexagonal-lattice PnC structure in the middle and the transducer electrodes on each side, and (**b**) cross sectional view of the same structure [8]

the frequency range of 119 MHz $< f <$ 150 MHz (i.e., a 23% band gap to midgap frequency ratio). This frequency range is in excellent agreement with the theoretical predictions shown in Fig. 5.16(right).

For more details on the discussion of this section, the readers are encouraged to refer to [8].

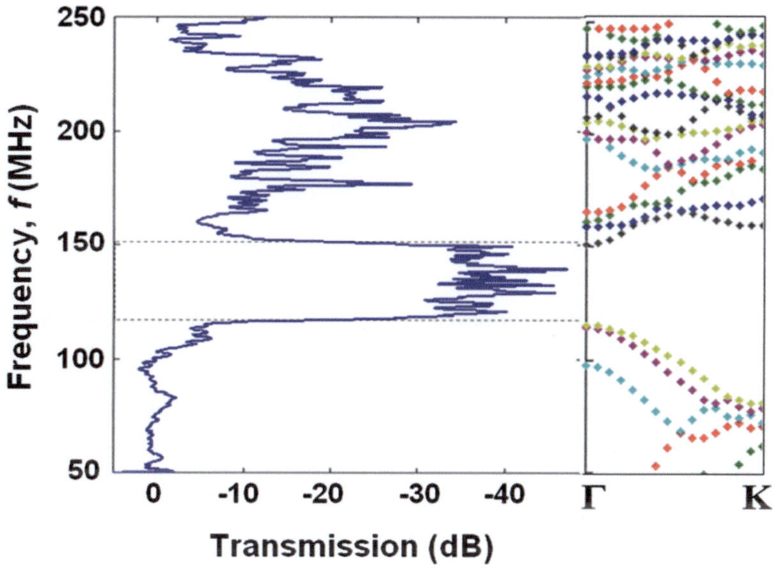

Fig. 5.16 (*Left*) Normalized average transmission of elastic waves propagating through the PnC structure in ΓK direction. More than 30 dB attenuation is observed in the range of 119 MHz $< f <$ 151 MHz, which is very well matched with the theoretical calculations using PWE method shown on the *right*

5.6 PnC Crystal Slab Devices and Band Gap Engineering

Because of the possibility of confining and guiding of mechanical energy with low loss using PnC slabs with PnBGs, PnC slabs are excellent candidates for the implementation of chip-scale integrated wireless or sensing devices, especially for high-frequency applications in which PnC feature sizes are small, resulting in compact structures. The availability of CMOS-compatible fabrication facilities makes SOI substrate a natural choice for such PnC slabs. However, to implement functionalities needed for integrated wireless applications, reliable frequency-selective structures must be realized. Such structures are usually formed using resonators with high Qs. Thus, the development of high-Q resonators operating at high resonance frequencies in PnC slabs is a fundamental step for the deployment of PnC-based integrated wireless and sensing systems such as oscillators, filters, and directional sources. Such resonators combined with waveguides can form a core set for implementing various functional signal processing devices such as filters, delay lines, and de/multiplexers. In what follows we will briefly discuss the design and implementation of a set of PnC slab devices such as resonators, waveguides, and their combination for forming more complex devices. These devices are all based on the void/Si PnC slab structures discussed above and are first functional devices of their kind. For more details on these structures, the reader is recommended to refer to the referenced articles.

5.6.1 Energy Confinement in MM Resonators in PnC Slabs

As explained before, the energy confinement in the out of plane direction due to the acoustic mismatch gives an unfair advantage to PnC slabs for confining mechanical energy and obtaining high quality factor resonators and waveguides. Because of the fundamental role of energy confinement in resonators, in this section we will discuss a variety of PnC slab-based MM resonators.

5.6.1.1 MM PnC Slab Resonators with Resonant Tunneling Excitation

The first demonstration of PnC slab resonators was reported in [48]. The structure of this resonator was made by removing a period of holes from the PnC structure to form a Fabry–Perot type cavity. The geometrical parameters are similar to those of the previous section that results in a complete PnBG with extents in the range of 119 MHz $< f <$ 150 MHz allowing for confining mechanical vibrations in a wide range. The cavity is surrounded by three periods (twelve rows) of holes on each side in the x direction and is considered very large compared to the wavelength in the y direction.

The structure was fabricated on an SOI wafer with two transducers on its two sides as illustrated in Fig. 5.17a. The fabrication procedure of the device was similar to what discussed in the previous sections resulting an SEM image of the fabricated structure shown in Fig. 5.17b.

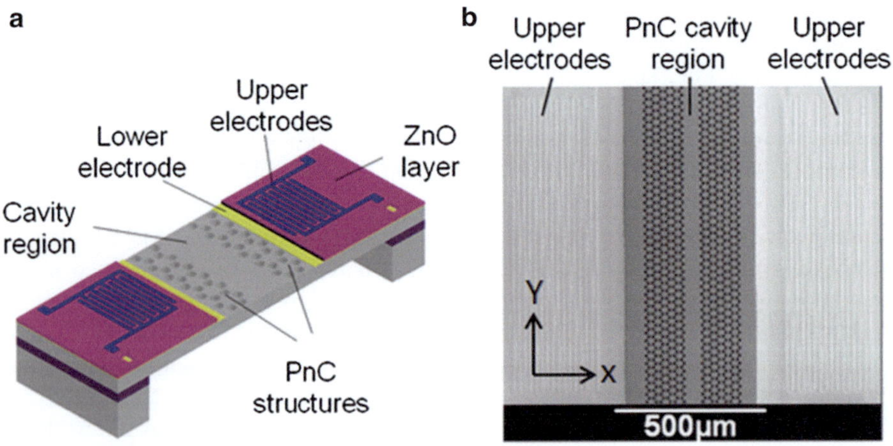

Fig. 5.17 (**a**) Schematic of the PnC slab resonator structure with excitation and detection transducers on its two sides. In this schematic, the cavity region is surrounded by four rows of holes (one period of the PnC) on each side. (**b**) Top SEM image of a fabricated PnC slab resonator with the transducer electrodes on each side. The cavity region is surrounded by twelve rows (three periods) of holes on each side

Using IDTs, with appropriate frequency respond, flexural plate waves are launched towards the cavity and the two flexural vibration modes of the resonator are excited through the resonant tunneling phenomenon. The resonance frequency of the first flexural mode is approximately in the middle of the complete PnBG at 126 MHz, while the resonance frequency of the second flexural mode is at 149.5 MHz, which matches very well with FE simulations.

Similar to resonant tunneling effect in semiconductors [49], resonant peaks will appear in the transmission of acoustic waves through this structure at the frequencies of resonance. The result of the normalized transmission through the PnC resonator (with two layers of PnC on each side) is shown in Fig. 5.18. The anticipated PnBG for flexural acoustic waves is shaded in Fig. 5.18.

As shown in the original [48], by increasing the number of PnC layers, the Q can be significantly improved (e.g., fivefold for one additional layer on each side of the cavity), which puts this PnC slab resonator among the outstanding state of the art micromechanical resonators. This simple PnC slab resonator showed the effectiveness of PnC slab structures in confining the elastic energy at high frequencies.

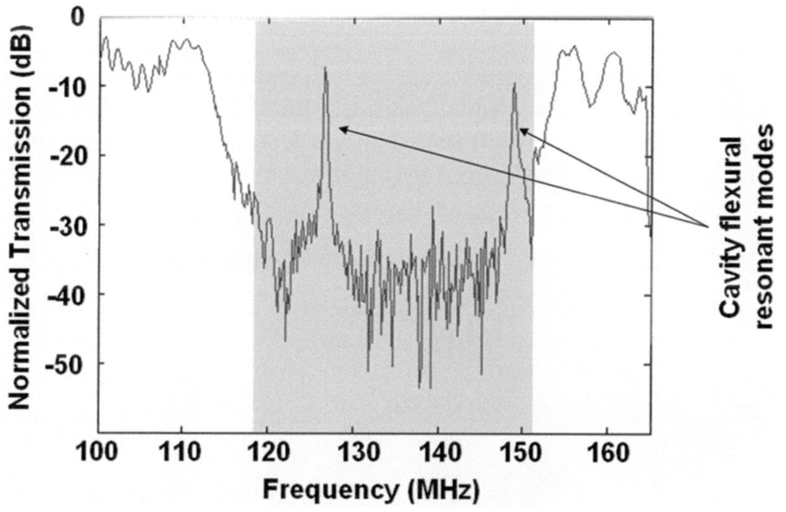

Fig. 5.18 Normalized transmission of flexural acoustic modes passing through the structure shown (*right*) but with only two periods of holes on each side. The calculated PnBG for flexural waves is *shaded* in this figure. As expected, the two resonant flexural modes of the cavity appear as sharp peaks in the transmission spectrum within the PnBG

5.6.2 Support Loss Suppression of MM Resonators Using PnBG Structures

It should be noted the properties of the resonant modes can be significantly improved by confining the cavities entirely by the PnC structures. In the resonators discussed in the previous section, no reflectors are fabricated to confine the acoustic energy in the *y* direction. By completely enclosing the cavity by the PnC structures with complete PnBGs (or in a combination with other types of reflectors), high-Q and compact MM resonators should be obtained. In such a scheme, the resonant region can be almost completely isolated from the substrate at the frequencies of resonance due to the complete PnBG of the PnC. The mechanical support and electrical signal to the cavity would be provided through the structure of the PnC surrounding the cavity. Therefore, MM resonators free of support loss [50] can be obtained using PnC structures with complete PnBGs. In this section, it is shown that with a PnC-based support loss-free resonator, higher Qs compared to the conventional resonators maybe obtained.

5.6.2.1 Support Loss in MM Resonators

High-Q MM resonators are often made by suspending a resonating mass in air or vacuum with some anchor (or supporting) structures that provide mechanical support to the resonating mass as shown in Fig. 5.19. Due to the large impedance mismatch between the solid structure and air or vacuum, most of the energy is stored

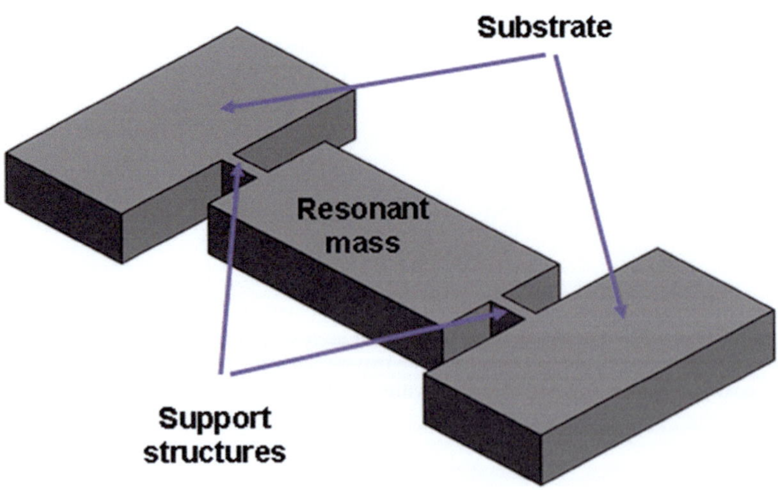

Fig. 5.19 An illustration of a typical MM resonator made by suspending a resonating structure to obtain high Q resonances. The structure is suspended via some supporting structures that mechanically hold the structure

within the resonating structure except an important portion of energy which is lost through the anchors. The loss of mechanical energy through the anchors or support structures is called the *support loss* [50], which is an important factor in determining the Q of the resonators, especially at high frequencies. Therefore, suppression of support loss is an important step toward MM resonators with higher Qs and better performance.

Although efforts have been constantly made to reduce the support loss from the MM resonators by techniques such as placement of the support structures at the resonant mode nodes, such techniques cannot completely suppress the support loss and the few number of support structures in such schemes results in lower levels of mechanical support to the structure. As we will address, the use of appropriate PnC structures with complete PnBGs can result in MM resonators that can be inherently support loss suppressed while being adequately mechanically supported.

5.6.2.2 A Support Loss Suppressed PnC Slab Resonator

To obtain the characteristics and the possibility of suppressing support loss in MM resonators, a typical two-port MM resonator is fabricated. A top-view scanning electron microscope (SEM) image of the fabricated structure is shown in Fig. 5.20a. In order to de-couple the PnC slab from the substrate, the SOI wafer is patterned on the back using optical lithography and deep plasma etching.

Figure 5.20b–d shows the transmission profile between the two ports of the device, where two resonant peaks appear in the transmission spectrum at 106.3 MHz, and 122.4 MHz. FE simulations show that these two modes are associated with a flexural and an extensional vibration mode, respectively. Qs of 870 and 1,200 are obtained for the flexural and longitudinal modes, respectively.

To characterize the role of the supporting structures in determining the Q of the resonator, a similar resonant structure but with PnC structures to provide the mechanical support is designed and fabricated on the same chip as the conventional resonator. The structure and the geometrical parameters of the PnC slab structure is similar to the ones discussed in the previous sections. By having the piezoelectric transducers right on the resonant area of the resonator, the number of PnC periods placed around the resonant region can be increased to provide enough isolation from the surroundings. This can eliminate the reduction in the Q due to the support (coupling) loss to the surroundings as compared to the resonant tunneling excitation method [48].

Figure 5.21a shows a SEM image of the fabricated PnC slab resonator. The transmission profiles of the modes propagating from one port to the other are shown in Fig. 5.21b, c for the flexural and longitudinal vibrations, respectively. As can be seen in this figure, the resonant modes show an increased Q of approximately 6,000 which is approximately six times the Q of the modes of the fabricated conventional MM resonator. To ensure the accuracy of the Q calculations a BVD model was fit to the admittance of one of the ports. Although the Q has been greatly improved, as can be seen in Fig. 5.21b, c, multiple resonant modes are supported by the structure

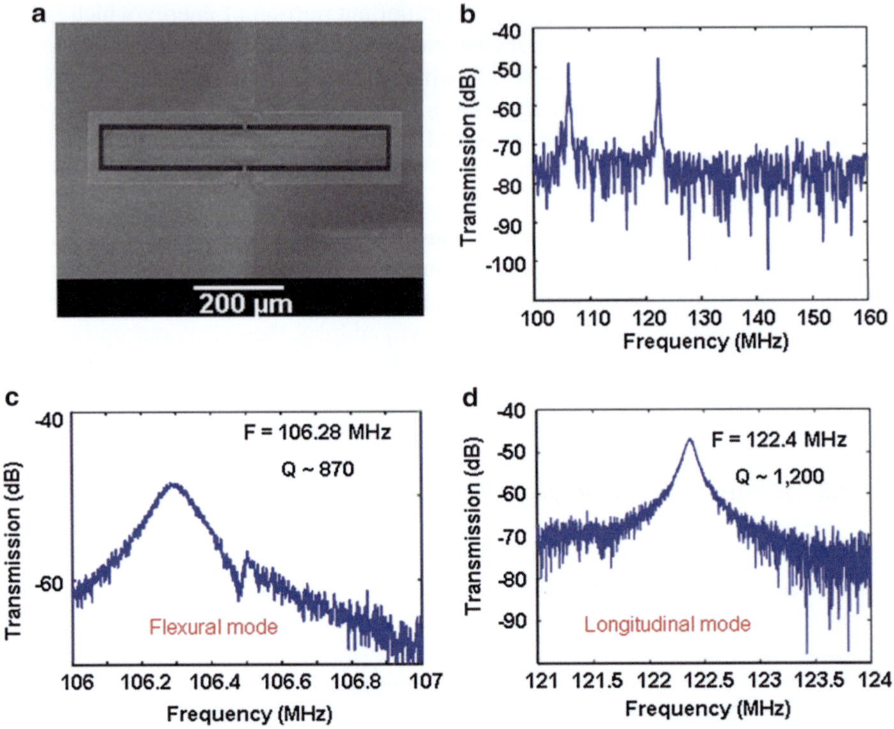

Fig. 5.20 (**a**) Scanning electron microscope (SEM) image of a conventional MM resonator. (**b**) The transmission spectrum between the two ports of the resonator in the range of interest. (**c**) High-resolution graph of the characteristics of the flexural resonant mode. (**d**) High-resolution graph of the characteristics of the longitudinal mode

in each frequency region. It can be deduced that this is mainly owing to the size of the resonator in the y direction, which is multiple times the wavelength of the resonant mode. One reason that can cause this problem is the Fabry–Perot reflectors that are corrugated in the case of the PnC resonator. As such spurious modes are not usually desirable for communications and sensing applications, a different method of fully confining the resonator with PnCs should be utilized. Such an approach will be discussed in what follows using PnC slab waveguides.

5.6.3 PnC Slab Waveguides

Wave guiding is one of the most fundamental functionalities needed for manipulating waves and energy flow. Along with resonators as energy storage elements, waveguides are used for development of a variety of functional devices such as delay lines and filters. Therefore, there have been efforts and studies regarding PnC

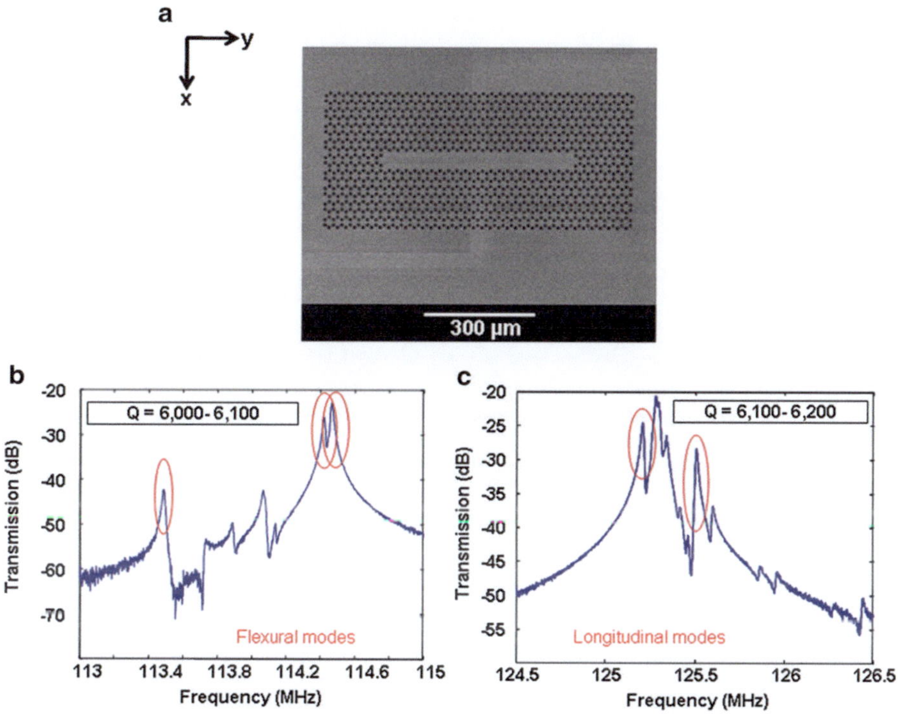

Fig. 5.21 (**a**) SEM image of the fabricated PnC-surrounded resonator structure. (**b**) First-order flexural resonant modes of the structure in the x direction. Multiple modes are excited because of the corrugated PnC structure. (**c**) First-order extensional modes present in the PnC structure

slab waveguides [9, 26, 31, 51, 52] since appropriate PnBGs have been developed. Here we briefly discuss the development of an innovative PnC slab waveguide that beside other applications can also be used to develop high quality resonators and is well suited for developing complex waveguide-resonator coupled devices.

5.6.3.1 An Efficient PnC Slab Waveguide

It is usually required to have a waveguide with lower number of supported propagating modes so that the level of inter-modal interference and cross talk within the waveguide is minimized. In order that only a small number of modes fall into the guiding region of the waveguide, it is better to perturb the structure of PnC as little as it gives us the intended guided mode. For example, instead of completely removing rows of holes to form a waveguide as reported before, the waveguide should be created by reducing the size of rows of holes. A schematic of such a PnC waveguide is shown in Fig. 5.22. To obtain appropriate dispersion diagrams, in the defect-free part of the structure, the normalized radius of the holes in the PnC slab

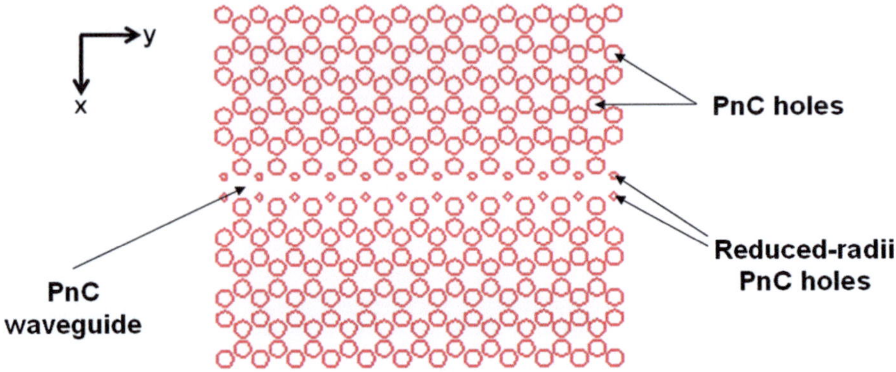

Fig. 5.22 Schematic of a PnC waveguide made by reducing the radii of two rows of holes

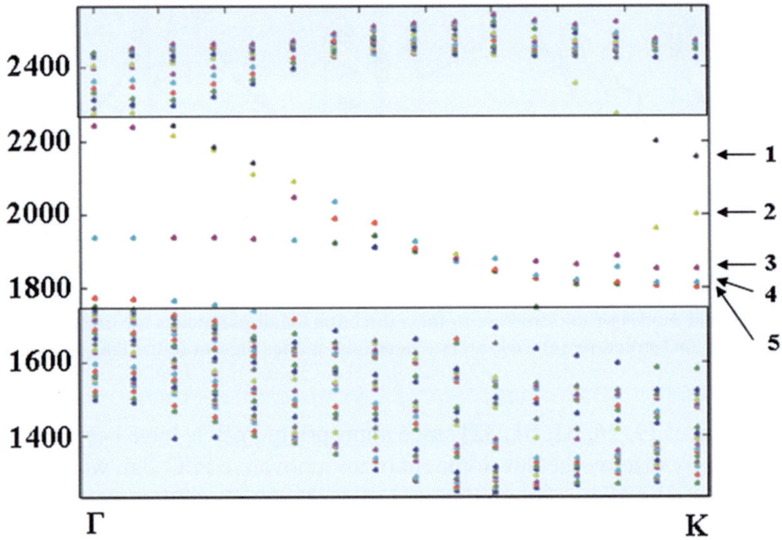

Fig. 5.23 Dispersion diagram of the PnC waveguide shown in Fig. 5.22 with five guided modes numbered from 1 to 5. The normalized radius of the PnC holes is $r/a = 0.43$ while the normalized radius of the two rows of holes forming the line defect is $r_1/a = 0.2$

structure is $r/a = 0.43$ and the normalized thickness of the slab is $d/a = 1$, where a is the distance between the centers of the closest holes and d is the thickness of the slab. The normalized radius of the two rows of the holes forming the defect is designed to be $r_1/a = 0.2$ to accommodate a close-to single-mode operation for the PnC waveguide while preserving symmetry. The dispersion curve associated with this PnC waveguide is shown in Fig. 5.23.

As can be seen in this figure, the number of guided modes in the band structure is significantly reduced compared to the waveguides reported previously [51].

5.6.4 High-Q Waveguide-Based PnC Resonators

As discussed, the support-loss-free MM resonators discussed in the previous sections suffered from multiple resonant modes with their frequencies in the vicinity of that of the principal resonant mode. This is often not a desirable phenomenon and structures with single-resonance characteristics with a larger spurious-free spectral range are preferred for the envisioned applications in both communications and sensing. To mitigate this problem, recently a PnC slab resonator based on PnC slab waveguides has been developed [53]. In this new type of resonator, a few periods of the waveguide in the guiding direction are chosen and the two ends of a waveguide are closed by the PnC structure so that the guided modes can be completely confined within the formed cavity. To interrogate the resonator modes, direct piezoelectric transduction [54] was utilized.

Figure 5.24 shows the schematic and SEM image of the designed waveguide-based resonator. The resonator is composed of 10 periods of the PnC slab waveguide and operates at the second extensional mode indicated in Fig. 5.23.

The amplitude of the admittance of the BVD model of one of the ports of the resonator, for the [100]-directed device is shown in Fig. 5.25. As can be seen in Fig. 5.25, unlike the profile of the modes in the Fabry–Perot-based resonator of Fig. 5.21, all the range of interest (which covers the PnBG) is free of the spurious modes.

The performance of this resonator can be further improved by using materials of better quality. As an example, a similar structure with AlN excitation has led to resonators with quality factors of approximately 15,000 and better characteristics impedance [55].

The fantastic characteristics of this type of MM resonator can also be the basis for implementing complex functionalities for on-chip phononic signal processing devices such as filters and multiplexers/demultiplexers as a result of better waveguide-coupling possibilities. The possibility of multiplexing/demultiplexing using such devices has recently been demonstrated [56] (see Fig. 5.26).

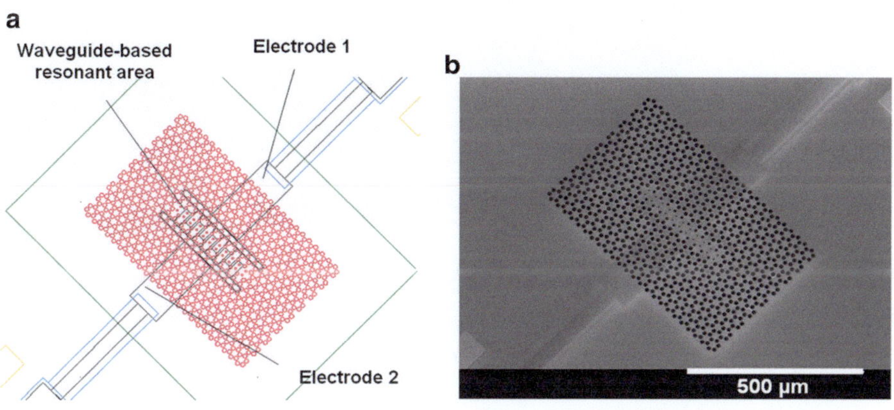

Fig. 5.24 (**a**) Schematic and (**b**) SEM image of the fabricated waveguide-based MM resonator

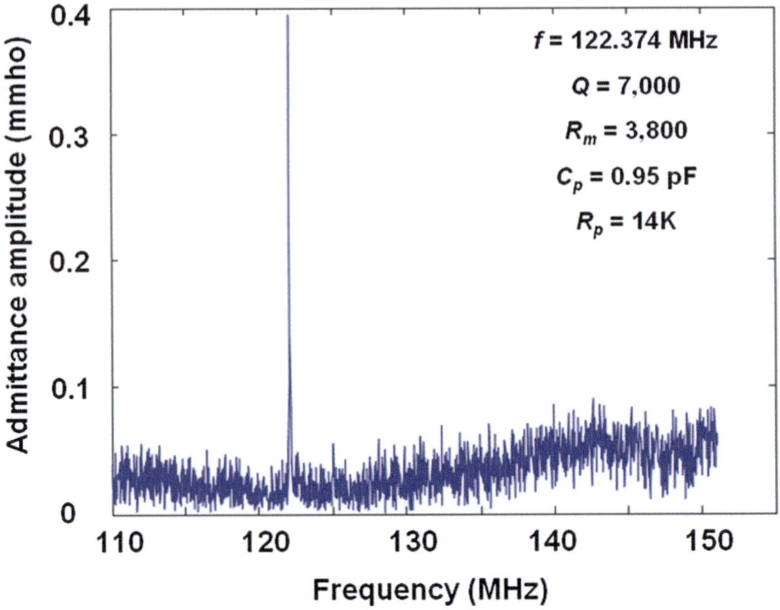

Fig. 5.25 The amplitude of the admittance of the waveguide-based resonator structure of Fig. 5.24

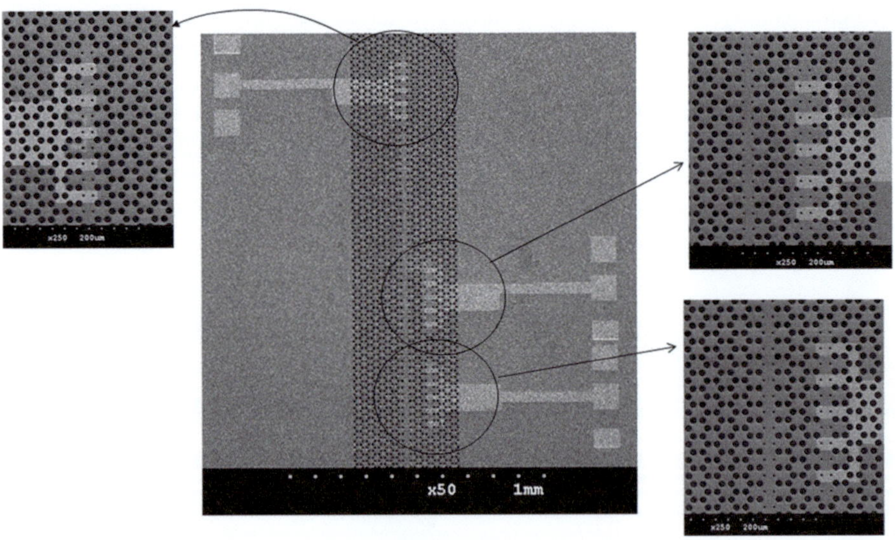

Fig. 5.26 SEM images of the fabricated structure showing the device as well as individual resonator ports magnified

5.7 Further Research Trends and Perspectives of PnC Slab Structures

5.7.1 Dispersive PnC Slab Structures

Besides the applications of PnC slabs in utilizing PnBGs for confining and controlling the flow of elastic waves leading to resonators, waveguides, and the devices based on their combination, the strong modification of the dispersion characteristics of the waves propagating in the slab can be quite interesting even without the emergence of PnBGs. Due to the low loss characteristics of PnC slabs, there has been interesting studies in developing different types of devices based on dispersive PnCs. These studies include observation of negative refraction [57, 58] and design and analysis of flat lenses using gradient index PnC slabs [59]. Such PnC slabs can be used for focusing and guiding of acoustic waves to the waveguides and resonators.

5.7.2 Optomechanical Crystal Slabs

Optomechanical interactions in structures where simultaneous confinement of photons and phonons (elastic vibrations) occurs can be quite rewarding. Phenomena such as quantum back action, enhanced Brillouin scattering, and optical pumping are all interesting phenomena that can have unique applications in communications and sensing. The PnC slabs can be a great potential platform for utilizing optomechanical interactions as it can provide high-Q photonic and phononic resonators with small mode volume. The small mode volume structures means higher levels of field concentration and therefore higher levels of photon/phonon interaction. This provides a great advantage for PnC slabs compared to other types of optomechanical devices. Recently, it has been shown that simultaneous photonic and phononic band gaps can be obtained using PnC slab structures [10, 60–62] which shows the possibility of high-Q confinement in such periodic slab. These periodic slabs, also named phoxonic crystal slabs to cover both photons and phonons, are currently under active study, but yet need more design and analysis to reach their potential applications. For instance, photon/phonon interactions in different forms of PnC slabs as well as their corresponding cavities and waveguides have shown promising results [11, 12, 63, 64]. Further it has been theoretically shown that it is possible to obtain simultaneous phononic and photonic band gaps for the membranes with stubs as well [65]. Demonstration of the performance of these band gaps in terms of optical loss is still to be assessed.

5.7.3 Thermal Phonons Control in PnC Slabs

An emerging application of PnCs is the control of heat transfer and thermal phonons. Since the nature of thermal phonons and acoustic phonons is essentially the same, with a PnC with a PnBG at the right range of frequencies, it is possible to suppress and control the propagation of thermal phonons and therefore, heat. The possibility of controlling heat transfer has important applications in the fields of bolometric, calorimetric, thermal isolation, thermoionic, and thermoelectric devices. PnC slabs made by nanometric periodicity have been a major theme in development of such devices. Especially, due to the possibility of reducing thermal conductivity independently from the electrical conductivity, there have been several recent efforts in PnC slab devices for thermal control which show about an order of magnitude reduction in thermal conductivity [66–70]. Such promising results draw the possibility of performance devices for applications in various emerging fields such as energy generation and microelectronic cooling.

References

1. S.G. Joshi, S.M. Ieee, B.D. Zaitsev, Reflection of plate acoustic waves produced by a periodic array of mechanical load strips or grooves. IEEE Trans. Ultrason. Ferroelectr. Freq. Control **49**(12), 1730–1734 (2002)
2. B.A. Auld, Waves and vibrations in periodic piezoelectric composite materials. Mater. Sci. Eng. A **122**, 65–70 (1989)
3. M.M. Sigalas, E.N. Economou, Elastic and acoustic wave band-structure. J. Sound Vib. **158**(2), 377–382 (1992)
4. L. Ye, G. Cody, M. Zhou, P. Sheng, Observation of bending wave localization and quasi mobility edge in 2 dimensions. Phys. Rev. Lett. **69**(21), 3080–3082 (1992)
5. A. Khelif, B. Aoubiza, S. Mohammadi, A. Adibi, V. Laude, Complete band gaps in two-dimensional phononic crystal slabs. Phys. Rev. E **74**(4), 46610–46615 (2006)
6. B. Bonello, C. Charles, F. Ganot, Lamb waves in plates covered by a two-dimensional phononic film. Appl. Phys. Lett. **90**(2), 021909 (2007)
7. S. Mohammadi, A.A. Eftekhar, A. Khelif, H. Moubchir, R. Westafer, W.D. Hunt, A. Adibi, Complete phononic bandgaps and bandgap maps in two-dimensional silicon phononic crystal plates. Electron. Lett. **43**, 898–899 (2007)
8. S. Mohammadi, A.A. Eftekhar, A. Khelif, W.D. Hunt, A. Adibi, Evidence of large high-frequency complete phononic band gaps in silicon phononic crystal plates. Appl. Phys. Lett. **92**(22), 221905-1-3 (2008)
9. J.O. Vasseur et al., Waveguiding in two-dimensional piezoelectric phononic crystal plates. J. Appl. Phys. **101**(11), 114904 (2007)
10. S. Mohammadi, A.A. Eftekhar, A. Khelif, A. Adibi, Simultaneous two-dimensional phononic and photonic band gaps in opto-mechanical crystal slabs. Opt. Express **18**(9), 9164–9172 (2010)
11. V. Laude et al., Simultaneous guidance of slow photons and slow acoustic phonons in silicon phoxonic crystal slabs. Opt. Express **19**(10), 9690–9698 (2011)
12. A.H. Safavi-Naeini, O. Painter, Design of optomechanical cavities and waveguides on a simultaneous bandgap phononic–photonic crystal slab. Opt. Express **18**(14), 14926–14943 (2010)

13. I. El-Kady, R.H. Olsson, J.G. Fleming, Phononic band-gap crystals for radio frequency communications. Appl. Phys. Lett. **92**(23), 233504 (2008)
14. C. Hsu, T.-T. Wu, Efficient formulation for band-structure calculations of two-dimensional phononic-crystal plates. Phys. Rev. B **74**(14) (2006)
15. K.H. Fung, Z. Liu, C.T. Chan, Transmission properties of locally resonant sonic materials with finite slab thickness. Z. Kristallogr. **220**(9–10), 871–876 (2005)
16. D. Yu, G. Wang, Y. Liu, J. Wen, J. Qui, Flexural vibration band gaps in thin plates with two-dimensional binary locally resonant structures. Chin. Phys. **15**(2) (2006)
17. J.-C. Hsu, T.-T. Wu, Lamb waves in binary locally resonant phononic plates with two-dimensional lattices. Appl. Phys. Lett. **90**(20), 201904 (2007)
18. B. Djafari-Rouhani, Y. Pennec, H. Larabi, Band structure and wave guiding in a phononic crystal constituted by a periodic array of dots deposited on a homogeneous plate. Proc. SPIE **7223**, 72230F-72230F-10 (2009)
19. M. Oudich, M.B. Assouar, Z. Hou, Propagation of acoustic waves and waveguiding in a two-dimensional locally resonant phononic crystal plate. Appl. Phys. Lett. **97**(19), 193503 (2010)
20. J.-C. Hsu, Local resonances-induced low-frequency band gaps in two-dimensional phononic crystal slabs with periodic stepped resonators. J. Phys. D: Appl. Phys. **44**(5), 055401 (2011)
21. S. Mohammadi, A. Khelif, R. Westafer, E. Massey, W.D. Hunt, A. Adibi, Full band-gap silicon phononic crystals for surface acoustic waves, in *Proc. 2006 ASME International Mechanical Engineering Congress and Exposition, IMECE2006*, November, 2006
22. S. Benchabane, A. Khelif, J.Y. Rauch, L. Robert, V. Laude, Evidence for complete surface wave band gap in a piezoelectric phononic crystal. Phys. Rev. E **73**, 65601–65612 (2006)
23. J.H. Sun, T.T. Wu, Propagation of surface acoustic waves through sharply bent two-dimensional phononic crystal waveguides using a finite-difference time-domain method. Phys. Rev. B **74**, 174305–174307 (2006)
24. T.-T. Wu, W.-S. Wang, J.-H. Sun, J.-C. Hsu, Y.-Y. Chen, Utilization of phononic-crystal reflective gratings in a layered surface acoustic wave device. Appl. Phys. Lett. **94** (2009)
25. S. Benchabane, O. Gaiffe, G. Ulliac, R. Salut, Y. Achaoui, V. Laude, Observation of surface-guided waves in holey hypersonic phononic crystal. Appl. Phys. Lett. **98**, 171908 (2011)
26. R.H. Olsson, I.F. El-Kady, M.F. Su, M.R. Tuck, J.G. Fleming, Microfabricated VHF acoustic crystals and waveguides. Sens. Actuators A: Phys. **145**, 87–93 (2008)
27. C.M. Reinke, M.F. Su, R.H. Olsson, I. El-Kady, Realization of optimal bandgaps in solid–solid, solid–air, and hybrid solid–air–solid phononic crystal slabs. Appl. Phys. Lett. **98**, 061912 (2011)
28. Y. Pennec, B. Djafari-Rouhani, H. Larabi, J. Vasseur, A. Hladky-Hennion, Low-frequency gaps in a phononic crystal constituted of cylindrical dots deposited on a thin homogeneous plate. Phys. Rev. B **78**(10), 1–8 (2008)
29. Y. Pennec, J.O. Vasseur, B. Djafari-Rouhani, L. Dobrzyński, P.A. Deymier, Two-dimensional phononic crystals: examples and applications. Surf. Sci. Rep. **65**(8), 229–291 (2010)
30. S. Halkjær, O. Sigmund, J.S. Jensen, Maximizing band gaps in plate structures. Struct. Multidiscip. Opt. **32**(4), 263–275 (2006)
31. F.-L. Hsiao, A. Khelif, H. Moubchir, A. Choujaa, C.-C. Chen, V. Laude, Waveguiding inside the complete band gap of a phononic crystal slab. Phys. Rev. E **76**(5), 1–6 (2007)
32. N.K. Kuo, C.J. Zuo, G. Piazza, Microscale inverse acoustic band gap structure in aluminum nitride. Appl. Phys. Lett. **95**, 93501–93503 (2009)
33. A. Taflove, S. Hagness, *Computational Electrodynamics: The Finite-Difference Time-Domain Method* (Artech House, Norwood, 1995)
34. M.M. Sigalas, N. García, Theoretical study of three dimensional elastic band gaps with the finite-difference time-domain method. J. Appl. Phys. **87**(6), 3122 (2000)
35. P.-F. Hsieh, T.-T. Wu, J.-H. Sun, Three-dimensional phononic band gap calculations using the FDTD method and a PC cluster system. IEEE Trans. Ultrason. Ferroelectr. Freq. Control **53**(1), 148–158 (2006)
36. S. Mohammadi, *Phononic Band Gap Micro/Nano-Mechanical Structures for Communications and Sensing Applications* (Georgia Institute of Technology, Atlanta, 2010)

37. J. Lourtioz, H. Benisty, V. Berger, J. Gerard, D. Maystre, A. Tchelnokov, Photonic crystals: towards nanoscale photonic devices (2005)
38. M. Wilm, S. Ballandras, V. Laude, T. Pastureaud, A full 3D plane-wave-expansion model for 1–3 piezoelectric composite structures. J. Acoust. Soc. Am. **112**(3), 943–952 (2002)
39. J.-C. Hsu, T.-T. Wu, Calculations of Lamb wave band gaps and dispersions for piezoelectric phononic plates using Mindlin's theory-based plane wave expansion method. IEEE Trans. Ultrason. Ferroelectr. Freq. Control **55**(2), 431–441 (2008)
40. COMSOL multi-physics modeling guide, v. 3.3, August 2006
41. I.E. Psarobas, N. Stefanou, A. Modinos, Scattering of elastic waves by periodic arrays of spherical bodies. Phys. Rev. B **62**(1), 278–291 (2000)
42. C.Y. Qiu, Z.Y. Liu, J. Mei, M.Z. Ke, The layer multiple-scattering method for calculating transmission coefficients of 2D phononic crystals. Solid State Commun. **134**(11), 765–770 (2005)
43. S. Noda, T. Baba, O. Industry, *Roadmap on Photonic Crystals* (Springer, Berlin, 2003)
44. X. Zhang, T. Jackson, E. Lafond, P. Deymier, J. Vasseur, Evidence of surface acoustic wave band gaps in the phononic crystals created on thin plates. Appl. Phys. Lett. **88**(4), 041911 (2006)
45. T. Gorishnyy, C.K. Ullal, M. Maldovan, G. Fytas, E.L. Thomas, Hypersonic phononic crystals. Phys. Rev. Lett. **94**(11) (2005)
46. D. Profunser, O. Wright, O. Matsuda, Imaging ripples on phononic crystals reveals acoustic band structure and Bloch harmonics. Phys. Rev. Lett. **97**(5), 1–4 (2006)
47. K.E. Petersen, Silicon as a mechanical material. Proc. IEEE **70**(5), 420–457 (1982)
48. S. Mohammadi, A.A. Eftekhar, W.D. Hunt, A. Adibi, High-Q micromechanical resonators in a two-dimensional phononic crystal slab. Appl. Phys. Lett. **94**(5), 051906-051901-051903 (2009)
49. L.L. Chang, L. Esaki, R. Tsu, Resonant tunneling in semiconductor double barriers. Appl. Phys. Lett. **24**(12), 593–595 (1974)
50. Z.L. Hao, A. Erbil, F. Ayazi, An analytical model for support loss in micromachined beam resonators with in-plane flexural vibrations. Sens. Actuators A: Phys. **109**(1–2), 156–164 (2003)
51. S. Mohammadi, A.A. Eftekhar, W.D. Hunt, A. Adibi, Demonstration of large complete phononic band gaps and waveguiding in high-frequency silicon phononic crystal slabs, in *2008 IEEE International Frequency Control Symposium*, vol. 43, no. 21066, pp. 768–772, May 2008
52. T.-C. Wu, T.-T. Wu, J.-C. Hsu, Waveguiding and frequency selection of Lamb waves in a plate with a periodic stubbed surface. Phys. Rev. B **79**(10), 1–6 (2009)
53. S. Mohammadi, A.A. Eftekhar, A. Adibi, Support loss-free micro/nano-mechanical resonators using phononic crystal slab waveguides, in *IEEE Frequency Control Symposium*, 2010, pp. 521–523
54. S. Mohammadi, A.A. Eftekhar, R. Pourabolghasem, A. Adibi, Simultaneous high-Q confinement and selective direct piezoelectric excitation of flexural and extensional lateral vibrations in a silicon phononic crystal slab resonator. Sens. Actuators A: Phys. **167**(2), 524–530 (2011)
55. S. Mohammadi, A. Adibi, Waveguide-based phononic crystal micro/nano-mechanical high-Q resonators. J. Microelectromech. Syst. (to appear)
56. S. Mohammadi, A.A. Eftekhar, A. Adibi, Resonator/waveguide coupling in phononic crystals for demultiplexing and filtering applications, in *Ultrasonics Symposium (IUS)*, 2010 IEEE, pp. 155–157
57. J. Pierre, O. Boyko, L. Belliard, J.O. Vasseur, B. Bonello, Negative refraction of zero order flexural Lamb waves through a two-dimensional phononic crystal. Appl. Phys. Lett. **97**(12), 121919 (2010)
58. M.K. Lee, P.S. Ma, I.K. Lee, H.W. Kim, Y.Y. Kim, Negative refraction experiments with guided shear-horizontal waves in thin phononic crystal plates. Appl. Phys. Lett. **98**(1), 011909 (2011)
59. T.-T. Wu, Y.-T. Chen, J.-H. Sun, S.-C.S. Lin, T.J. Huang, Focusing of the lowest antisymmetric Lamb wave in a gradient-index phononic crystal plate. Appl. Phys. Lett. **98**(17) (2011)

60. S. Mohammadi, A.A. Eftekhar, A. Adibi, Large simultaneous band gaps for photonic and phononic crystal slabs, in *2008 Conference on Lasers and Electro-Optics*, vol. 1, no. c, pp. 1–2, May 2008
61. Y. Pennec, B. Rouhani, E. El Boudouti, C. Li, Y. El Hassouani, J. Vasseur, N. Papanikolaou, S. Benchabane, V. Laude, A. Martinez, Simultaneous existence of phononic and photonic band gaps in periodic crystal slabs. Opt. Express **18**, 14301–14310 (2010)
62. B. Djafari Rouhani et al., Band gap engineering in simultaneous phononic and photonic crystal slabs. Appl. Phys. A **103**(3), 735–739 (2010)
63. M. Eichenfield, R. Camacho, J. Chan, K.J. Vahala, O. Painter, A picogram- and nanometre-scale photonic-crystal optomechanical cavity. Nature **459**(7246), 550–555 (2009)
64. A.H. Safavi-Naeini, O. Painter, Proposal for an optomechanical traveling wave phonon–photon translator. New J. Phys. **13**(1), 013017 (2011)
65. Y. El Hassouani et al., Dual phononic and photonic band gaps in a periodic array of pillars deposited on a thin plate. Phys. Rev. B **82**(15), 1–7 (2010)
66. R. Venkatasubramanian, E. Siivola, T. Colpitts, B. O'Quinn, Thin-film thermoelectric devices with high room-temperature figures of merit. Nature **413**(6856), 597–602 (2001)
67. R. Yang, G. Chen, Thermal conductivity modeling of periodic two-dimensional nanocomposites. Phys. Rev. B **69**(19), 1–10 (2004)
68. J.-K. Yu, S. Mitrovic, D. Tham, J. Varghese, J.R. Heath, Reduction of thermal conductivity in phononic nanomesh structures. Nat. Nanotechnol. **5**(10), 718–721 (2010)
69. P.E. Hopkins, L.M. Phinney, P.T. Rakich, R.H. Olsson, I. El-Kady, Phonon considerations in the reduction of thermal conductivity in phononic crystals. Appl. Phys. A **103**(3), 575–579 (2010)
70. P.E. Hopkins et al., Reduction in the thermal conductivity of single crystalline silicon by phononic crystal patterning. Nano Lett. **11**(1), 107–112 (2011)

Chapter 6
Surface Acoustic Waves in Phononic Crystals

Tsung-Tsong Wu, Jin-Chen Hsu, Jia-Hong Sun, and Sarah Benchabane

6.1 Introduction

Over the past two decades, propagation of acoustic waves in periodic structures comprised of multi-components has received much attention because of renewed physical properties and potential applications in a variety of fields, such as noise and vibration isolation, frequency filters in wireless communication, super lens design, etc. These composite materials, called phononic crystals (PCs) [1, 2], give rise to forbidden gaps for acoustic waves which are analogous to the band gaps for electromagnetic waves in photonic crystals. Major mechanisms leading to the forbidden gaps are Bragg scattering and localized resonances (LR) [3]. The former opens up the Bragg gap at the Brillouin-zone boundaries, and the band-gap frequency corresponds to the wavelength in the order of the structural period, i.e. of the lattice constant, and relates to the lattice symmetry. On the other hand, localized

T.-T. Wu
Institute of Applied Mechanics, National Taiwan University , No. 1, Section 4, Roosevelt Road, Da'an District, Taipei 106, Taiwan
e-mail: wutt@ndt.iam.ntu.edu.tw

J.-C. Hsu
Applied Mechanics Department, National Yunlin University of Science and Technology, No. 123, Section 3, Daxue Road, Douliu, Yunlin 604, Taiwan
e-mail: uem@yuntech.edu.tw

J.-H. Sun
Applied Mechanics Department, National Taiwan University, No. 1, Section 4, Roosevelt Road, Da'an District, Taipei 106, Taiwan
e-mail: jhsun@mail.cgu.edu.tw

S. Benchabane (✉)
Institut FEMTO-ST, Université de Franche-Comté, CNRS, ENSMM, UTBM, 15b Av des Montboucons, Besançon 25030, France
e-mail: sarah.benchabane@femto-st.fr

A. Khelif, A. Adibi (eds.), *Phononic Crystals*, DOI 10.1007/978-1-4614-9393-8_6

resonance creates resonant gaps dictated by the frequency of resonance associated with scattering units and depends less on the lattice symmetry, orderliness, and periodicity of the structure.

In the literature, theoretical calculations of bulk acoustic waves (BAWs) in phononic band structures of two-dimensional (2D) periodic composites were reported using the plane wave expansion (PWE) method [1–14], multiple scattering theory [15–18], and finite difference time domain (FDTD) method [19–21]. On the other hand, the experimental evidence for the existence of absolute acoustic band gaps (BAW modes) has also been reported [19, 22–27]. The existence of band gap, transmission properties, and local resonance of 2D phononic structures in the low frequency range were demonstrated. A comprehensive review of bulk waves in PCs can be found in a review article by Sigalas et al. [28].

Investigations into the surface acoustic wave (SAW) properties of PCs in which the periodic modulation occurs on the traction-free surface did not take place until the late 1990s [29–36]. The theoretical studies of SAWs on a square and hexagonal superlattice consisting of cubic (AlAs/GaAs) and isotropic (Al/polymer) materials were reported [29, 30] and extended to the case of general anisotropy a couple of years later by Wu et al. [31]. The case of SAW modes in 2D piezoelectric PC was then reported subsequently [32]. The surface mode of a square-lattice Y-cut lithium niobate PC with a large full band gap (a fractional bandwidth of 34 %) was found [33]. In [34], existence of the Bleustein–Gulyaev (BG) piezoelectric SAW in a 2D piezoelectric PC (ZnO(100)/CdS(100)) was reported. The authors showed that the BG surface waves have higher acoustic wave velocities, high electromechanical coupling coefficients, and larger band gap width than those of the Rayleigh surface waves and pseudosurface waves. In a recent article [35], the concept of the localization factor was introduced to study the Rayleigh surface wave propagation and localization in disordered piezoelectric PCs and the authors found that the larger the randomness degree, the stronger the degree of wave localization.

In the late 1990s, experimental studies of SAW in 2D PCs were concentrated on cases with frequency in the range of a few MHz or lower. The existence of surface states for sonic propagation in elastic band gap finite periodic systems was first demonstrated by Torres et al. [36] both theoretically and experimentally. In their experimental work, surface modes as well as localization phenomena in linear and point defects were demonstrated by drilling millimeter cylinders filled up with Hg in an Al matrix. At about the same time, experiments on the attenuation of Rayleigh surface waves by a semi-infinite 2D elastic band gap crystal made up of marble with centimeter size were reported by Meseguer et al. [37]. Vines et al. [38] conducted an experimental study on the surface waves generated by a line-focus acoustic lens at the water-loaded surfaces of a number of 2D superlattices that intersect the surface normally. Highly anisotropic transmission of SAW with frequency in the low MHz range was observed by continuously scanning the wave vector angle. Recently, further experimental demonstrations of the band gaps of surface modes in 2D PCs in the MHz range using laser ultrasonics have also been reported [39, 40].

Although demonstration of the band gap properties of SAWs in PCs at the millimeter scale has been conducted since the late 1990s, studies at the micrometer scale, which may find applications in the radio frequency (RF) communications or

MEMS devices, had not started until the mid of last decade. In an effort toward the integration of PC and SAW frequency filters or oscillators, Wu et al. [41] utilized silicon micromachining to fabricate air/silicon square lattice PC with layered slanted finger interdigital transducers (SFIT) attached. The transmission of high-frequency SAW through 6 layers of PC with more than 30 dB attenuation was observed in the band gap between 183 and 215 MHz. In a subsequent paper, Benchabane et al. [42] demonstrated experimentally the existence of a complete SAW band gap in a 2D square lattice piezoelectric PC etched in lithium niobate. The SAW was generated by normal interdigital transducer (IDT) and a complete band gap extending from 203 to 226 MHz was demonstrated experimentally. Scattering and propagation of SAWs in PCs revealed by optical methods were given by Kokkonen et al. [43] and Profunser et al. [44]. Recently, a design that combines two-port SAW devices and PCs acting as reflective gratings was demonstrated [45]. The design consists of a layered ZnO/Si SAW device and a square lattice PC composed of cylindrical holes in silicon. With 15-layer PC cylinders, experimental insertion loss shows a 7 dB improvement at the central frequency of 212 MHz. As compared with the conventional metallic reflective gratings, the major advantage of PC is that the size of the CMOS compatible layered SAW filter can be reduced significantly.

In this chapter, investigation of SAW in 2D phononic band gap structures is presented. The content is organized as follows. In Sect. 6.2, the theoretical formulations of the PWE and the FDTD methods for SAW in periodic structures are briefly summarized. In Sect. 6.3, calculated results of SAW propagation and related phenomena are presented. Then, in Sect. 6.4, utilizing the properties of phononic band gaps, waveguides constructed with the band gap structures for SAW are investigated. The numerical results show that the acoustic wave energy can be efficiently guided or confined with the proposed structures. In Sects. 6.5 and 6.6, experiments are conducted to demonstrate the phononic band gaps and potential applications to SAW resonator devices. Conclusions are finally given in Sect. 6.7.

6.2 Theoretical Formulations

6.2.1 PWE Method

6.2.1.1 General Equations

Consider a PC composed of a 2D periodic array (on the $x_1 - x_2$ plane) of material A embedded in a background material B. Both materials A and B are piezoelectric or dielectric. In the quasi-static approximation, the governing field equations of piezoelectricity can be expressed as

$$\frac{\partial T_{ij}}{\partial x_i} = \rho \frac{\partial^2 u_j}{\partial t^2}, \tag{6.1}$$

$$\frac{\partial D_i}{\partial x_i} = 0, \; i,j = 1,2,3, \tag{6.2}$$

where $\rho(\mathbf{x})$ is the position-dependent mass density, $u_j(\mathbf{r}, t)$ is the displacement vector, $\mathbf{r} = (x_1, x_2, x_3) = (\mathbf{x}, x_3)$ is the position vector, and $T_{ij}(\mathbf{r}, t)$ and $D_i(\mathbf{r}, t)$ are the stress and electric displacement, respectively. The piezoelectric constitutive equations with the displacement $u_j(\mathbf{r}, t)$ and the electric potential $\varphi(\mathbf{r}, t)$ as variables are given by

$$T_{ij} = c_{ijkl}\frac{\partial u_k}{\partial x_l} + e_{lij}\frac{\partial \varphi}{\partial x_l}, \tag{6.3}$$

$$D_i = \varepsilon_{il}\frac{\partial \varphi}{\partial x_l} - e_{ikl}\frac{\partial u_k}{\partial x_l}, \; i,j,k,l = 1,2,3, \tag{6.4}$$

where $c_{ijkl}(\mathbf{x})$, $e_{lij}(\mathbf{x})$, and $\varepsilon_{il}(\mathbf{x})$ are the position-dependent elastic stiffness constants, piezoelectric constants, and permittivity, respectively. Due to the spatial periodicity, the material constants can be expanded in Fourier series with respect to the 2D reciprocal lattice vectors. On utilizing the Bloch–Floquet theorem and expanding the displacement vector and electric potential in Fourier series, a generalized eigenvalue problem with respect to the x_3 component of the wave vector k_3 can be derived:

$$\left(\mathbf{P} + k_3\mathbf{Q} + k_3^2\mathbf{R}\right)\mathbf{U} = \mathbf{0}. \tag{6.5}$$

The dimensions of the matrices **P**, **Q**, **R**, and **U** are dependent on the number of the reciprocal lattice vectors used in the Fourier series expansions. As N reciprocal lattice vectors are used, **P**, **Q**, and **R** are $4N \times 4N$ square matrices, and **U** is a $4N \times 1$ column matrices. Equation (6.5) with an arbitrarily guessed frequency ω yields a set of eigenvalues ${k_3}^{(m)}$, $(m = 1\text{–}8N)$. The displacement field and electric potential of a SAW exponentially decay into the structure ($x_3 < 0$) away from the surface. Thus, $4N$ eigenvalues with the following characteristics are picked out from the $8N$ eigenvalues:

$$\begin{cases} \mathrm{Re}\left(k_3^{(m)}\right) > 0 & \text{if } \mathrm{Im}\left(k_3^{(m)}\right) = 0, \\ \mathrm{Im}\left(k_3^{(m)}\right) < 0 & \text{otherwise.} \end{cases} \tag{6.6}$$

For SAWs propagating along the traction-free surface, the following mechanical boundary conditions are satisfied:

$$T_{i3} = c_{i3kl}\frac{\partial u_k}{\partial x_l} + e_{li3}\frac{\partial \varphi}{\partial x_l} = 0, \quad \text{at } x_3 = 0. \tag{6.7}$$

For a piezoelectric medium, the electrical boundary condition is imposed. Here, two important cases are distinguished:

- Open-circuit condition: If the adjacent medium is air, the normal component of the electric displacement across the surface is continuous, that is,

$$D_3 = \left(e_{3kl} \frac{\partial u_k}{\partial x_l} - \varepsilon_{3l} \frac{\partial \varphi}{\partial x_l} \right) = -\varepsilon_0 \frac{\partial \varphi_{\text{air}}}{\partial x_3}, \tag{6.8}$$

  giving the angular frequency $\omega = \omega_0$. The electric potential φ_{air} in air can be solved by the Laplace equation together with the boundary conditions $\varphi_{\text{air}} = \varphi$ at $x_3 = 0$, and $\varphi_{\text{air}} = 0$ as $x_3 \to \infty$.
- Short-circuit condition: If the surface is covered by a very thin metallic film, the electric potential at the surface is equal to zero, that is, $\varphi = 0$, giving the angular frequency $\omega = \omega_\infty$.

6.2.1.2 Case of Void Inclusions in a Solid Matrix

The PWE method as briefly stated above and detailed in Chap. 4 generally considers a PC with a solid–solid composition. In practice, however, a very convenient way to fabricate PCs consists in drilling holes in a solid matrix. This particularly holds in the case of periodical structures exhibiting features at the micron-scale. The resulting air/solid compositions in addition offer the highest possible impedance contrast between host and inclusion, hence contributing to the opening of wide band gaps. The case of inclusions made of air or vacuum, as propagation in air can be neglected when considering PCs in the ultrasonic regime, is therefore of definite interest.

In the literature, this issue is usually tackled by representing vacuum in a PWE model as an artificial medium with appropriate elastic constants and density. Laude et al. [33], for instance, proposed to fake vacuum by a material exhibiting zero elastic constants and mass density. In their paper, the case of a piezoelectric solid/void PC is specifically dealt with. Their starting point is to consider air or vacuum as a non-piezoelectric solid where wave propagation is governed by the equation of motion in its purely elastic form, i.e.:

$$T_{ij} = c_{ijkl} \frac{\partial u_k}{\partial x_l}, \tag{6.9}$$

$$\rho \frac{\partial u_j}{\partial t^2} = \frac{\partial T_{ij}}{\partial x_j}. \tag{6.10}$$

Setting c_{ijkl} to zero imposes $T_{ij} = c_{ijkl} \frac{\partial u_k}{\partial x_l} = 0$ independently of the displacements, while $\rho = 0$ ensures free displacements at the vacuum/solid interface. These solutions hence remain compatible with the definition of a free interface (unspecified displacement and zero-stress) while verifying the relation of continuity of the normal component of the stress tensor and of the elastic displacement field components at the interface, that is, $u_i^{\mathrm{air}} = u_i^{\mathrm{solid}}$ and $T_{ij}^{\mathrm{air}} n_j = T_{ij}^{\mathrm{solid}} n_j$. A simple pseudo-solid is hence defined, that will fake a vacuum in solid–solid PWE computations.

An alternative initially proposed by Tanaka et al. [20] and later refined by Manzanares-Martínez et al. [46] and Vasseur et al. [47], consists in representing vacuum as an artificial isotropic solid material characterized by physical parameters allowing for a fast decay of the elastic displacement field in the holes. The choice of these parameters is obviously governed by the boundary conditions at the vacuum–solid interface that impose $\rho c_{\mathrm{t,l}}^2 \to 0$, where ρ is the mass density and c_{t} and c_{l} are the transverse and longitudinal wave velocities, respectively. A natural choice consists in imposing a very low mass density ρ^{air} (of the order of 10^{-4} kg/m^3) and longitudinal and transverse velocities much larger than the usual sound of speed in solids (typically $c_{\mathrm{l}} = c_{\mathrm{t}} = 10^5$ m/s) to limit propagation of waves to the surrounding, actual solid while remaining with finite values. Under such conditions, the obtained pseudo-solid presents non-zero elastic constants C_{11} and C_{44} around 10^6 N/m^2, four orders of magnitude lower than those encountered in real solids.

These two possible implementations offer stable numerical solutions by ensuring at all time that the Fourier coefficients describing the material constants of both matrix and inclusions are well-defined.

6.2.2 Finite-Difference Time-Domain Method

The FDTD method was introduced in 2000 for the study of PCs. Sigalas and Garcia [48] calculated the transmission of acoustic waves passing a PC structure. Kafesaki et al. [49] demonstrated the wave propagation and transmission inside a PC waveguide. Further, Tanaka et al. [20] applied the Bloch theorem to FDTD method and analyzed the acoustic dispersion. A directed periodic boundary condition was suggested by Hsieh et al. [50] to study the band structures of PCs. So far, the FDTD method has been well-developed to calculate the dispersion, transmission, and propagation of BAWs inside PC structures. Further application of the FDTD method to SAWs was suggested by Sun and Wu [51]. The principle of FDTD method is mentioned in the following.

In a linear elastic material, the constitutive law and equation of motion of elastic materials are expressed as Eqs. (6.9) and (6.10). These equations describe the property of an infinitesimal element of an anisotropic material in general. They are applicable to the inhomogeneous structure of PCs by arranging the density and elastic constant periodically. Further, with staggered grids, the differential equations (6.9) and (6.10) are transferred into difference equations based on the Taylor's expansion to develop the 3D heterogeneous finite difference formulation. Then these equations are calculated recursively to study the wave propagation in PCs.

The Bloch's theorem is introduced to treat the periodic boundary condition of a unit cell of PCs. The displacement and stress fields of PCs can be expressed by a plane wave and a periodic function as:

$$u_i(\mathbf{x}, t) = e^{\mathrm{i}\mathbf{k}\cdot\mathbf{x}} U_i(\mathbf{x}, t), \tag{6.11}$$

$$T_{ij}(\mathbf{x}, t) = e^{\mathrm{i}\mathbf{k}\cdot\mathbf{x}} S_{ij}(\mathbf{x}, t), \tag{6.12}$$

where **k** is a wave vector. $U_i(\mathbf{x}, t)$ and $S_{ij}(\mathbf{x}, t)$ are periodic functions satisfying

$$U_i(\mathbf{x} + \mathbf{a}, t) = U_i(\mathbf{x}, t), \tag{6.13}$$

$$S_{ij}(\mathbf{x} + \mathbf{a}, t) = S_{ij}(\mathbf{x}, t), \tag{6.14}$$

with a lattice translation vector **a**. In [20], the equation of motion and constitutive law were transformed using the periodic functions to satisfy the conditions (6.13)–(6.14). Alternatively, we combined Eqs. (6.11)–(6.12) and (6.13)–(6.14) to write a periodic boundary condition (PBC) of displacement u_i and stress T_{ij} directly as [50]:

$$u_i(\mathbf{x} + \mathbf{a}, t) = e^{\mathrm{i}\mathbf{k}\cdot\mathbf{a}} u_i(\mathbf{x}, t), \tag{6.15}$$

$$T_{ij}(\mathbf{x} + \mathbf{a}, t) = e^{\mathrm{i}\mathbf{k}\cdot\mathbf{a}} T_{ij}(\mathbf{x}, t). \tag{6.16}$$

With the PBC, both the 2D and 3D PC cases can be analyzed by calculating a unit cell. In the analyses of dispersions, a small disturbance in a random position of the unit cell is set as the initial condition. Thus all possible wave modes are transported inside the considered PC, and the displacement is recorded and expanded into Fourier series. Then the eigenfrequencies as a function of the wave vector **k** are indicated by selecting the resonance peaks of the spectrum. This procedure allows finding acoustic modes which satisfy the periodic boundary condition in PCs.

To solve the SAW modes of PCs, extra boundary conditions are needed to develop SAWs inside a PC unit cell. In this study, we set a free surface boundary condition and use the PML absorbing boundary condition to deal with the reflection from the numerical boundary.

Berenger [52] introduced the concept of PML to reduce the electromagnetic wave reflection from the boundary, and PML have also been developed in the case of elastic wave propagation [53, 54]. In the present study, a 3D PML program applicable to orthogonal materials is adopted to serve the non-reflection boundary condition. A stretched coordinate is defined with a complex variable

$$e_i = a_i + \mathrm{i}\frac{\Omega_i}{\omega} \tag{6.17}$$

to derive the code of the PML region [53]. In Eq. (6.17), the real part a_i is the scale factor, and Ω_i/ω is the imaginary part with the attenuation factor Ω_i and the circular frequency ω. Then the differential operation in the stretched coordinate is defined and applied to the equation of motion and constitutive law. After introducing the plane wave solutions into the equations, the numerical attenuation is achieved via the factor Ω_i. Besides, a non-reflection condition at the interface between PML region and inner space is obtained by setting corresponding material constants and the unity scale factor. In the elastodynamic equations for the stretched coordinate, displacement and stress components are taken via spatial partial differential operation in all directions; thus, the variables are split into three components to realize the difference equations. Then actual values are obtained from the summation of splitting components. Finally, the PML equations can be written as follows:

$$\rho\frac{\partial u_{i/j}}{\partial t^2}+\rho\Omega_j\frac{\partial u_{i/j}}{\partial t}=\frac{\partial T_{ij}}{\partial x_j}, \tag{6.18}$$

$$\frac{\partial T_{ij/m}}{\partial t}+\Omega_m T_{ij/m}=\frac{c_{ijkl}}{2}\frac{\partial}{\partial t}\left(\frac{\partial u_k}{\partial x_l}\delta_{ml}+\frac{\partial u_l}{\partial x_k}\delta_{mk}\right). \tag{6.19}$$

In the above equations, $u_{i/j}$ and $T_{ij/m}$ are the splitting displacement and splitting stress which satisfy $u_i = u_{i/1} + u_{i/2} + u_{i/3}$ and $T_{ij} = T_{ij/1} + T_{ij/2} + T_{ij/3}$. δ_{ij} is the delta function and $\delta_{ij} = 1$ when $i = j$, else $\delta_{ij} = 0$.

After transforming equations (6.18) and (6.19) into the difference formula, the PML is arranged outside the space boundaries as a buffer zone with a matched acoustic impedance to suppress reflection. It performs a numerical attenuation so that waves decay rapidly inside the region. With this absorbing boundary condition, reflection is reduced to less than 1 % and the PML boundary is used in both the calculations of dispersion relation and transmission.

6.3 SAWs in Phononic Crystals

6.3.1 *Frequency Band Structures*

An example of a frequency band structure of a 2D PC is presented in Fig. 6.1. The structure is a 2D square lattice of circular cylinders of radius r_0 embedded periodically in a background material. The lattice spacing is a, and the reciprocal lattice vectors are $\mathbf{G} = 2\pi(n_1/a,\ n_2/a)$ with integers n_1 and n_2. Figure 6.1 is calculated by using the PWE method with 25 Fourier components in the series expansion (i.e., $-2 \leqq n_1,\ n_2 \leqq 2$). The filling fraction $F = \pi r_0{}^2/a^2$ is 0.6. The cylinders are assumed to be made of bismuth germanium oxide ($Bi_{12}GeO_{20}$), and the background material is silica (SiO_2). The three fourfold axes of $Bi_{12}GeO_{20}$ are aligned with the x_1-, x_2-, and x_3-axis of the PC, respectively. The material constants used in the calculations are listed in Table 6.1. In Fig. 6.1, bulk and surface wave

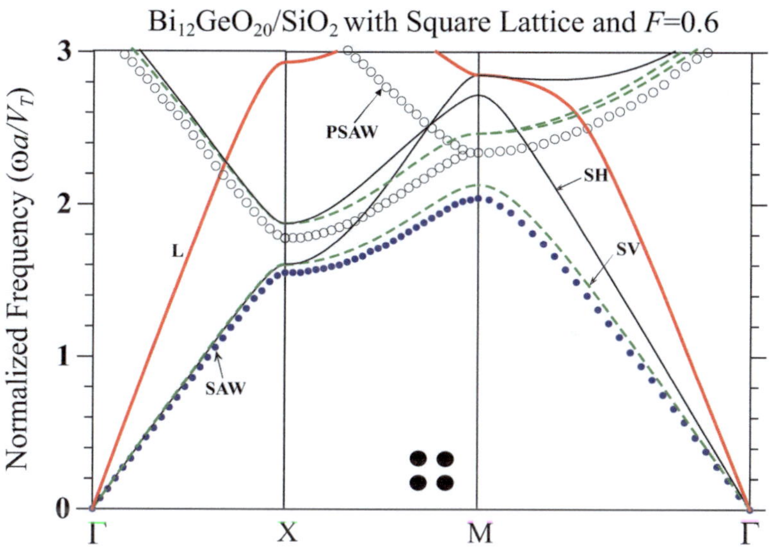

Fig. 6.1 Frequency band structures of the bulk, SAW and PSAW modes in the $Bi_{12}GeO_{20}/SiO_2$ piezoelectric PC with square lattice. $F = 0.6$

Table 6.1 Material constants used in the calculation

Material constants	$Bi_{12}GeO_{20}$	SiO_2	ZnO	CdS
c_{11} (10^{10} N/m^2)	12.8	7.85	20.97	8.56
c_{12}	3.05	1.61	12.11	5.32
c_{13}			10.51	4.62
c_{33}			21.09	9.36
c_{44}	2.55		4.25	1.49
ρ (kg/m^3)	9,230	2,203	5,676	4,824
e_{14} (C/m^2)	0.99			
e_{15}			−0.59	−0.21
e_{22}			−0.61	−0.24
e_{31}			1.14	0.44
ε_{11} (10^{-11} F/m)	34.2		7.38	7.99
ε_{22}			7.83	8.44

modes are presented. The thick and thin solid lines represent the quasi-longitudinal (denoted by L) and quasi-shear-horizontal modes (denoted by SH) polarized on the $x_1 - x_2$ plane, respectively. The dashed lines represent the piezoelectric shear-vertical modes (denoted by SV). The surface modes are represented by the solid dots and the open circles, respectively, for SAWs and pseudosurface acoustic waves (PSAWs), where the PSAWs leak energy into the structure.

Figure 6.1 shows that the SAW branch exists along the boundary Γ–X–M of the Brillouin zone and slightly below the lowest branch of the SV modes. After reaching the Brillouin zone boundary, the folded branch of SAWs turns into the PSAWs.

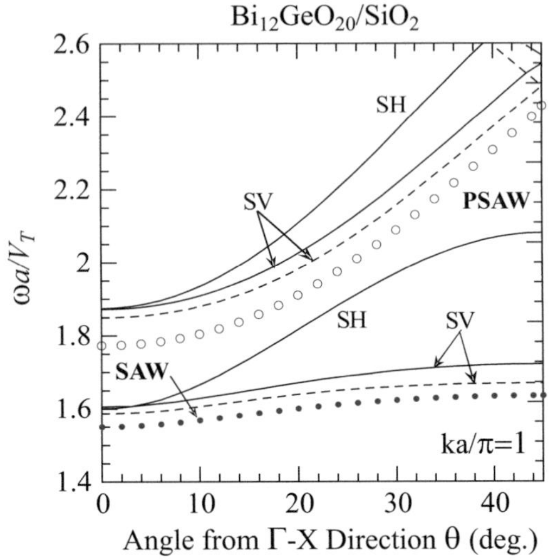

Fig. 6.2 Angular dependence of the frequencies of the bulk waves, surface wave, and pseudosurface wave at $ka/\pi = 1$. The *dashed lines* denote the branches of SV mode with the piezoelectric effect ignored

The PSAWs have a characteristic that they are attenuated by radiating the acoustic energy into the bulk of the structure. More detail about PSAW characteristics will be given in Sect. 6.3.2. It is noted that in this case, only the SH wave is piezoelectric. The L and SH modes are not affected by the piezoelectricity since the only nonzero piezoelectric constants of $Bi_{12}GeO_{20}$ are $e_{14} = e_{25} = e_{36}$. In this regard, Eq. (6.5) can be decoupled into the mixed in-plane modes (i.e., L and SH vibrations are coupled), which are piezoelectric inactive, and piezoelectric SV modes with polarization of the displacement along the x_3 direction.

6.3.2 *Band Gaps and SAW Characteristics*

Figure 6.2 shows the angular dependence of dispersion relations at the magnitude of the reduced wave vector $ka/\pi = 1$, where $k = \sqrt{k_1^2 + k_2^2}$. The PSAW branch exists in between the fundamental branches and the folded branches of bulk shear modes, and its frequency increases as the propagation angle θ increases. Figure 6.3 shows the SAW frequency ω_{SAW} and PSAW frequency ω_{PSAW}, as well as the magnitude of the frequency gap $\Delta\omega = \omega_{\mathrm{PSAW}} - \omega_{\mathrm{SAW}}$ at the symmetry point X as a function of filling fraction F. Both the SAW and PSAW frequencies at the lower and upper band-gap edges decrease monotonically with the increase in F; the decrease in the frequency is reasonable because the filling material $Bi_{12}GeO_{20}$ has a lower SAW velocity propagating along its crystallographic X-axis than that of SiO_2. Therefore, increasing the filling fraction of $Bi_{12}GeO_{20}$ accordingly decreases the SAW and PSAW frequencies. The maximum gap width appears at $F = 0.183$. The value of maximum gap width is 0.5 in normalized frequency. When the filling

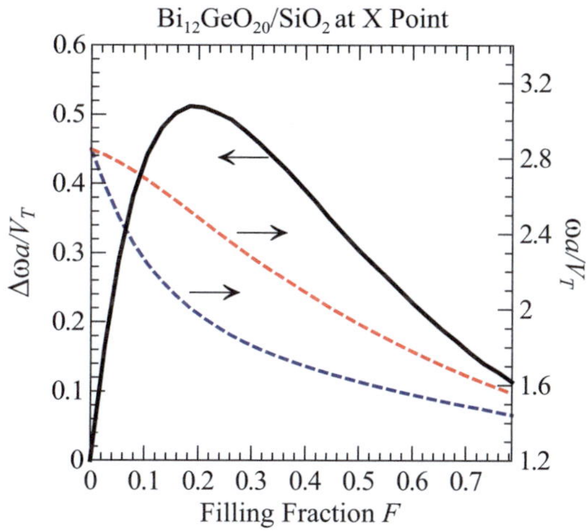

Fig. 6.3 Width of the frequency gap $\Delta\omega = \omega_{PSAW} - \omega_{SAW}$ in between the SAW and PSAW (*solid line*) at the symmetry point X as a function of filling fraction. The frequencies ω_{PSAW} at the upper edge and ω_{SAW} at the lower edge of the gap are also plotted (*dashed lines*)

fraction decreases to zero, the phase velocity of SAW reduces to the SAW velocity in homogeneous SiO_2.

Figure 6.2 shows that the folded branch of surface mode exists inside the bulk bands and turns into a PSAW branch. The reason can be understood thanks to Fig. 6.4 which shows the $k_1 - k_3$ and $k - k_3$ sections of the equifrequency surfaces for the bulk waves, respectively. Here, k represents the magnitude of the Bloch wave vector along Γ–M direction. Figure 6.4a shows that, at the given value of normalized frequency equal to 1.4, the wave number of the SAW, denoted by k_{SAW}, exists along k_1-axis and outside the three equifrequency surfaces of the bulk wave modes. Instead, Fig. 6.4b shows a different case. At the normalized frequency equal to 1.8, which is in the frequency gap of the bulk shear modes at X point, a real wave vector $\mathbf{K} = (k_{SAW} = k_{PSAW}, k_3)$ of a PSAW can be found by drawing a dashed line, from the point $k_1 = k_{PSAW}$, parallel to the k_3-axis and intersecting the curves of the bulk shear modes. The corresponding group-velocity directions are plotted by the outward normal of the surfaces that point inside the bulk of the medium. This means that the surface mode couples to bulk shear modes, and thus that it propagates with attenuation by radiating acoustic energy into the medium. This coupling is attributed to the simultaneous anisotropic effect of the constituents and geometry of the structure. Because of the coupling, the folded branch of the surface modes in the frequency band structure of Fig. 6.1b turns into a PSAW branch. Along the Γ–M direction, Fig. 6.5 also shows a similar phenomenon.

The coupling of a surface mode to the bulk-wave bands to become a PSAW can also be revealed by the eigenvalues solved by Eq. (6.5); that is, the energy radiating into the bulk of medium leads to a part of the $4N$ eigenvalues being real, and a real eigenvalue means a non-decayed out-of-plane propagation. As the general solution of surface modes contains a part of the partial waves constructed by these real eigenvalues and their corresponding eigenvectors, the surface mode has a leaky property of propagation.

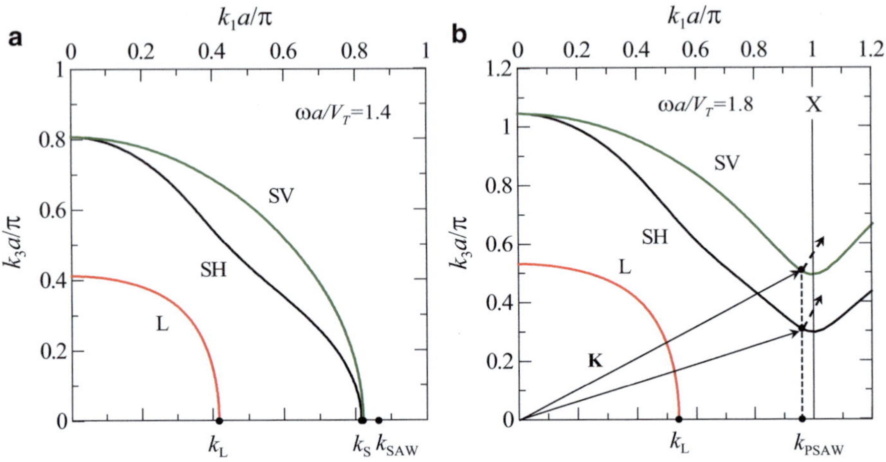

Fig. 6.4 The section of equifrequency surfaces of the bulk waves by the $k_1 - k_3$ plane. (**a**) Normalized frequency equal to 1.4. (**b**) Normalized frequency equal to 1.8

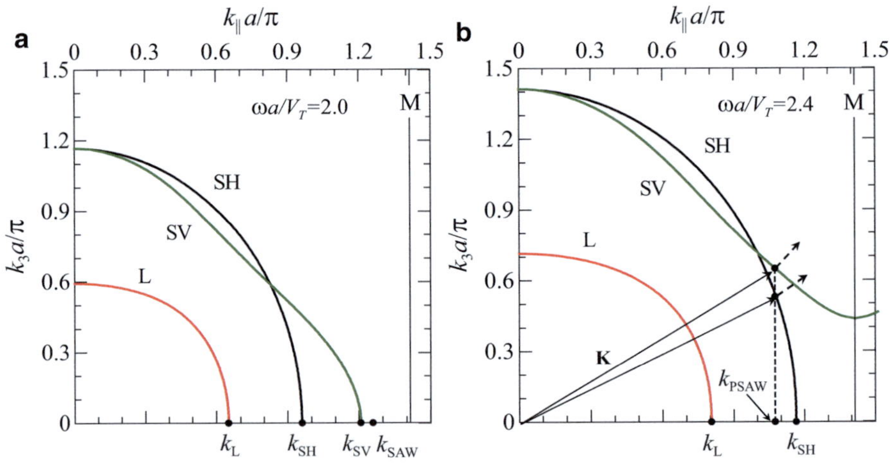

Fig. 6.5 The section of constant-frequency surfaces of the bulk waves by the $k - k_3$ plane, where k is the magnitude of the wave vector along Γ–M direction. (**a**) Normalized frequency equal to 2.0. (**b**) Normalized frequency equal to 2.4

6.3.3 Bleustein–Gulyaev Waves

In a homogeneous piezoelectric crystal, a transverse SAW with no counterpart in a non-piezoelectric material was discovered by Bleustein and Gulyaev in 1968. They showed that the non-vanishing piezoelectric coupling factor in a class of transverse isotropic piezoelectric crystal (e.g., crystal symmetry class 6 mm) can result in a surface wave with a very simple mechanical displacement and electric potential fields that decay exponentially into the medium. In early 1969, Shimizu et al. theoretically and experimentally demonstrated the existence of such a unique wave in a poled lead zirconate titanate (PZT) ceramic independent of the knowledge of Bleustein and Gulyaev's papers. Since then, the follow-up researches and related applications have made the Bleustein–Gulyaev wave theory as one of the cornerstones for the modern signal processing and electro-acoustic technology. In this section, Bleustein–Gulyaev (BG) waves in a homogeneous piezoelectric crystal and in a 2D piezoelectric PC are discussed, respectively.

First, let us consider the homogeneous case. The equations governing the BG wave with purely transverse polarization u_2 and propagating along the x_2-direction in a semi-infinite homogeneous piezoelectric crystal whose surface outward normal is in the x_3-direction are

$$c_{i22l}\frac{\partial^2 u_2}{\partial x_i \partial x_l} + e_{l2j}\frac{\partial^2 \varphi}{\partial x_i \partial x_l} = \rho \frac{\partial^2 u_2}{\partial t^2}, \tag{6.20}$$

$$e_{i2l}\frac{\partial^2 u_2}{\partial x_i \partial x_l} - \varepsilon_{il}\frac{\partial^2 \varphi}{\partial x_i \partial x_l} = 0, \tag{6.21}$$

$$\frac{\partial^2 \varphi_{\text{air}}}{\partial x_i \partial x_i} = 0, \tag{6.22}$$

with the mechanical boundary conditions at the surface

$$T_{i3} = c_{i32l}\frac{\partial u_2}{\partial x_l} + e_{li3}\frac{\partial \varphi}{\partial x_l} = 0, \quad x_3 = 0, \tag{6.23}$$

and the electrical boundary conditions, either

$$\begin{aligned} e_{32l}\tfrac{\partial u_2}{\partial x_l} - \varepsilon_{3l}\tfrac{\partial \varphi}{\partial x_l} &= -\varepsilon_0 \tfrac{\partial \varphi_{\text{air}}}{\partial x_3}, & x_3 &= 0, \\ \varphi &= \varphi_{\text{air}}, & x_3 &= 0, \\ \varphi_{\text{air}} &= 0, & x_3 &\to \infty, \end{aligned} \tag{6.24}$$

for open-circuit surface, or

$$\begin{aligned} \varphi &= 0, & x_3 &= 0, \\ \varphi_{\text{air}} &= 0, & x_3 &\to \infty, \end{aligned} \tag{6.25}$$

for short-circuit surface. The solutions for u_2 and ϕ can be written as

$$\begin{aligned} u_2 &= A_2 \cdot e^{\beta x_3} \cdot e^{ikx_1 - i\omega t}, \\ \varphi &= A_4 \cdot e^{\beta x_3} \cdot e^{ikx_1 - i\omega t}, \qquad \text{Re}(\beta) > 0 \end{aligned} \tag{6.26}$$

where β is a decay constant. Substituting Eq. (6.26) into the governing equations and boundary conditions accordingly solves the BG wave velocity, decay constant, and the electromechanical coupling coefficient.

As an example, consider a case for the existence of the BG wave: a hexagonal crystal of 6 mm class including the [010] propagation direction in the (100) plane. The physical properties transformed from the crystallographic coordinates are given by

$$\begin{pmatrix} c_{11} & c_{13} & c_{12} & 0 & 0 & 0 \\ c_{13} & c_{33} & c_{13} & 0 & 0 & 0 \\ c_{12} & c_{13} & c_{11} & 0 & 0 & 0 \\ 0 & 0 & 0 & c_{44} & 0 & 0 \\ 0 & 0 & 0 & 0 & \frac{c_{11}-c_{12}}{2} & 0 \\ 0 & 0 & 0 & 0 & 0 & c_{44} \end{pmatrix}, \tag{6.27}$$

$$\begin{pmatrix} 0 & 0 & 0 & 0 & 0 & e_{15} \\ e_{31} & e_{33} & e_{31} & 0 & 0 & 0 \\ 0 & 0 & 0 & e_{15} & 0 & 0 \end{pmatrix}, \tag{6.28}$$

$$\begin{pmatrix} \varepsilon_{11} & 0 & 0 \\ 0 & \varepsilon_{33} & 0 \\ 0 & 0 & \varepsilon_{11} \end{pmatrix}. \tag{6.29}$$

Analytically, BG wave velocity and decay constant for this configuration are given by

$$V_0 = V_T \sqrt{1 - \frac{K_T^4}{(1 + \varepsilon_{11}/\varepsilon_0)^2}}, \tag{6.30}$$

$$\beta = \frac{K_T^2}{(1 + \varepsilon_{11}/\varepsilon_0)}, \tag{6.31}$$

for open-circuit surface, and

$$V_\infty = V_T \sqrt{1 - K_T^4}, \tag{6.32}$$

$$\beta = K_T^2, \tag{6.33}$$

for short-circuit surface, where

$$V_T = \sqrt{\frac{c_{44} + \left(e_{15}^2/\varepsilon_{11}\right)}{\rho}}, \tag{6.34}$$

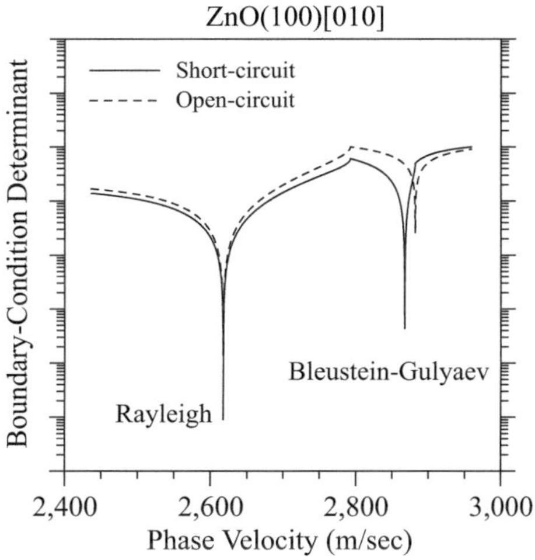

Fig. 6.6 The boundary-condition determinant vs. phase velocity. The sharp minima indicate the eigen-solutions of surface waves, where the chosen orientation of ZnO substrate is [010] direction on (100) plane

$$K_T = \frac{e_{15}}{\sqrt{c_{44}\varepsilon_{11} + e_{15}^2}}, \tag{6.35}$$

and V_T and K_T are the piezoelectric bulk transverse wave velocity and the bulk transverse-wave electromechanical coupling factor, respectively. Equations (6.30)–(6.35) tell that the decay is more rapid, and the difference in velocity from the bulk transverse wave velocity for BG wave increases when the piezoelectric constant is large. For the case of zinc oxide (ZnO) that belongs to class 6 mm, Fig. 6.6 shows the values of boundary-condition determinant versus the phase velocity $V = \omega/k$. The solid line denotes the short-circuit condition, and the dashed line denotes the open-circuit condition. There are two sharp local minima that represent the roots of the boundary-condition determinant for each electrical condition, respectively. The one with lower phase velocity indicates the phase velocity of Rayleigh wave, and the other one indicates the phase velocity of BG wave. For Rayleigh surface wave, there is no difference in phase velocities between the open-circuit and short-circuit conditions. For BG surface wave, the difference in phase velocities between the two electrical conditions can be clearly differentiated because the BG waves are always piezoelectric; otherwise, the BG wave degenerates into a non-piezoelectric bulk SH wave. The phase velocity of BG wave under short-circuit condition is slower than that under open-circuit condition, and they are 2,869.79 m/s and 2,884.04 m/s, respectively. This always happens because the short-circuit surface eliminates the tangential electric field, and thus, partially suppresses the piezoelectricity of the material. The electromechanical coupling for BG wave in ZnO is 9.88 %.

Figures 6.7 and 6.8 show the depth dependence of the displacement and electric potential fields for Rayleigh wave and BG wave under short-circuit surface condition, respectively. In Fig. 6.7, the Rayleigh wave is polarized in the sagittal plane with zero u_2 and φ, so the wave is a non-piezoelectric wave. This means

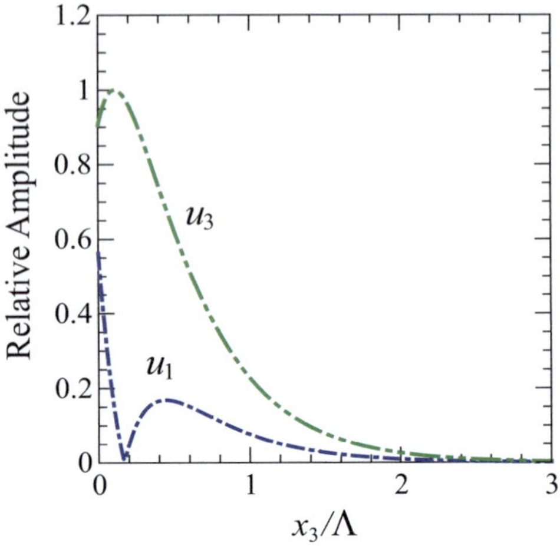

Fig. 6.7 The depth dependence of displacement amplitudes of Rayleigh surface wave propagating along [010] direction on (100) plane of ZnO

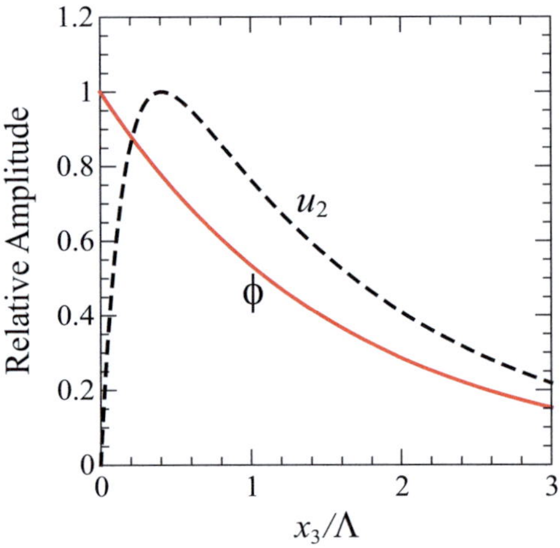

Fig. 6.8 The depth dependence of displacement and electric potential amplitudes of BG surface wave propagating along [010] direction on (100) plane of ZnO (short-circuit condition)

that, in this propagation direction and plane orientation, the mechanical terms are decoupled from the electrical terms for sagittal-plane vibrations. In Fig. 6.8, the piezoelectric BG wave polarized transversely accompanies with an electric potential field. Moreover, the BG wave penetrates much deeper than the Rayleigh wave into the medium.

As an example of piezoelectric PC, let us consider a square array of circular ZnO cylinders, arranged with lattice constant a, embedded periodically in a cadmium sulfide (CdS) background material. In the arrangement of the constituents of the piezoelectric PC, the crystallographic plane (100) of ZnO and CdS is parallel to the $x_1 - x_2$ plane, and their sixfold symmetry axis (i.e., the crystallographic Z-axis) coincides with the x_2-axis of the PC so that, according to BG wave theory, the

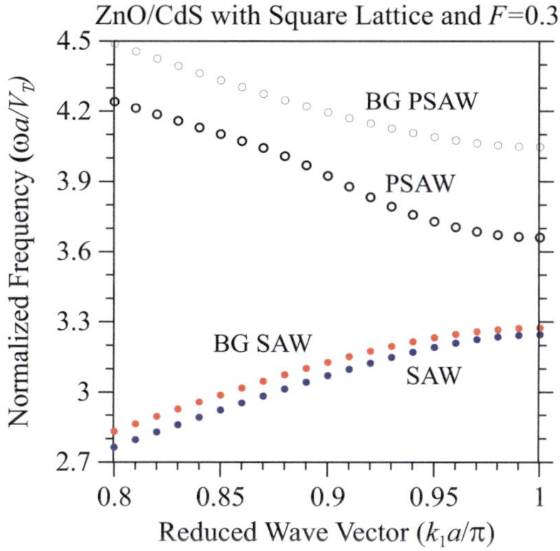

Fig. 6.9 Dispersion relations of SAW (*blue dots*), BG SAW (*red dots*), PSAW (*black circles*), and BG PSAW wave (*red circles*) under short-circuit condition in the frequency range around the band gap and near the point X

piezoelectric effect can compensate for the mechanical displacement field to satisfy the stress-free boundary condition, as well as the SH displacement could decay into the medium.

Figure 6.9 shows the dispersion relations of surfaces modes under short-circuit boundary condition in the frequency range around the band gap near the X point. The filling fraction F is equal to 0.3. In the figure, another fundamental branch of surface wave solution (red dots) exists in addition to the common surface wave for this configuration. Further calculation will show that this surface wave branch is mainly composed of SH displacement, i.e., it is a BG wave in the 2D piezoelectric PC. This wave mode degenerates into the piezoelectric bulk wave as the piezoelectricity is ignored. Moreover, the folded branch of the BG wave has also been found, the folded branch of BG waves is higher than the PSAW branch. This folded BG branch is a pseudosurface-type wave (denoted by BG PSAW) with very small attenuation. The band gap width of the BG waves in the ΓX direction is 0.773, in the unit of normalized frequency.

To reveal the influence of the electrical boundary conditions on the BG surface wave in the piezoelectric PC, a detailed analysis at the symmetry point X is conducted. Figure 6.10 shows the boundary-condition determinants of the fundamental and the folded branches of surface modes as a function of frequency, respectively. In both figures, solid lines denote the short-circuit boundary condition, and dashed lines denote the open-circuit condition. In the case of the fundamental branch (Fig. 6.10a), there are two sharp minima for each boundary-condition determinant. The first one with lower frequency indicates eigenfrequency of the common SAW, while the other one with higher frequency denotes the BG surface wave. For the common SAW, the frequency difference between the open-circuit and short-circuit conditions is very small. In other words, the electromechanical coupling coefficient almost vanishes for the common SAW in this configuration. For the BG waves,

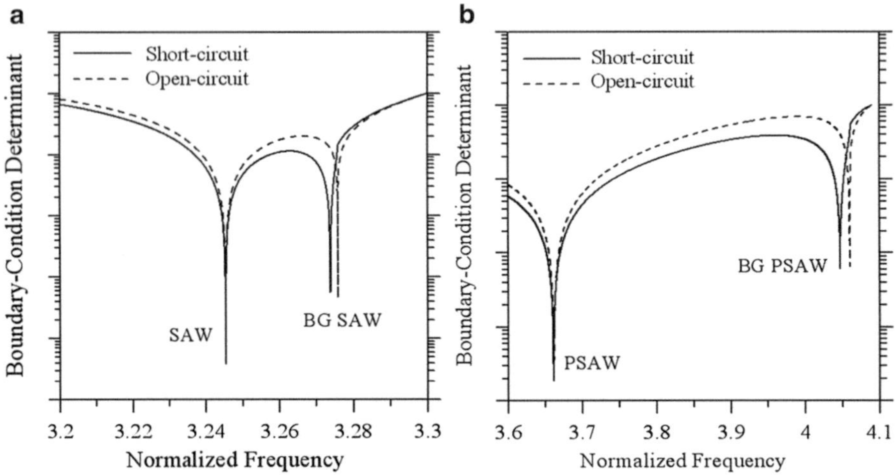

Fig. 6.10 The boundary-condition determinant vs. normalized frequency. (**a**) The frequency range is around the fundamental branches of SAW. (**b**) The frequency range is around the folded branches of SAW at the point X

there is a clear difference in frequency with the same wave vector between the two boundary conditions, and the resulted electromechanical coupling coefficient is about 0.12 %. Figure 6.10b shows results for the folded branches, and the sharp minima indicate the frequencies of PSAWs. The electromechanical coupling coefficient of the folded BG PSAW is 0.63 %.

The depth dependence of the displacement and electric potential fields of the BG waves at the center of the zinc oxide circular cylinder at symmetry point X in the piezoelectric PC are shown in Figs. 6.11 and 6.12. Figure 6.11a, b shows the profiles of the displacement and electric potential amplitudes in relative magnitudes of the fundamental modes of the BG waves under open-circuit and short-circuit conditions, respectively. The results show that, compared with the SH component, the other two components are negligible. In addition, the decay of the displacements and the electric potential is much more rapid for the short-circuit condition than that for the open-circuit condition. From the dispersion relations, the phase velocities of fundamental BG surface modes with reduced wave vector $k_1 a/\pi$ at the symmetry point X are 1,832.09 and 1,830.97 m/s for open-circuit condition and electrical shorted condition, respectively. Figure 6.12a, b shows the profiles of relative amplitudes of the displacement and electric potential of the folded BG waves at point X under short- and open-circuit conditions, respectively. It is observed that the folded BG waves decay more rapidly into the medium than the fundamental modes. However, in this branch a non-vanishing longitudinal component exists. In other words, the BG wave in this 2D piezoelectric PC contains displacement components other than the SH component, which differs from the BG waves in homogeneous piezoelectric materials. This phenomenon is caused by the scattering effect due to the periodic cylindrical fillers in the structure and can be explained by two reasons:

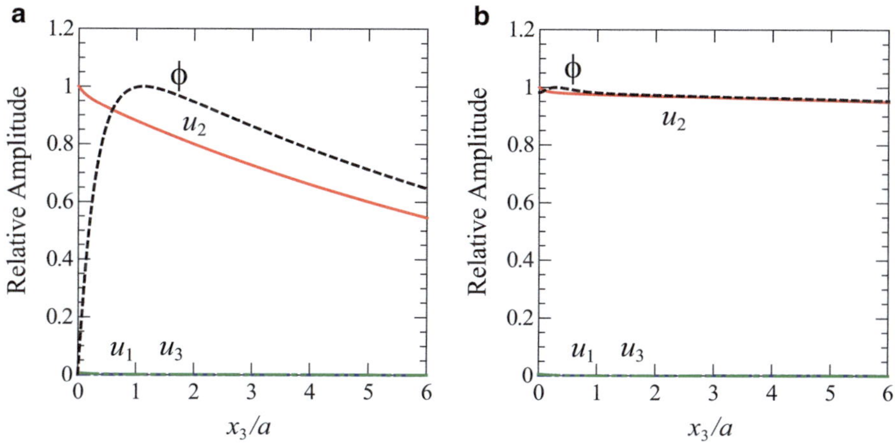

Fig. 6.11 Displacement and electrical potential fields of BG surface wave at the center of the ZnO cylinder vs. the distance from the surface. (**a**) Under short circuit condition. (**b**) Under open-circuit condition

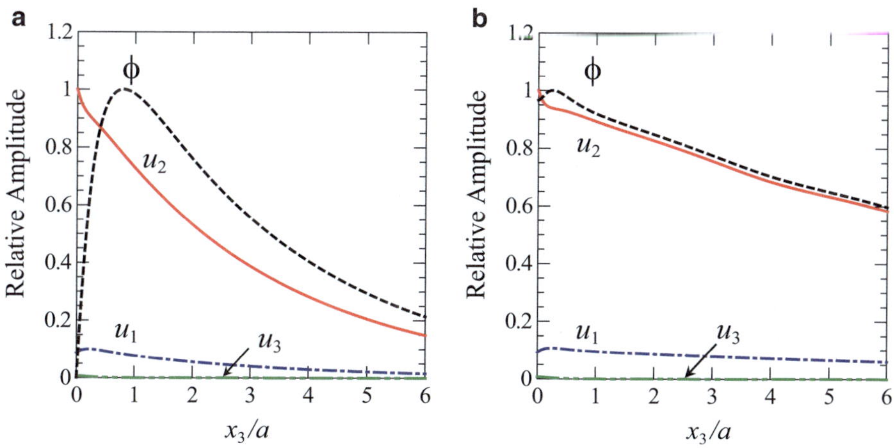

Fig. 6.12 Displacement and electrical potential fields of BG PSAW at the center of the ZnO cylinder vs. the distance from the surface. (**a**) Under short-circuit condition. (**b**) Under open-circuit condition

1. The SH vibration, u_2, is coupled with the sagittal-plane vibrations, u_1 and u_3, through the elastic terms in Eq. (6.5).
2. The electric potential is coupled with the mechanical vibration, not only with the SH vibration but also with the sagittal-plane vibrations, through the piezoelectric and dielectric terms in Eq. (6.5).

These two effects (i.e., the coupling from scattering effect of the periodic structure as well as the anisotropic property of the constituents) do not appear in

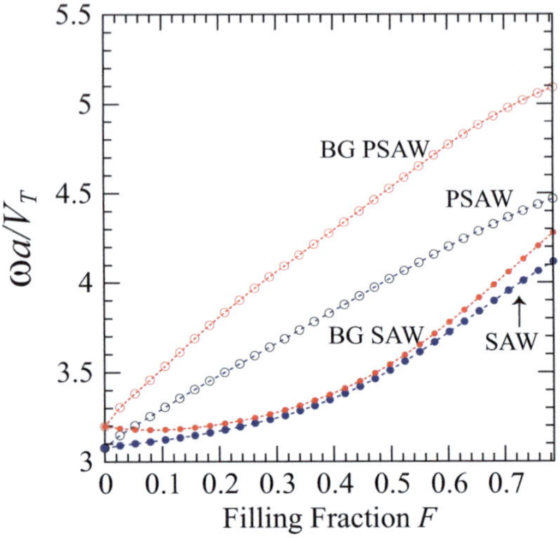

Fig. 6.13 The band edges of the frequency gaps in between the SAW and PSAW, as well as BG SAW and BG PSAW at the symmetry point X as a function of filling fraction

the homogeneous media belonging to 6 mm crystal where the SH vibration with piezoelectricity is decoupled from the sagittal-plane vibrations (and the sagittal-plane vibrations are piezoelectric inactive) for acoustic waves propagating in the [010] direction on (100) plane of the materials belonging to 6 mm symmetry class. The coupling of u_1 with u_2 is stronger than u_3; therefore, the longitudinal displacement is larger than the SV displacement which is very small. The phase velocities of folded mode of the BG waves for open- and short-circuit conditions with reduced wave vector at symmetry point X, calculated by using ω/k, are 2,270.60 and 1,163.43 m/s, respectively.

Figure 6.13 shows the band edges of the frequency gaps of the surface modes and their normalized gap widths at the point X as a function of filling fraction F. The figures show that the BG wave (open circles) band gaps in the Γ–X direction are larger than the gaps in between the common SAW and the PSAW. The maximum band gap width appears at $F = 0.576$ in this material configuration.

6.4 Phononic Crystal Waveguides for SAW

This section deals with the PC waveguides for SAW. First, a steel/epoxy PC is introduced and the band gap is investigated. According to the band gap property, line defects in the PC are defined as acoustic waveguides and the guided SAW propagating inside the structure are analyzed [51].

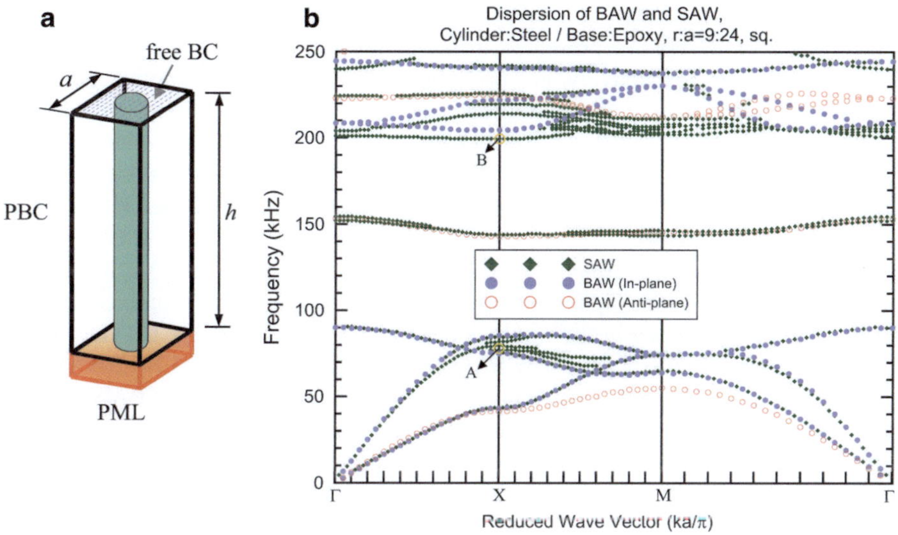

Fig. 6.14 (**a**) The 3D unit cell of PC employed to calculate SAWs. (**b**) Dispersion of acoustic waves in the square lattice steel/epoxy PC with a free surface [51]

6.4.1 Complete Band Gaps of Steel/Epoxy Phononic Crystals

In this study, a square lattice PC consisting of steel cylinders embedded in an epoxy half space was employed. The lattice constant a of the PC is 8 mm and the radius r of steel cylinder is 3 mm, and thus the filling ratio F equals to 0.442. The density and elastic constant C_{11} and C_{44} of steel are assumed as 7,900 kg/m^3, 280.2 GPa and 82.9 GPa, respectively, and those for epoxy are 1,180 kg/m^3, 7.61 GPa and 1.59 GPa. Thus the impedance of the cylinders is more than 15 times larger than the one of the matrix.

To investigate the band gap, the dispersion of acoustic waves inside PCs was calculated first. The PC structure has a traction free surface ($x_3 = 0$) to allow SAW propagation. Although the SAW propagates within the $x_1 - x_2$ plane perpendicular to the cylinders, the SAW propagation is indeed a 3D problem. Thus we defined a 3D unit cell as shown in Fig. 6.14a to calculate the characteristic SAW modes. The dimension of this unit cell is chosen as $1a \times 1a$ of the cross section on $x_1 - x_2$ plane and the height h is $9a$ to develop SAWs. Then a free surface boundary is on the surface at $x_3 = 0$, and a PML region is under the bottom surface to apply non-reflection boundary. Finally, the PBC is defined on the other four surfaces of the unit cell. In the FDTD method, each unit cell is divided into 24×24 grids and the time step interval is 20 ns to satisfy the numerical stability conditions.

In this PC, the structure is infinite along the x_3-direction, and the SAWs propagate in the $x_1 - x_2$ plane on the free surface. Meanwhile, BAWs accompany SAW and propagate in the $x_1 - x_2$ plane. The polarization of BAW modes are decoupled as the in-plane mode ($x_1 - x_2$ plane) and the anti-plane mode. In Fig. 6.14b, the eigenmodes of BAWs are presented in solid and hollow circles for in-plane and anti-plane modes, respectively. For the in-plane modes, there is a noticeable wide complete band gap from 90 to 204 kHz and several narrow complete band gaps at higher range 231–237 kHz, 245–255 kHz. From the curves of anti-plane modes, band gaps are located at 55–143, 153–212, 225–250, and 270–276 kHz. Considering both modes together, the complete band gaps are at 90–143, 153–204, 231–237, and 245–250 kHz.

Then we used the 3D unit cell with PBC, free surface and PML to calculate the dispersion of SAWs. The eigenmodes are obtained and marked as rhombus symbols in Fig. 6.14b. As shown in the figure, extra modes are detected by the sensors on the free surface. For example, new bands appear at 7–79.5 kHz, 199–201 kHz in ΓX direction. Also, some selected peaks overlap the circle symbol of BAW modes, such as the first and second bands of in-plane and anti-plane transverse waves in ΓX direction. In the higher frequency range, there are numerous peaks but most peak values are very small.

To investigate the modes of acoustic wave detected in the 3D PC unit cell, the displacement distributions of the eigenmodes are calculated. The calculating setup is the same as Fig. 6.14a and a wave source generating a wave package of chosen frequency was defined. Because both the wave vector and frequency are assigned, the specific eigenmode is excited and the displacement field is recorded to help identify the properties of modes. The displacement distributions of the eigenmodes are plotted in a 3D vector plot.

First, two eigenmodes which differ from those of BAW are calculated. The eigenmodes of wave vector $\mathbf{k} = (\pi/a, 0)$ with frequency $f = 77$ kHz (point A in Fig. 6.14b) and $f = 199$ kHz (point B in Fig. 6.14b) are shown in Fig. 6.15a, b. In the vector plots, the direction of the cones indicates the polarization of displacement vector and the size reflects the magnitude. The unit cell has grid lines in the location of multiples of the lattice constant. These modes of wave vector $\mathbf{k} = (\pi/a, 0)$ propagate along the x_1-direction in Fig. 6.15. The wave length of first band with $|\mathbf{k}| = \pi/a$ is $2a$ and thus only a half of the periodic displacement field is shown is one unit cell. The wave in Fig. 6.15a has a main polarization in the sagittal plane, the $x_1 - x_3$ plane, and the wave in Fig. 6.15b has a main polarization in the $x_1 - x_2$ plane. The result shows confined displacement fields—the amplitude decays rapidly as the depth increases and presents a typical property of SAWs. An interesting phenomenon is that the band which the mode in Fig. 6.15a belongs to has a higher velocity than the transverse wave of BAW modes. That means the surface wave is a leakage type, i.e. a PSAW.

In addition, the peaks overlapped on the BAW modes were also analyzed. For example, the eigenmode of wave vector $\mathbf{k} = (\pi/a, 0)$ with frequency $f = 43$ kHz was calculated and the distribution of displacement field showed a x_2-direction polarization and the amplitude does not decay rapidly within nine lattice constants. This reflects the property of transverse waves, the same with the mode of in-plane transverse bulk waves. Analysis of another mode of wave vector $\mathbf{k} = (\pi/2a, 0)$ with

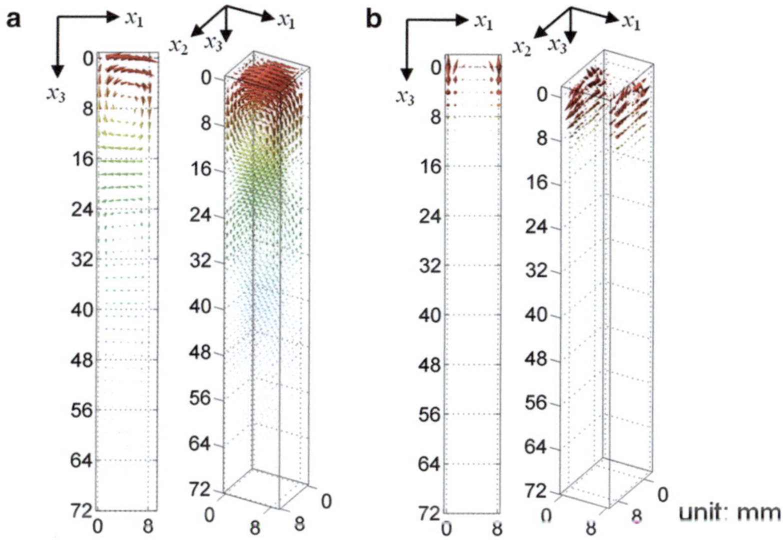

Fig. 6.15 The 3D displacement fields of eigenmodes. (**a**) $\mathbf{k} = (\pi/a, 0)$, $f = 77$ kHz (point A of Fig. 6.14b); (**b**) $\mathbf{k} = (\pi/a, 0)$, $f = 199$ kHz (point B of Fig. 6.14b) [51]

frequency $f = 84$ kHz also shows a rotation field which remains within the depth of the unit cell. Thus, the peaks overlapped on the modes of BAW present the property of BAW.

The analysis shows that the 2D steel/epoxy square PC with a free surface allows PSAW instead of SAW. The dispersion curves show an important fact that there exists a complete band gap for both BAW and PSAW (SAW) in this PC. The complete band gaps obtained by both results of 2D and 3D unit cell locate from 90 to 143 kHz and 154 to 199 kHz. This is the basis to design a PC waveguide of surface waves.

6.4.2 SAWs Inside a Phononic Crystal Waveguide

A PC waveguide is constructed by arranging adjacent point defects in a PC structure. These defects form a continuous region which is free of inclusions and acoustic waves are expected to propagate within. A PC waveguide of SAWs is formed based on the complete band gap of SAW. The acoustic waves in the waveguide cannot penetrate the periodic structure and thus the guided modes result from the existence of defects appearing in the range of complete band gaps.

For the steel/epoxy PC presented in the last section, the complete frequency ranges of band gaps are 90–143 and 154–199 kHz.

To understand the property of SAW inside PC waveguides, the supercell technique is used to analyze the dispersion of a waveguide. We consider a waveguide

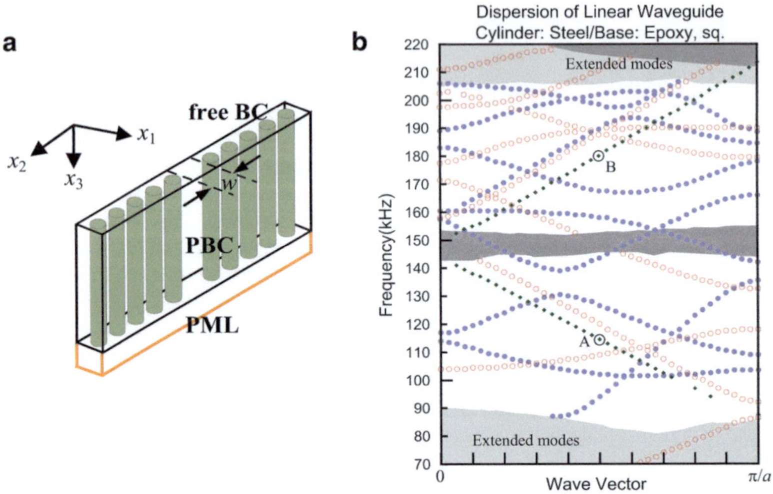

Fig. 6.16 (**a**) The supercell used to analyze the PC waveguide; (**b**) Dispersion diagram of BAW and SAW eigenmodes in the steel/epoxy PC waveguide [51]

with a 10 mm width w (the distance between two neighboring cylinders on both sides) as shown in Fig. 6.16a. A 3D supercell consisting of $11 \times 1 \times 9$ unit cells is defined, and then the waveguide is constructed of ten periodic arrayed cylinders and a dislocation in the center with the PBC surrounding it. Similarly, an initial disturbance is defined inside the waveguide, and thus acoustic waves are generated and the displacement field is recorded. Then the eigenmodes are selected from the local maximum peaks of spectra as the procedure of analyzing a unit cell. By setting corresponding boundaries, both the dispersion calculated by 2D and 3D supercells are obtained and plotted in Fig. 6.16b. We focused on the frequency of 70–220 kHz to observe defect modes in the range of complete band gaps. The extended bands outside the complete band gaps are omitted and the region is marked in gray. As shown in the result, new defect modes of acoustic waves appear in the range of complete band gaps. The eigenmodes of BAW are marked as solid and hollow circles and those obtained from the 3D supercell are rhombus symbols.

Comparing the dispersion for the PC waveguides to the case of perfect PC, new defect modes can be obtained. In the 3D supercell, sensors on the free surface of PC waveguide receive the signal of defect modes of both BAW and SAW. The signal of BAW modes has a small peak value and make the spectrum shows a complex pattern. Thus, in the 3D waveguide calculation, the BAW modes are removed and only the SAW mode is shown in Fig. 6.16b. Inside the first complete band gap of 90–143 kHz, a band extends from wave vector $\mathbf{k} = (\pi/a, 0)$ with frequency $f = 87.5$ kHz to $\mathbf{k} = (0.05\pi/a, 0)$ with $f = 141$ kHz, and another band appears from $\mathbf{k} = (0.75\pi/a, 0)$ with $f = 154$ kHz to $\mathbf{k} = (0.9\pi/a, 0)$ with $f = 206$ kHz in the second complete band gap of 154–199 kHz.

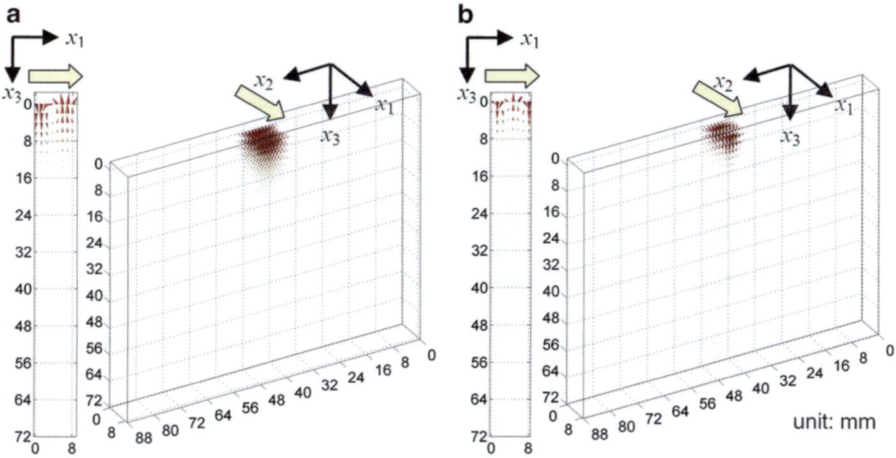

Fig. 6.17 The 3D displacement fields of defect modes inside the PC waveguide. (**a**) $\mathbf{k} = (0.5\pi/a, 0), f = 114.5$ kHz (point A of Fig. 6.16b); (**b**) $\mathbf{k} = (0.5\pi/a, 0), f = 180$ kHz (point B of Fig. 6.16b) [51]

The displacement fields of the two defect modes are plotted in 3D vector diagrams. The eigenmodes of $\mathbf{k} = (0.5\pi/a, 0)$ with $f = 114.5$ kHz (point A in Fig. 6.16b) and $\mathbf{k} = (0.5\pi/a, 0)$ with $f = 180$ kHz (point B in Fig. 6.16b) are calculated and shown in Fig. 6.17a, b, respectively. Since the dispersion diagram is presented in a reduced zone scheme, the actual wave vectors of these two modes in folded bands are $\mathbf{k} = (1.5\pi/a, 0)$ and $\mathbf{k} = (2.5\pi/a, 0)$ with wavelengths of $1.33a$ and $0.8a$, and the consistent result is obtained from the displacement fields of $x_1 - x_3$ plane. The guided waves propagate along the x_1-direction, and the displacement fields have a confined distribution within the region of waveguides. The maximum amplitude occurs on the free surface and the amplitude decays rapidly in a deeper location. Thus these two bands have a typical property of SAWs.

A further discussion of the new defect modes in PC waveguide is presented as follows. In the calculation of eigenmode displacement fields, a Gaussian weighted wave packet with specified frequency f is set as the source and input finite energy into the supercell structure to excite the selected eigenmode. Then, the sensor on the free surface detects a monotonous displacement variation and the vertical component U_3 recorded after the wave packet is excited. The result shows that the eigenmode has decaying amplitude. This means an energy leakage, partial acoustic waves propagates downward into the half space, and thus the defect modes in this PC waveguide are actual PSAWs. Finally, converting the period of signal into propagation distance, the attenuation coefficient is obtained as -0.049 dB/a and -0.06 dB/a for these eigenmodes, respectively.

SAW propagation inside a straight PC waveguide was further calculated. Figure 6.18 shows a top view of the free surface and the actual structure extends vertically to the surface. The periodic circles present the boundaries of steel

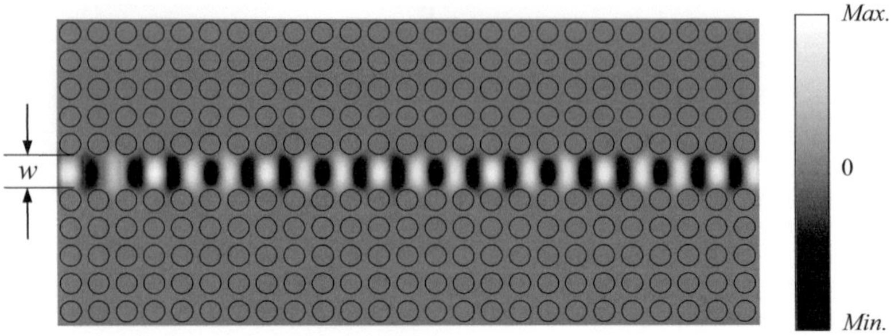

Fig. 6.18 The x_3-direction displacement field of the 114.5 kHz PSAW inside a linear PC waveguide with a width of 10 mm [51]

cylinders and the epoxy matrix. The waveguide has a width of 10 mm by removing one row of adjoining cylinder inclusions along ΓX-direction, and the length of waveguide is $25a$. An acoustic wave is generated in the inlet of waveguide with five line sources to simulate an IDT. The PSAW defect mode of wavelength $\lambda = 1.33a$ with frequency $f = 114.5$ kHz is generated and the wave propagation pattern is demonstrated. The displacement field of x_3-direction component U_3 is presented in a gray scale to show the surface wave propagation. The scale is fixed in this study for the convenience to compare the amplitude in different cases. The result shows that the PSAW is confined well within the waveguide and only slight energy leaks into the periodic PC area. In addition, although the PSAW has energy leakage, the PSAW propagates along the waveguide and carries most energy.

In acoustic waveguides, a bend section is an elementary structure to change the wave propagation direction. Here the PC waveguide with a sharp bend is also calculated to estimate the transmission efficiency. The first case is a simple junction of two ΓX-direction waveguide. The structure is a $17\ a \times 17\ a$ area and two $11\ a$ waveguides are connected as shown in Fig. 6.19a. The same PSAW of 114.5 kHz defect mode is launched into the waveguide from the inlet in the lower-left corner and the U_3 displacement field on the surface is shown in the same gray scale. The displacement pattern shows that PSAW is still confined within the bending waveguide structure, but most energy of the incident wave is reflected at the corner of the bend. In the case of a linear straight waveguide in Fig. 6.18, the amplitude of the 114.5 kHz PSAW in the outlet is chosen as a reference value which results from the same monotonous wave source. With the identical wave source, the maximum amplitude on the surface of the vertical waveguide branch is about 59 % of the reference amplitude in the bending waveguide. Another phenomenon is that BAW modes are also excited as the turned-wave, and thus considerable energy is taken away and the amplitude in the vertical branch decays rapidly. Thus a simple joint bend is not an efficient design.

Another bend structure is shown in Fig. 6.19b. It contains two linear waveguide in ΓX-direction and one in ΓM-direction, thus the wave turns 45° twice to the left direction. The U_3 displacement pattern of 114.5 kHz PSAW defect mode also

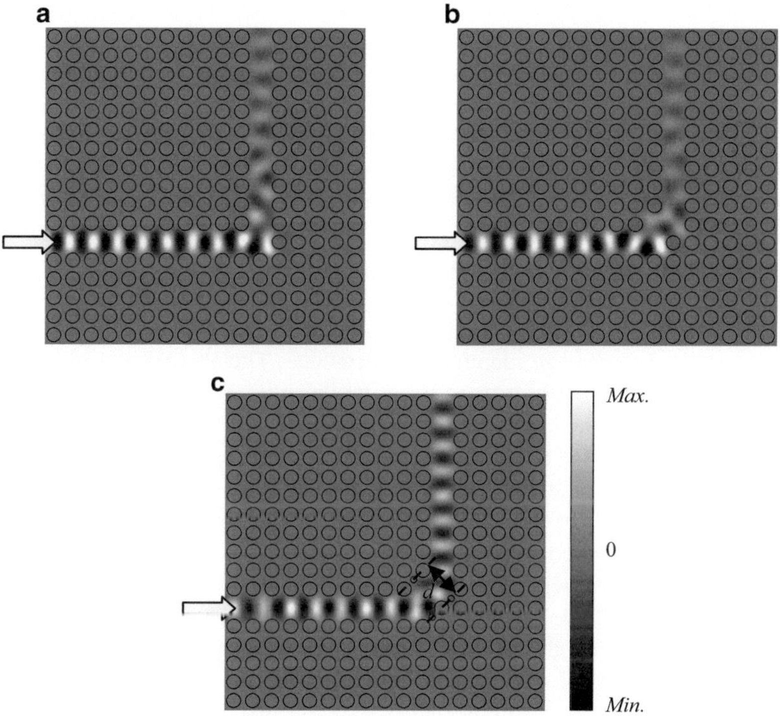

Fig. 6.19 The x_3-direction displacement field of the 114.5 kHz PSAW inside (**a**) a sharp bending waveguide, (**b**) a three-segment bending waveguide, and (**c**) an improved PC bending waveguide with scatterers [51]

shows a strong reflection of the incident wave. The amplitude of turned-wave is about 52 % of the reference amplitude in ΓM-direction segment and 26 % in the vertical ΓX-direction segment. Finally, the mode of turned-wave is rebuilt to a symmetric PSAW mode. In these cases, the wavelengths of PSAW are almost the same as a lattice constant, thus there is a strong scattering phenomenon when acoustic waves encounter inclusions at their propagating direction. In addition, the anti-symmetry conjunction section in Fig. 6.19b also enhances the scattering and the mode conversion. Therefore energy disperses while the excited BAW propagates downward, and thus the surface amplitude decays when the PSAW encounters the boundary of the bending waveguides.

To raise the transmission of bending PC waveguide, an improved bend structure is introduced. In the study of photonic crystals, the scatters are used to enhance transmission through waveguide bends in two-dimensional photonic crystal slabs. The idea is adopted to raise the transmission of PSAW in PC waveguides. The design is based on the prototype shown in Fig. 6.19c and the cylinders of smaller radius 1.3 mm are inserted into ΓM-direction segment. With the inserted scatterers, the conjunction section has a better symmetric geometry property. This also changes the dispersion curve, as the result shown in the photonic crystal cases, and result in eigenmodes that propagate in the conjunction section with less loss. In Fig. 6.19c, the distance d between the center of scatters is 16.97 mm ($3\sqrt{2}a/2$). Then the U_3 displacement pattern of PSAW of 114.5 kHz is calculated and shown. Obviously, the PSAW passing the corner remains the same eigenmode and therefore the amplitude is raised to 72 % of the reference amplitude.

Further calculation shows that the transmission of the guided PSAW is affected by the scatters obviously. If the scatters are moved toward the center of the waveguide, the dispersion relation is altered. In the case of the distance d of 11.31 mm ($\sqrt{2}a$), the amplitude of turned-PSAW at 114.5 kHz becomes 48 % of the reference amplitude. Another important fact is that waves of different frequency in the structure do not have identical performance. The PSAW of 135 kHz in the bend structure of Fig. 6.19c shifts to BAW mode and the mode decays very rapidly in the turned waveguide, but has a 62 % amplitude ratio in the structure with d of 11.31 mm. Therefore, the transmission efficient is affected by the scatters and also a function of frequency. Indeed, setting scatters properly in the PC waveguides can raise the transmission and change the propagation direction in a small region.

In brief, we introduced steel/epoxy PC with complete band gaps for both bulk and SAWs. Further, PC waveguides are designed based on the complete band gap property. The defect mode inside the PC waveguide has an energy leakage because the 3D structure allows the wave propagating downward to the half space. The eigenmodes of the 2D PC waveguide with a free surface is thus identified as PSAW. Although the PSAW has a slight energy leakage inside the 2D waveguide, it is possible to have a waveguide with a sharp bend. Adding scatters inside the PC waveguides can raise the transmission and the design depends on the frequency of PSAW.

6.5 Experiment on the SAW Band Gap

In this section, the experimental studies of SAW in the 2D PCs are presented. Using the MEMS process, micro-scale PC were realized and the band gaps were observed in the hundreds MHz frequency range. In the following, the PC fabricated on silicon wafers [41] and lithium niobate wafers [42] are introduced.

6.5.1 Band Gap for SAWs in Silicon-Based Phononic Crystals

To demonstrate the frequency band gap of SAWs in a 2D silicon PC, a square array of circular holes in a silicon substrate was realized [41]. High-frequency wideband slanted finger interdigital transducers (SFITs) on the silicon substrate were used to generate and detect the SAW signals. Since silicon is not a piezoelectric material, the layered structure SFIT/ZnO/silicon as shown in Fig. 6.20a was considered in the

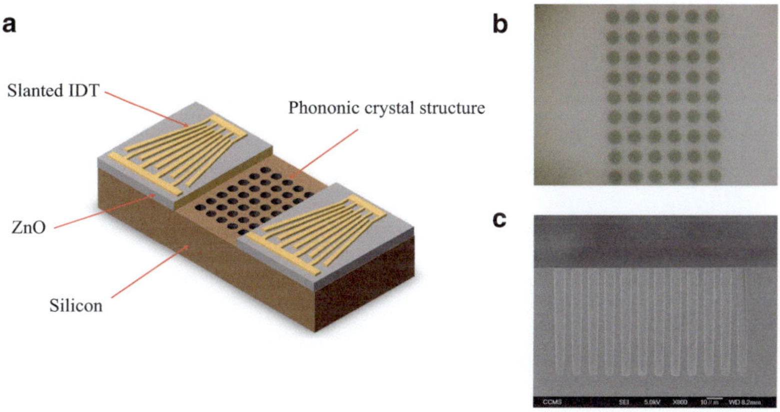

Fig. 6.20 (**a**) The schematic of the experimental setup for measuring the band gap of SAW. Top view (**b**) and cross section (**c**) of the 2D micro silicon PC [41]

Table 6.2 Design parameters of the SFIT/ZnO/silicon-layered structure

	ZnO (2 μm)/Si			
Substrate design no.	1	2	3	4
Center frequency	200	195	215	185
Bandwidth	~35	~24	~28	~28
λ_{min} (μm)	17	18	16	19
λ_{max} (μm)	25	24	22	26
Input pairs (pairs)	40	40	40	40
Output pairs (pairs)	32	32	32	32
Metal thickness (A)	1,600	1,600	1,600	1,600
Propagation length (μm)	631.25	606	489.5	656.5
Aperture (μm)	2,500	2,400	2,200	2,600
Max. tile angle (°)	3.66	2.86	3.12	3.08

device, where the ZnO layer is piezoelectric, and the metal SFITs are on the top of the ZnO layer so that SAWs can be excited electrically. The design parameters of the SFITs and the geometry of the layered structure depend upon the band-gap frequency of the 2D PC. The lattice constant of the PC was 10 μm, and the radius of the circular holes r was 3.5 μm which corresponds to a filling ratio F of 0.385. Analysis with the PWE method for this structure showed that the band-gap frequencies along the ΓX direction are from 183 to 215 MHz.

According to the calculated band-gap frequencies, the design parameters of the SFITs are listed in Table 6.2. In design 1, the excited SAW frequency range by the layered SFITs covers the band gap of the PC. Therefore, the frequency band-gap width of the PC can be measured by the device. In Fig. 6.20a, an array of cylindrical holes between the two SFITs were micromachined to act as the PC.

The fabricated 2D silicon-based PC consisting of six rows of air cylinders is shown in Fig. 6.20b. Figure 6.20c is the cross section view of the structure. The

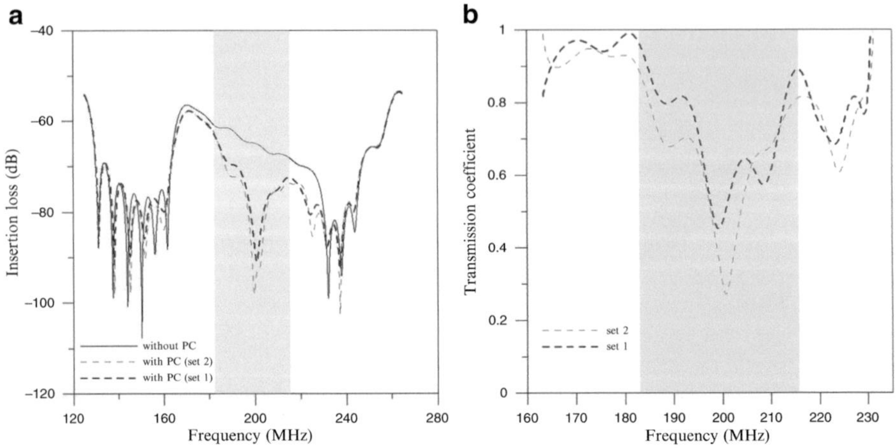

Fig. 6.21 (**a**) The band gaps for SAW in the silicon PCs are revealed by the layered SFIT of design 1 in the range of pass-band frequency. The *shadow region* represents the band gap of fabricated PCs with six rows of air cylinders. (**b**) The acoustic transmission coefficient [41]

depth of the 2D cylindrical holes is about 80 μm which is more than two times of the wave length. Figure 6.21a shows the measured frequency band gaps of the air/silicon PCs by using the SFITs of design 1. The shaded region denotes the theoretical band gap of the fabricated PCs. The dashed curves and solid curve denote the measured frequency responses of the layered structures with and without the PC slab, respectively. The two dashed curves are the results of different samples with the same design. The insertion losses of SAW propagated in the layered structure with PCs are clearly higher than that without the PC for frequency located in the band gap. This corresponds to the fact that most of the SAW energy in the band-gap range is blocked by the phononic structures. The ratio of the insertion loss with phononic structures to that without phononic structures gives the transmission coefficients. The transmission spectrum is shown in Fig. 6.21b. The transmission spectrum also clearly shows that the SAWs are attenuated greatly in the forbidden band.

6.5.2 *Piezoelectric Phononic Crystals*

The combination of SAWs and piezoelectric monocrystalline solids such as quartz, lithium tantalate ($LiTaO_3$) or lithium niobate ($LiNbO_3$), amongst others, occupy a prominent position in the field of wireless telecommunications and signal processing. Monocrystalline substrates indeed offer piezoelectric and electro-mechanical coupling properties that remain unmatched by the currently available piezoelectric thin films. The demonstration of phononic band gaps for SAWs in such materials hence offers particularly rich applicative prospects while constituting a very good field for more fundamental investigations. Very efficient electrical SAW sources and detectors can be found in interdigital transducers and single-mode operation is made possible by the very nature of the generated surface waves. These very appealing properties are however counter-balanced by the difficulty of processing these materials that are quite often complex oxides using standard micromachining technologies. Fabrication then stands as a challenge, while design is not made easier: the strong anisotropy of acoustic wave propagation inherent to piezoelectric materials, combined with the quasi-systematic mixing of shear and longitudinal polarizations, put tighter constraints on the geometrical parameters of the periodical structure itself.

The most obvious configuration from a conceptual as well as from a fabrication point of view that can be implemented to proceed with the demonstration of a frequency band gap in a piezoelectric crystal consists in considering a square lattice of air inclusions obtained by drilling holes in a piezoelectric solid. Such a demonstration has in particular been achieved in lithium niobate [42]. $LiNbO_3$ benefits from rich optical properties that add up to its acoustic characteristics to extend the application field of PCs made out of this material to the field of integrated acousto-optics and dual photonic and phononic structures that will be dealt with later in Chap. 9.

The PWE theory can, for instance, be used for the design of such a PC. Figure 6.22a displays the computation of the density of surface states in the case of a perfect, infinite 2D PC for $LiNbO_3$ in the Y-crystallographic orientation plotted along the closed path Γ–X–M–Y–Γ in the first irreducible Brillouin zone. The considered holes have a circular cross section with a diameter $d = 0.9a$, where a is the lattice constant. The filling fraction then equals 63 %. Seven harmonics are used in each direction in the PWE computations, resulting in a total of $N = 49$ harmonics. It can be seen that a full band gap (i.e. a band gap for any direction of propagation and polarization) exists for waves propagating in-plane from $f.a = 1{,}935$ to 2,745 m/s. The fractional bandwidth is then larger than 34 %. Qualitatively, it can be observed that although in principle anisotropy makes it more difficult to open a full band gap than with isotropic materials, the free boundaries of void inclusions are very efficient scatters for elastic waves of any polarization. Another interesting observation is that the frequency band gap width and position were found to be exactly coincident with those found via PWE for in-plane propagating bulk waves [33]. Figure 6.22b reports some finite element method (FEM) simulations of bulk waves propagating in the same periodical structure that confirm this initial observation. The displayed propagation directions have here been kept to the only ones that can actually be investigated experimentally.

The corresponding PC was then fabricated in a 500 μm thick Y-cut $LiNbO_3$ substrate. To relieve the technological constraints in this initial experiment and to be able to use optical lithography and collective fabrication methods for the patterning of the phononic structure, the operating frequency was set to 200 MHz, hence setting the lattice parameter to 10 μm and the hole diameter to about 9 μm. The FEM theory consequently predicts a complete band gap spanning from 190 to 250 MHz. It can

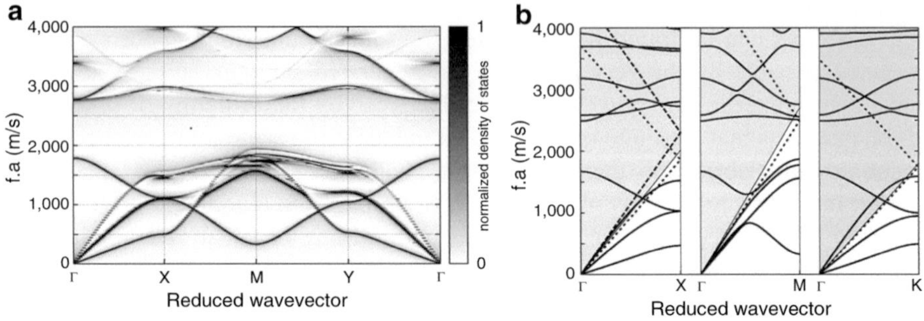

Fig. 6.22 (**a**) Band structure for surface waves along the Γ–X–M–Y–Γ path of the first Brillouin zone for a PC made of a square-lattice of air holes etched in a Y-cut lithium niobate matrix. The filling fraction has been set to 64 %. (**b**) Theoretical band diagrams for bulk waves in the same crystal. The dispersion relations on a free surface are indicated for the Rayleigh (*short dashed line*) and the leaky (*long dashed line*) surface waves. The *grayed regions* indicate the sound cone

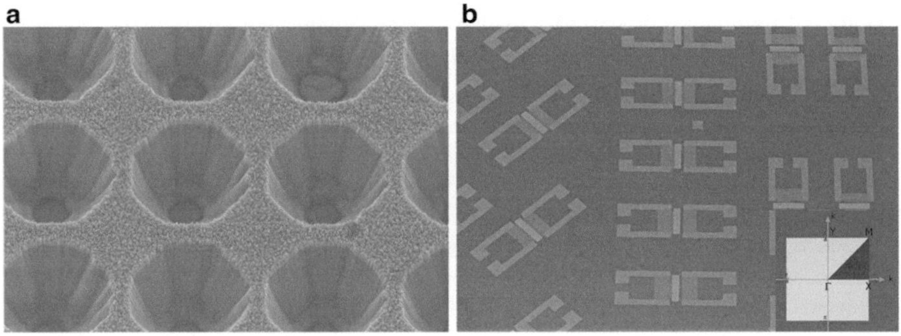

Fig. 6.23 (**a**) Scanning electron microscope photograph of the array of 10 μm deep, 9 μm diameter holes etched in a lithium niobate substrate constituting the PC, before removal of the etching mask. (**b**) Optical microscope image of the series of devices used to demonstrate the band gap phenomenon over the whole Brillouin zone

be noted in Fig. 6.23a that the obtained holes exhibit a rather low sidewall slope angle, around 72°, which limits the depth to about 11 μm and therefore the aspect ratio to 1.2.

Interdigital transducers in a delay-line configuration were used to launch and detect surface waves propagating through the crystal. In order to keep full control of the wave vector direction in a highly anisotropic material such as $LiNbO_3$ and to make sure that Rayleigh waves exclusively were generated, a series of eight ten-finger-pair IDTs of varying mechanical period, corresponding to emission wavelengths ranging from 12.2 to 26 μm were used and preferred over the slanted IDTs of Sect. 6.5.1. The frequency range of interest was then covered by juxtaposing neighboring responses. The IDTs were oriented for elastic wave propagation along the ΓX, ΓM, and ΓY directions of the first Brillouin zone, as illustrated in Fig. 6.23b.

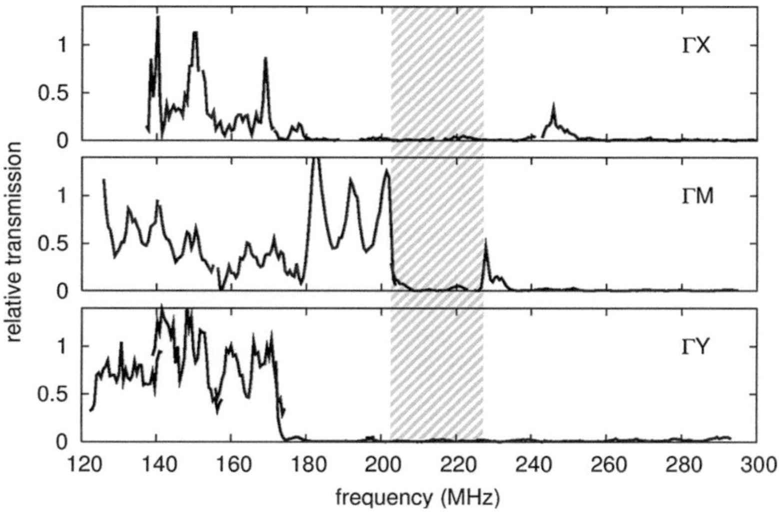

Fig. 6.24 Relative transmission, defined as the ratio of the transmittances with and without a PC. The hatched area delimits the complete band gap that extends from 203 to 226 MHz

Figure 6.24 displays the measured electrical normalized transmission, computed as the ratio of the transmission scattering parameters with and without the PC for each set of delay lines. If a full band gap is definitely observed, with the crystal causing a 20 dB loss in the transmission value, no transmitted signal was measured at frequencies lying above the band gap. This phenomenon can be accounted for by considering the actual geometry of the fabricated SAW PC: since the surface wave penetration depth in the substrate is at least of the order of the hole depth, the experimental structure can be seen as a stratified medium, with a void/lithium niobate PC slab lying on a semi-infinite lithium niobate substrate. This, combined with the conicity of the holes tends to favor out-of-plane scattering, leading to coupling with bulk modes of the substrate. A *sound line*, by analogy with the *light line* of photonic crystal slabs, can then be defined, that corresponds to the dispersion relation of the slowest bulk acoustic mode in the substrate, i.e. of the slowest shear wave. Surface modes lying inside the sound cone are, possibly, highly lossy radiation modes. Only modes below the sound line are guaranteed to be guided by the surface, which agrees well with the experimental results. Another origin of the signal extinction at higher frequencies can also be found in elastic wave diffraction by the PC that can indeed be considered as a 2D diffraction grating [43]. Clearly, electrical characterizations through RF probe testing cannot provide us with a means of observing this kind of phenomena. Additional characterization tools giving access to more direct information on wave propagation such as optical mapping of the displacement fields at the surface are then required to reach a higher level of understanding of both scattering and diffraction phenomena in PCs. These techniques will be more thoroughly accounted for in Chaps. 7 and 8.

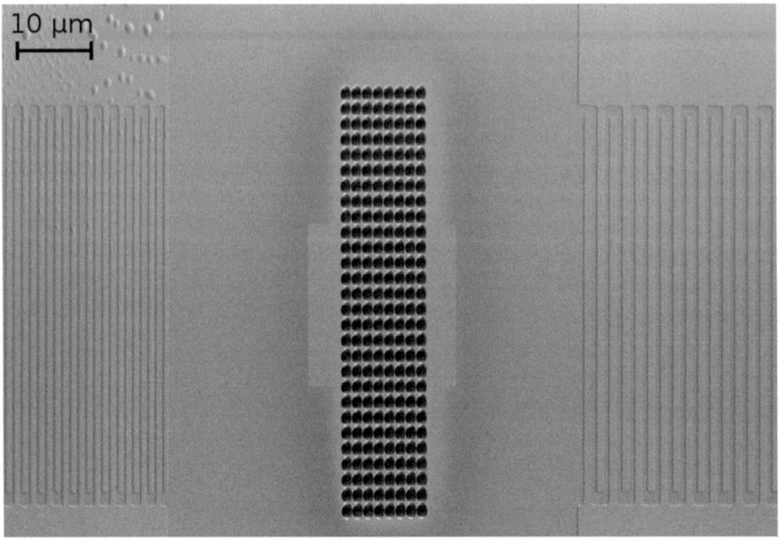

Fig. 6.25 Scanning electron microscope image of a hypersonic PC device. The pitch of the structure is about 2.2 μm and the holes are around 2 μm in diameter, resulting in a filling fraction of about 64 %. A pair of broadband chirped transducers is used for electrical emission and detection of the elastic waves

The sound cones are superimposed on the band diagrams in Fig. 6.22b and correspond to the grayed areas. This kind of representation for both the sound and light cones is widely adopted by both the photonic and phononic community, though it can somehow be considered as misleading. It is indeed quite tempting to fully discard the dispersion properties of modes lying within the sound cone. Yet, stating that the modes lying beyond the sound line are likely to radiate to the bulk does not necessarily exclude the existence of surface or pseudosurface guided waves. This was, for example, demonstrated experimentally through the fabrication and characterization of the structure reported in Fig. 6.25. The PC considered here is a scaled-down version, by a factor of 5, of the previous one. It consists again in a square-lattice crystal with a filling fraction close to 64 %. This time, however, the period of the structure is about 2.2 μm and the $LiNbO_3$ substrate was chosen in the X-crystallographic orientation. A critical difference lies in the patterning method used for the 2 μm diameter air holes that were milled by focused ion beam (FIB). This fabrication technique indeed allows obtaining a nearly vertical hole profile, with with a sidewall slope angle around 85, which is a significant improvement compared to 72° previously reported.

The main drawback of the FIB fabrication method lies in the time required to etch a single one of the phononic structures presented here that was of the order of 9 hours with the equipment at hand (Orsay Physics LEO FIB 440). For this reason, the fabrication of multiple occurrences of identical PCs was this time not conceivable. The standard interdigital transducers were then replaced by chirped

IDTs exhibiting a linear variation in the electrical period along their length and allowing for a broadband emission over more than an octave. Only a couple of identical PCs, surrounded by two different sets of transducers, were then required for a full characterization of the expected band gap, as opposed to eight pairs of transducers in the previous case. If using chirped IDTs instead of SFITs still allows setting the $\boldsymbol{k}$-vector direction, the detection bandwidth along a given propagation direction is limited by possible interferences between pure Rayleigh waves and slow, shear leaky-surface waves in some cases. In X-cut $LiNbO_3$ for example, wide band electrical detection is only possible in the (*XZ*) propagation direction, while the achievable bandwidth along (*XY*) is limited by a transverse leaky wave of velocity $v = 4{,}100$ m/s ($v = 3{,}680$ m/s for the Rayleigh wave).

The raw transmission data for the reference delay lines and for the PCs, along with the normalized transmittance are reported in Fig. 6.26. A very good overlap between the two signals can be observed at frequencies below 620 MHz. Beyond this point, and much similarly to what has been reported in the lower frequency SAW $LiNbO_3$ PC, the electrical response is strongly attenuated, with an extinction ratio of the order of 13 dB. Partial re-transmission occurs from a frequency of about 1 GHz and there is once again no undisputable evidence of surface modes above the band gap, although a close study of the reflection scattering parameters seemed to indicate that the frequency band gap should be exited at a frequency of about 1 GHz [55]. This point was further confirmed by optical measurements performed via heterodyne interferometry. Maps of the amplitude wave profile across the PC were taken at frequencies supposedly lying before, within and after the expected band gap. The resulting averaged cross sections of the surface motion through the PC area are reported in Fig. 6.27. As expected, below the band gap, e.g. at 540 MHz, the elastic wave passes through the crystal and is almost unaltered. In contrast, a standing wave pattern is clearly observed at 660 MHz (band gap edge) and at 800 MHz (center), confirming that both frequencies lie within the band gap for surface guided modes. The wave penetration length inside the PC is higher at the band edge than at the central frequency and up to 25 % of the elastic energy is still transmitted at 660 MHz. At 800 MHz, the plot of the averaged cross-section of the surface motion reveals an exponential decrease of the wave amplitude inside the crystal, resulting in an output wave amplitude within the noise level. The most striking results are observed at higher frequencies: at 1.05 GHz, at which frequency the electrical transmittance is weak, the optical measurements clearly demonstrate that transmission occurs above the band gap. The average amplitude at the output was evaluated to be around 0.17 nm, versus 0.22 nm for the incident wave, resulting in an amplitude transmission close to 75 %. This result shows that configurations do exist where the sound line limit can be partly overcome: the sound cone designates a frequency region where leakage from surface to bulk modes can happen, but this does not imply that surface wave propagation is fully prohibited in this part of the spectrum.

The implementation of PCs for SAW in piezoelectric materials therefore remains at an early stage and both the physical description of such structures and the technological means used for their realizations need to be improved. The current

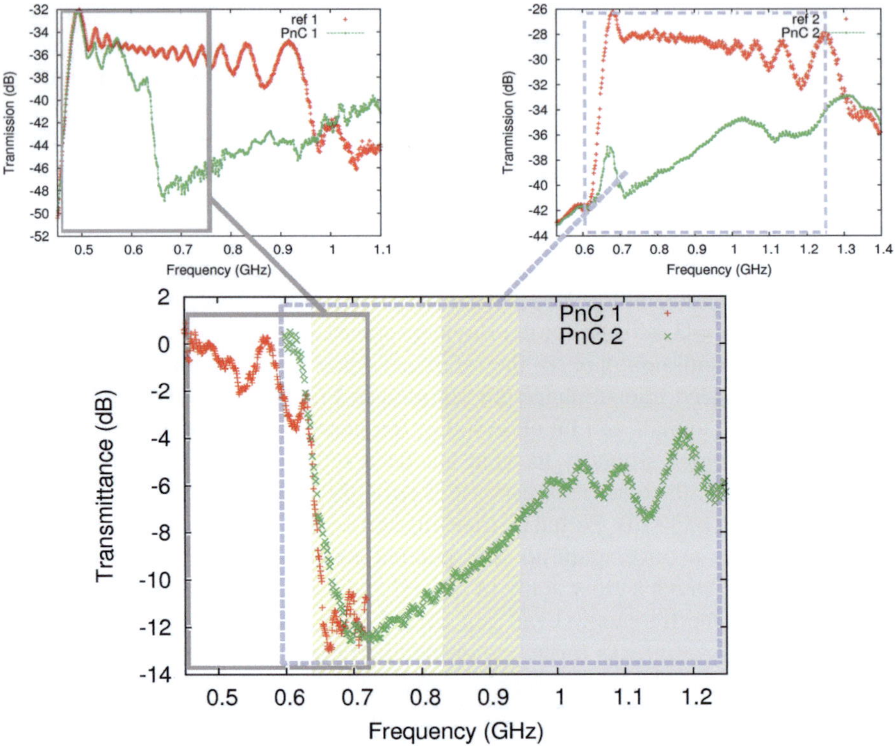

Fig. 6.26 Transmission responses of each PC, along with the response of the electro-acoustic delay lines used as references. The normalized transmittance versus frequency for surface guided modes propagating through the PCs is also plotted. The transmittance plot gathers the results obtained for the two phononic structures

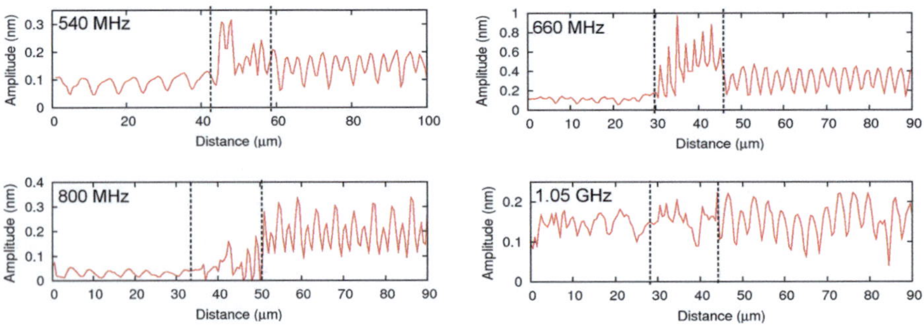

Fig. 6.27 Measured vertical displacements through the PC structure obtained by optical interferometry at different frequencies: 540 MHz (below the band gap), 660 MHz (bandgap edge), 800 MHz (within the bandgap), and 1.05 GHz (above the bandgap). The amplitude data are averaged in the *y*-direction. In all cases, the emitting transducer is on the right-hand side

theoretical, perfect, two-dimensional representation of these periodical structures cannot even find a remote real-life equivalent. Surface structuration using micro-nano-fabrication processes remain to date the most obvious choice of realization of these PCs and these techniques will inevitably lead to limited aspect ratio structures, particularly in piezoelectric materials that usually happen to be quite resilient to usual etching processes. Pure surface-guided modes will under such configuration be clearly difficult to maintain at frequencies lying above the sound line, and this is true that in practical applications, radiation to the bulk can prevent from taking full advantage of the PC capabilities over the entire frequency range of operation. Yet, the band gaps observed for surface-guided modes lying below the sound line remain perfectly valid. The frequency range outside the sound cone should hence be favored for the design of efficient waveguides, high-Q cavities or for the implementation of RF signal processing functions where electrical signal transmission is the essential vector of information. The demonstration of a clear transmission of the elastic waves outside of the band gap for modes lying within the sound cone however shows that losses experienced by high frequency surface waves can be partly overcome, provided that enhanced sidewall verticality and aspect ratio of the holes constituting the crystal can be improved. This may open interesting prospects for both fundamental and applicative studies involving elastic wave propagation on its own and its interaction with the host medium, as in the case, for instance, of elasto-optical interactions.

6.6 Application of SAW Band Gap to SAW Devices

6.6.1 Reflective Grating for SAWs Using Phononic Crystals

SAW devices consisting of IDTs and metal gratings have been widely used as resonators for decades. In order to have good reflection, hundreds of metal strips are usually employed, and it is volume consuming. PCs consist of periodic arranged media and perform acoustic band gaps. The experimental studies in the last section showed that PCs can block SAW propagation efficiently with only several lattice periods. In other words, PC can be designed as space-saving wave reflectors. In this section, the analysis and design of PC reflective grating for SAW devices are demonstrated.

The substrate of the SAW device is a silicon wafer, and cylindrical holes in the wafer were made from a square lattice PC. For an air/Si PC with square lattice, propagation of SAWs was analyzed in previous studies [56, 41]. With a filling fraction $F = 0.283$, there is a partial band gap for SAW along the ΓX direction [56]. Further, the band gap enlarges with the increase of filling fraction, and reaches the widest gap for the case of $F = 0.48$ [41]. To study the PC grating, the air/Si PC of $F = 0.283$ was adopted for the convenience of specimen fabrication. The dispersion curves were analyzed by the PWE method and the band structures of acoustic waves

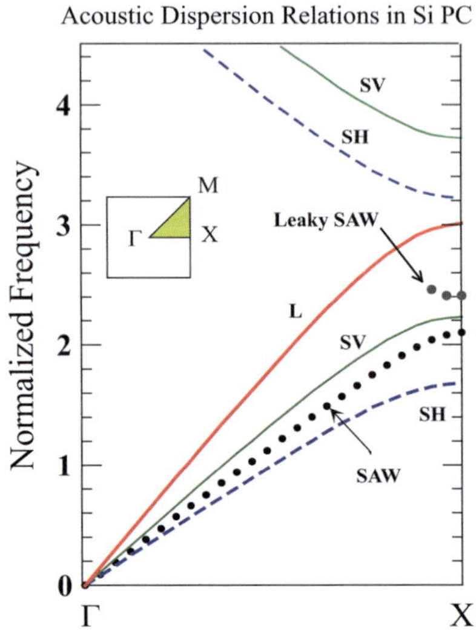

Fig. 6.28 Dispersion relations of SAW and BAW in the air/Si PC of $F = 0.283$. The denotations Γ, X, and M of the inset are the symmetry points of the first Brillouin zone in the **k** space. The normalized frequency is defined as ω_a/C_T, where ω is angular frequency, C_T transverse wave velocity [45]

are shown in Fig. 6.28. As marked in the figure, a partial band gap was obtained for SAW between the normalized frequencies 2.1 and 2.41. In this case, only partial band gap for bulk waves exists as well. We note that the SAW modes become a leaky SAW that radiates energy into the half space in the folded SAW band for the reduced frequency higher than 2.41 because it pierces into the bulk-wave bands and couples to the SV modes. However, the SAW modes of the lowest SAW band polarized in the sagittal plane and decoupled from the SH modes is still a true SAW. In the study, the lattice constant is 10 μm, thus the corresponding range of the partial band gap is 195–224 MHz. In the previous section, the existence of the band gap has been demonstrated experimentally.

In traditional SAW devices, hundreds of metal strips are usually adopted as reflective gratings to improve insertion losses of IDTs. The distance between the IDT and the grating has to be optimized to result in coherent reflective waves in the device. To apply PCs as a reflective grating, the location of effective reflective plane of the PC grating was investigated by using the FDTD method. A 210 MHz SAW is excited in a silicon half space and encounters an air/Si square lattice PC of lattice constant $a = 10$ μm and filling fraction $F = 0.283$. Because the frequency of the SAW is inside the band gap, the evanescent wave penetrating into the PC decays rapidly and most of the incident wave is reflected. With a continuous wave source, the reflected waves interact with the incident waves and form a standing wave pattern.

A surface displacement field of in-plane component U_1 is shown in Fig. 6.29. There are fifteen columns of cylinders in the calculation. The SAW is reflected by the PC and the incident one is attenuated inside the PC. Due to the circular boundaries of the cylinders, the reflective wave needs an extra delay distance to reconstruct the linear wavefront. As shown in Fig. 6.29a, l is defined as the distance

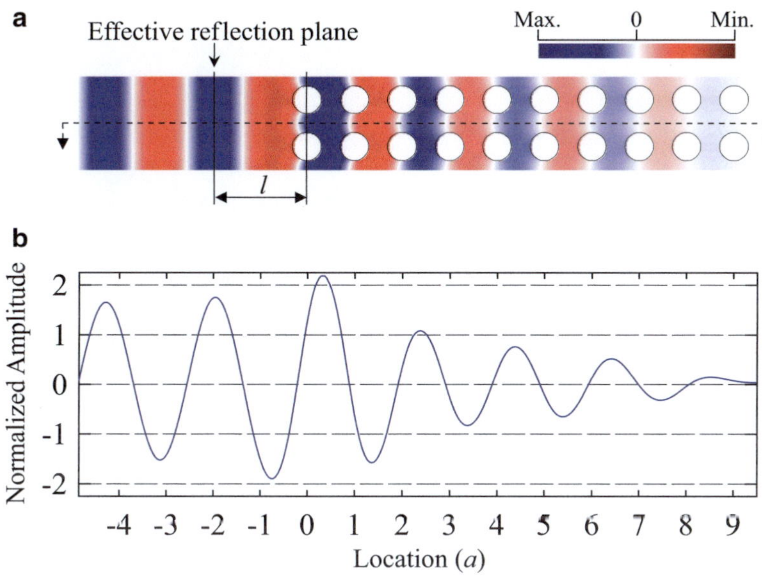

Fig. 6.29 The displacement field of in plane component U_1 shows SAW encountering the PC. (**a**) The picture shows an extra delay distance is needed to reconstruct the linear wavefront. (**b**) A profile along the *dashed line* in (**a**) shows the reflection and attenuation inside the PC. The amplitude is normalized to the U_1 of the case without the PC [45]

from the centerline of the first layer of cylinders to the location of a peak of U_1. The location where there is the peak value of U_1 is considered as the effective reflective plane of the PC grating. We note that the distance from the effective reflective plane to the location of wave sources (on the left) satisfies the condition of multiples of a half wavelength. For the SAW of 210 MHz in silicon, we chose l as about 1.32 λ, where λ is the wavelength of the 210 MHz SAW in silicon half space, i.e. about 23.39 μm. This delay distance is adopted as a reference to design the two-port resonant SAW devices in the following section.

6.6.2 SAW Resonator

After investigating the reflection of SAW on the PC grating, two-port layered SAW devices combining PC gratings are designed to investigate the reflection efficiency and the optimum delay distance between IDTs and PCs. A schematic diagram of the two-port layered SAW device is shown in Fig. 6.30a. The center frequency 210 MHz of the layered ZnO/Si SAW device was calculated based on the effective permittivity method [57]. The designed distance between the central line of the first two metal strips of IDT and the edge of ZnO film is 1.5 λ. There are two fifteen-layer PCs

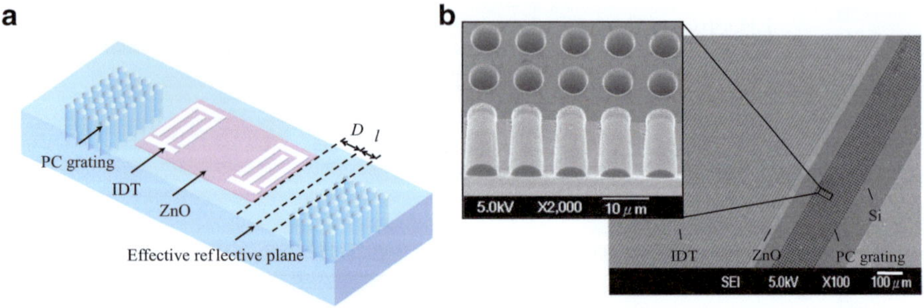

Fig. 6.30 (**a**) Schematic of a two-port layered SAW device using PC grating. (**b**) SEM photo of a layered SAW device with PC gratings [45]

outside the IDTs as reflective gratings. To study the effect of the delay distance D between the IDT and the PC grating, four different D were used; i.e., 0.875 λ, 1.0 λ, 1.125 λ, and 1.25 λ.

Herein a process realizing a two-port layered SAW device with PC gratings is briefly described. A 0.5 μm ZnO film was deposited on silicon by sputtering. Aluminum thin film of 150 nm thickness was evaporated on the ZnO film, and interdigital transducers were formed by conventional photolithography and lift-off process. The line width of IDTs is 5.5 μm, the aperture is 100 wavelengths, and IDT pair number is 70. PC of 15-layer cylinders was fabricated by using the ICP process. Figure 6.30b shows SEM images of the realized device. 15 rows of cylindrical holes acting as reflective gratings were placed on both sides of transducers. As the figure shows, the depth of the cylindrical holes is about 20 μm. Moreover, the width of the PC gratings is only about seven pairs of IDTs. Comparing to a traditional two-port SAW resonator with hundreds of metal strips, the size of phononic-crystal reflective gratings is reduced significantly. After fabricating the SAW devices, the transmission coefficient S_{21} was measured by an RF network analyzer. The frequency responses of the two-port SAW devices without PC gratings were measured as reference signals. Then the device was measured again after fabricating the square lattice PC.

To verify the design of the optimum delay distance between the IDTs and PC gratings, two-port SAW devices of four different delay distance D were realized. Figure 6.31 presents the results of the cases with $D = 1.0\,\lambda$ and $D = 1.25\,\lambda$. In these cases, the transmission shows the central frequency is about 212 MHz. The shift of central frequency may be caused by the difference of material constants of ZnO between single crystals used for calculations and the deposited polycrystalline thin film. Although the frequency is not exactly the same, it does not interfere the effect of PC gratings. With PC gratings, Fig. 6.31a shows a significant improvement of insertion loss. However, Fig. 6.31b has a sharp drop-off at the central frequency of the SAW device. Thus, the delay distance in Fig. 6.31b results in destructive wave interaction inside the SAW device. Consequently, with improper delay distance, the cases of $D = 0.875\,\lambda$ and 1.125 λ present distorted frequency responses, whose

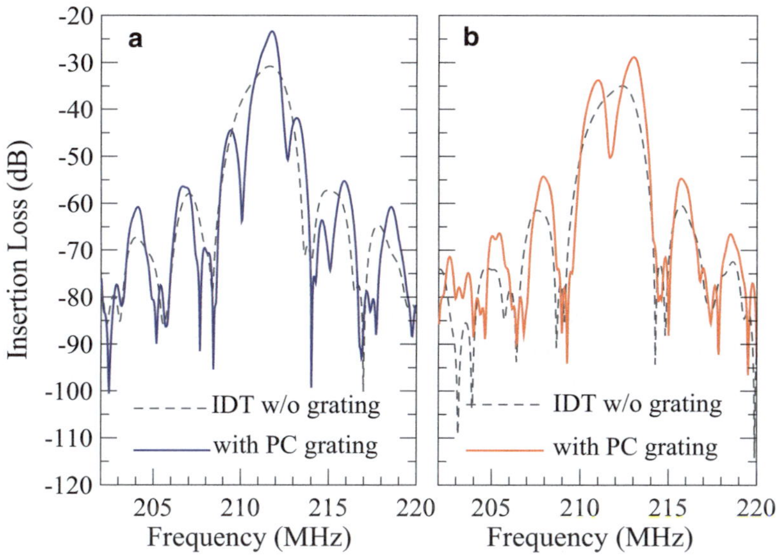

Fig. 6.31 Transmission of a two-port SAW device with PC gratings. (**a**) $D = 1.0\,\lambda$; (**b**) $D - 1.25\,\lambda$ [45]

peaks appear at higher and low frequencies. In addition, due to the wide band gap of the PC gratings, the waves of frequency ranging from 195 to 224 MHz were reflected. Several peaks and dips at different frequencies were observed in the transmission of Fig. 6.31, which result from constructive and destructive interactions affected by the delay distance. Indeed, the estimation of effective reflective plane plays an important role in designing the SAW devices with PC gratings.

To show the improvement of the PC grating compared to the metal reflector, a two-port SAW device with 300 metal reflectors was fabricated and measured. Since the PC grating was silicon-based, the metal reflectors were also fabricated on silicon. The measured insertion loss did not show obvious improvement with the existence of metal reflectors. This may be due to the fact that the metal reflectors deposited on silicon directly can only reflect SAW by the mechanical effect but electrical reflection.

In summary, a structure which combines a layered SAW device and PCs acting as gratings has been demonstrated. The results showed that the delay distance between the IDTs and PC grating is an important parameter and has to be designed carefully. By using PCs as wave reflectors, the experimental results show a 7 dB improvement in the central frequency. Moreover, the size of gratings is reduced significantly as compared to the traditional gratings with hundreds of metal strips.

6.7 Conclusions

In this chapter, we summarized recent theoretical and experimental studies of SAWs propagating in 2D phononic structures. The theoretical formulations of PWE method for SAWs in both non-piezoelectric and piezoelectric PC structures were briefly introduced, while the technique allowing to account for the case of void inclusions in a solid matrix was also addressed. The principle of FDTD method and useful boundary conditions were presented and employed to calculate the band structures of SAW in PCs. By using these numerical tools, acoustic waves propagating along the free surface of 2D PCs can be analyzed successfully. By analyzing the equifrequency surfaces of SAWs in general PCs, both the Rayleigh surface wave and pseudosurface waves can be identified clearly. For the case of a piezoelectric PC, the existence of BG SAWs was demonstrated and results show that the electrical boundary condition of piezoelectric media affects the phase velocities of BG SAW and PSAW. Nonetheless, the most important result is the demonstration of the existence of complete band gaps for SAW in 2D PCs with a free surface.

By utilizing the FDTD method with the supercell technique, salient features of SAW propagation in both straight and bent waveguides have been revealed. The results showed that it is quite straightforward to construct a straight SAW waveguide by removing a layer of PC from the 2D PC. However, to maintain proper transmission efficiency, the structure at the corner of a bent waveguide should be designed carefully. Results also indicated that surface waves inside straight waveguides are PSAWs which have slight energy leakage.

On the experimental side, both the silicon based and piezoelectric based substrates were studied in the past few years and the results demonstrated for the first time the existence of SAW band gap in micro-scale PCs [41, 42]. It is demonstrated that the transmission spectra could be obtained by employing IDTs to detect the penetrating SAW. In addition, based on the laser interferometer, the surface profile of the SAW field of micro-scale PC demonstrated further the band gap property of SAW in such a PC structures.

Finally, we note that the application of air/silicon PC in a two-port SAW resonator has been introduced recently. Through a detailed design of the band gap and effective reflecting distance, both numerical and experimental results showed that ten layers of PC are enough to serve as the reflectors of a SAW resonator. By using PCs as the reflective grating, the size of gratings can be reduced significantly as compared to the traditional gratings with hundreds of metal strips.

References

1. M.S. Kushwaha, P. Halevi, L. Dobrzynski, B. Djafari-Rouhani, Acoustic band structure of periodic elastic composites. Phys. Rev. Lett. **71**, 2022 (1993)
2. M.M. Sigalas, E.N. Economou, Elastic and acoustic wave band structure. J. Sound Vib. **158**, 377 (1992)

3. Z. Liu, C.T. Chan, P. Sheng, Three-component elastic wave band-gap material. Phys. Rev. B **65**, 165116 (2002)
4. M.S. Kushwaha, P. Halevi, Band-gap engineering in periodic elastic composites. Appl. Phys. Lett. **64**, 1085–1087 (1994)
5. M.S. Kushwaha, P. Halevi, G. Martinez, L. Dobrzynski, B. Djafari-Rouhani, Theory of acoustic band structure of periodic elastic composites. Phys. Rev. B **49**, 2313–2322 (1994)
6. J.O. Vasseur, B. Djafari-Rouhani, L. Dobrzynski, M.S. Kushwaha, P. Halevi, Complete acoustic band gaps in periodic fiber reinforced composite materials: the carbon/epoxy composite and some metallic systems. J. Phys. Condens. Matter **6**, 8759–8770 (1994)
7. M. Wilm, A. Khelif, S. Ballandras, V. Laude, Out-of-plane propagation of elastic waves in two-dimensional phononic band-gap materials. Phys. Rev. E **67**, 065602 (2003)
8. C. Goffaux, J.P. Vigneron, Theoretical study of a tunable phononic band gap system. Phys. Rev. B **64**, 075118 (2001)
9. F. Wu, Z. Liu, Y. Liu, Acoustic band gaps in 2D liquid phononic crystals of rectangular structure. J. Phys. D Appl. Phys. **35**, 162–165 (2002)
10. F. Wu, Z. Liu, Y. Liu, Acoustic band gaps created by rotating square rods in a two-dimensional lattice. Phys. Rev. E **66**, 046628 (2002)
11. X. Li, F. Wu, H. Hu, S. Zhong, Y. Liu, Large acoustic band gaps created by rotating square rods in two-dimensional periodic composites. J. Phys. D Appl. Phys. **36**, L15–L17 (2003)
12. M.M. Sigalas, E.N. Economou, Attenuation of multiple scattered sound. Europhys. Lett. **36**, 241–246 (1996)
13. M.S. Kushwaha, P. Halevi, Stop-bands for periodic metallic rods: Sculptures that can filter the noise. Appl. Phys. Lett. **70**, 3218–3220 (1997)
14. Y.-Z. Wang, F.-M. Li, K. Kishimoto, Y.-S. Wang, W.-H. Huang, Elastic wave band gaps in magnetoelectroelastic phononic crystals. Wave Motion **46**, 47–56 (2009)
15. M. Kafesaki, E.N. Economou, Multiple-scattering theory for three-dimensional periodic acoustic composites. Phys. Rev. B **60**, 11993 (1999)
16. I.E. Psarobas, N. Stefanou, Scattering of elastic waves by periodic arrays of spherical bodies. Phys. Rev. B **62**, 278 (2000)
17. Z. Liu, C.T. Chan, P. Sheng, Elastic wave scattering by periodic structures of spherical objects: Theory and experiment. Phys. Rev. B **62**, 2446 (2000)
18. J. Mei, Z. Liu, J. Shi, D. Tian, Theory for elastic wave scattering by a two-dimensional periodical array of cylinders: An ideal approach for band-structure calculations. Phys. Rev. B **67**, 245107 (2003)
19. D. Garcia-Pablos, M. Sigalas, F.R. Montero de Espinosa, M. Torres, M. Kafesaki, N. Garcia, Theory and experiments on elastic band gaps. Phys. Rev. Lett. **84**, 4349 (2000)
20. Y. Tanaka, Y. Tomoyasu, S. Tamura, Band structure of acoustic waves in phononic lattices: Two-dimensional composites with large acoustic mismatch. Phys. Rev. B **62**, 7387–7392 (2000)
21. A. Khelif, B. Djafari-Rouhani, V. Laude, M. Solal, Coupling characteristics of localized phonons in photonic crystal fibers. J. Appl. Phys. **94**, 7944–7946 (2003)
22. J.O. Vasseur, P.A. Deymier, G. Frantziskonis, G. Hong, B. Djafari-Rouhani, L. Dobrzynski, Experimental evidence for the existence of absolute acoustic band gaps in two-dimensional periodic composite media. J. Phys. Condens. Matter **10**, 6051–6064 (1998)
23. J.O. Vasseur, P.A. Deymier, B. Chenni, B. Djafari-Rouhani, L. Dobrzynski, D. Prevost, Experimental and theoretical evidence of absolute acoustic band gaps in two-dimensional solid phononic crystals. Phys. Rev. Lett. **86**, 3012–3015 (2001)
24. F.R. Montero de Espinosa, E. Jimenez, M. Torres, Ultrasonic band gap in a periodic two-dimensional composite. Phys. Rev. Lett. **80**, 1208–1211 (1998)
25. M. Torres, F.R. Montero de Espinosa, J.L. Aragón, Ultrasonic wedges for elastic wave bending and splitting without requiring a full band gap. Phys. Rev. Lett. **86**, 4282–4285 (2001)
26. P.S. Russell, E. Marin, A. Díez, Sonic band gaps in PCF preforms: enhancing the interaction of sound and light. Opt Express **11**, 2555–2560 (2003)

27. K.M. Ho, C.K. Cheng, Z. Yang, X.X. Zhang, P. Sheng, Broadband locally resonant sonic shields. Appl. Phys. Lett. **83**, 5566–5569 (2003)
28. M. Sigalas, M.S. Kushwaha, E.N. Economou, M. Kafesaki, I.E. Psarobas, W. Steurer, Classical vibrational modes in phononic lattices: theory and experiment. Z. Kristallogr. **220**, 765–809 (2005)
29. Y. Tanaka, S. Tamura, Surface acoustic waves in two-dimensional periodic elastic structures. Phys. Rev. B **58**, 7958 (1998)
30. Y. Tanaka, S. Tamura, Acoustic stop bands of surface and bulk modes in two-dimensional phononic lattices consisting of aluminum and a polymer. Phys. Rev. B **60**, 13294 (1999)
31. T.-T. Wu, Z.-G. Huang, S. Lin, Surface and bulk acoustic waves in two-dimensional phononic crystals consisting of materials with general anisotropy. Phys. Rev. B **69**, 094301 (2004)
32. T.-T. Wu, Z.-C. Hsu, Z.-G. Huang, Band gaps and the electromechanical coupling coefficient of a surface acoustic wave in a two-dimensional piezoelectric phononic crystals. Phys. Rev. B **71**, 064303 (2005)
33. V. Laude, M. Wilm, S. Benchabane, A. Khelif, Full band gap for surface acoustic waves in a piezoelectric phononic crystal. Phys. Rev. E **71**, 036607 (2005)
34. J.-C. Hsu, T.-T. Wu, Bleustein–Gulyaev–Shimizu surface acoustic waves in two-dimensional piezoelectric phononic crystals. IEEE Trans. Ultrason. Ferroelectr. Freq. Control **53**, 1169–1176 (2006)
35. Y.-Z. Wang, F.-M. Li, W.-H. Huang, Y.-S. Wang, The propagation and localization of Rayleigh waves in disordered piezoelectric phononic crystals. J. Mech. Phys. Solids **56**, 1578–1590 (2008)
36. M. Torres, F.R. Montero de Espinosa, D. Garcia-Pablos, N. Garcia, Sonic band gaps in finite elastic media surface states and localization phenomena in linear and point defects. Phys. Rev. Lett. **82**, 3504 (1999)
37. F. Meseguer, M. Holgado, D. Caballero, N. Benaches, J. Sanchez-Dehesa, C. López, J. Llinares, Rayleigh-wave attenuation by a semi-infinite two-dimensional elastic-band-gap crystal. Phys. Rev. B **59**, 12169–12172 (1999)
38. R.E. Vines, J.P. Wolfe, J.V. Every, Scanning phononic lattices with ultrasound. Phys. Rev. B **60**, 11871 (1999)
39. X. Zhang, T. Jackson, E. Lafond, Evidence of surface acoustic wave band gaps in the phononic crystals created on thin plates. Appl. Phys. Lett. **88**, 041911 (2006)
40. B. Bonello, C. Charles, F. Ganot, Velocity of a SAW propagating in a 2D phononic crystal. Ultrasonics **44**, 1259–1263 (2006)
41. T.-T. Wu, L.-C. Wu, Z.-G. Huang, Frequency band-gap measurement of two-dimensional air/silicon phononic crystals using layered slanted finger interdigital transducers. J. Appl. Phys. **97**, 094916 (2005)
42. S. Benchabane, A. Khelif, J.-Y. Rauch, L. Robert, V. Laude, Evidence for complete surface wave band gap in a piezoelectric phononic crystal. Phys. Rev. E **73**, 065601 (2006)
43. K. Kokkonen, M. Kaivola, S. Benchabane, A. Khelif, V. Laude, Scattering of surface acoustic waves by a phononic crystal revealed by heterodyne interferometry. Appl. Phys. Lett. **91**, 083517 (2007)
44. D.M. Profunser, E. Muramoto, O. Matsuda, O.B. Wright, U. Lang, Dynamic visualization of surface acoustic waves on a two-dimensional phononic crystal. Phys. Rev. B **80**, 014301 (2009)
45. T.-T. Wu, W.-S. Wang, J.-H. Sun, J.-C. Hsu, Y.-Y. Chen, Utilization of phononic-crystal reflective gratings in a layered surface acoustic wave device. Appl. Phys. Lett. **94**, 101913 (2009)
46. B. Manzanares-Martínez, F. Ramos-Mendieta, Surface elastic waves in solid composites of two-dimensional periodicity. Phys. Rev. B **68**, 134303 (2003)
47. J.O. Vasseur, P.A. Deymier, B. Djafari-Rouhani, Y. Pennec, A.-C. Hladky-Hennion, Absolute forbidden bands and waveguiding in two-dimensional phononic crystal plates. Phys. Rev. B **77**, 085415 (2008)
48. M.M. Sigalas, N. Garcia, Theoretical study of three dimensional elastic band gaps with the finite-difference time-domain method. J. Appl. Phys. **87**(6), 3122–3125 (2000)

49. M. Kafesaki, M.M. Sigalas, N. Garcia, Frequency modulation in the transmittivity of wave guides in elastic-wave band-gap materials. Phys. Rev. Lett. **85**(19), 4044–4047 (2000)
50. P.-F. Hsieh, T.-T. Wu, J.-H. Sun, Three-dimensional phononic band gap calculations using the FDTD method and a PC cluster system. IEEE Trans. Ultrason. Ferroelectr. Freq. Control **53**(1), 148–158 (2006)
51. J.-H. Sun, T.-T. Wu, Propagation of surface acoustic waves through sharply bent two-dimensional phononic crystal waveguides using a finite-difference time-domain method. Phys. Rev. B **74**, 174305 (2006)
52. J. Berenger, A perfectly matched layer for the absorption of electromagnetic waves. J. Comput. Phys. **144**, 185–200 (1994)
53. W.C. Chew, Q.H. Liu, Perfectly matched layers for elastodynamics: a new absorbing boundary condition. J. Comput. Acoustics **4**(4), 341–359 (1996)
54. F. Chagla, C. Cabani, P.M. Smith, Perfectly matched layer for FDTD computations in piezoelectric crystals. Proc. IEEE Ultrason. Symp., pp. 517–520 (2004)
55. S. Benchabane, O. Gaiffe, R. Salut, G. Ulliac, Y. Achaoui, V. Laude, Observation of surface-guided waves in holey hypersonic phononic crystal. Appl. Phys. Lett. **98**, 171908 (2011)
56. T.-T. Wu, Z.-G. Huang, S.Y. Liu, Surface acoustic wave band gaps in micro-machined air/silicon phononic structures – theoretical calculation and experiment. Zeitschrift für Kristallographie **220**, 841 (2005)
57. T.-T. Wu, Y.-Y. Chen, Exact analysis of dispersive SAW devices on ZnO/diamond/Si-layered structures. IEEE Trans. Ultrason. Ferroelectr. Freq. Control **49**, 142 (2002)

Chapter 7
Optical Characterization of Phononic Crystals in the Time Domain

Osamu Matsuda and Oliver B. Wright

7.1 Introduction

For the characterization of phononic crystals and their derivative devices, the measurement of the acoustic field evolution in space is an attractive goal. Such measurements allow one to retrieve fundamental properties of phononic crystals such as dispersion relations, phononic stop bands and acoustic eigenmode distributions or, in the case of phononic crystal devices in particular, the acoustic leakage characteristics. Optical techniques are well suited for application to such acoustic field measurements, from audio frequencies up to the gigahertz range, and can be classified into two categories, time domain, the subject of the present chapter, and frequency domain [1–9], the subject of the next chapter.

Time-resolved optical imaging of acoustic waves in solids has been demonstrated using photoelastic techniques [10–13], beam deflection [14, 15], holography [16], and interferometry [17–24]. Time-domain imaging of acoustic wave propagation in phononic crystals was first demonstrated in millimeter-scale structures and liquid/solid systems without the use of optics by means of focused transducers at MHz frequencies [25–27]. More recently the versatility of time-domain optical interferometry for measurements on micron-scale solid and solid/air phononic crystals has been demonstrated up to ∼1 GHz [28–32]. It is this latter technique that we shall describe in detail in this chapter.

We first describe the ultrafast optical technique used for time-domain imaging of the acoustic field on the surface of microscopic phononic crystals in two spatial dimensions in the range ∼ 100 MHz–1 GHz. The imaged area is typically

O. Matsuda • O.B. Wright (✉)
Faculty of Engineering, Division of Applied Physics, Hokkaido University,
060–8628 Sapporo, Japan
e-mail: omatsuda@eng.hokudai.ac.jp; assp@kino-ap.eng.hokudai.ac.jp

A. Khelif, A. Adibi (eds.), *Phononic Crystals*, DOI 10.1007/978-1-4614-9393-8_7

$\sim$100 μm × 100 μm, and the lateral spatial resolution, limited by optical diffraction, is $\sim$1 μm. We present applications for samples consisting of one- and two-dimensional phononic crystals exhibiting phononic stop bands.

7.2 Experimental Setup for Scanning Laser Interferometry in the Time Domain

The optical pump-probe technique combined with a scanning interferometer allows ultrafast stroboscopic imaging of surface displacements. Section 7.2.1 deals with this time-resolved measurement technique, Sect. 7.2.2 with the detection of picometer surface displacements, and Sect. 7.2.3 with the optical scanning system for imaging surface motion.

7.2.1 Optical Pump-Probe Technique

In this section we give a description of a typical optical pump-probe apparatus for use in interferometric imaging, in particular for application to surface acoustic wave (SAW) measurement [20, 24]. In the basic setup, optical reflectivity or phase changes induced by the absorption of light pulses, known as pump pulses, are measured using delayed light pulses known as probe pulses. By varying the delay time between the pump and probe pulses, transient changes in the time domain can be accessed stroboscopically.

The application of the optical pump-probe technique to the measurement of acoustic phenomena was proposed about 25 years ago, resulting in the field of laser picosecond acoustics [33, 34]. The pump light pulses generate acoustic waves in the sample, and the probe light pulses focused to the same point on the sample detect the modulation of the optical properties by the propagating acoustic waves as a function of the delay time. This technique is appropriate for measuring bulk acoustic waves that propagate in the through-thickness direction. The temporal resolution is mainly governed by the optical pulse duration, typically less than 1 ps. It was later shown that the optical pump-probe technique could be extended to the two-dimensional imaging of SAWs [20]. This is done by scanning the position of the probe light spot on the sample surface, as described in Sect. 7.2.3.

Figure 7.1 shows a typical experimental setup for the measurement. A mode-locked Ti-sapphire laser is used for the light source. It generates an optical pulse train with wavelength 830 nm, pulse duration $\sim$ 100 fs, and pulse repetition rate 80 MHz. Part of this pulse train is frequency doubled with an optical nonlinear crystal (second harmonic generation crystal) such as BBO (β-BaB_2O_4) to yield the wavelength 415 nm. The 415 nm light pulses are used as the pump light pulses, whereas the 830 nm light pulses are used as the probe light pulses.

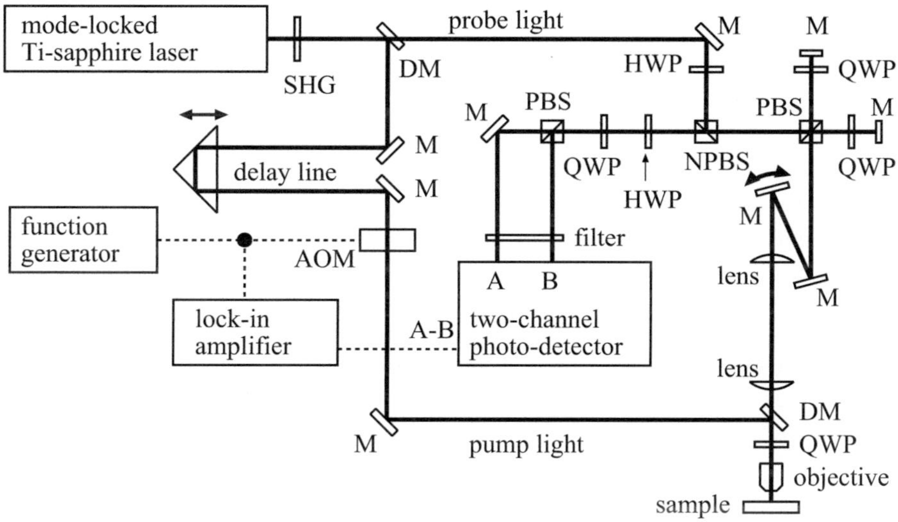

Fig. 7.1 Schematic diagram of a typical setup for time-resolved two-dimensional SAW imaging. *SHG*: second harmonic generation crystal, *AOM*: acousto-optic modulator, *HWP*: half wave plate, *QWP*: quarter wave plate, *PBS*: polarizing beam splitter, *NPBS*: non-polarizing beam splitter, *DM*: dichroic mirror, and *M*: mirror

The pump light pulses are focused with a high-power objective lens to a spot of diameter ~1 μm onto the surface of the sample, chosen to be absorbing at the pump wavelength. The typical pump light fluence is ~ 1 mJ cm^{-2} per pulse. An absorbed pump light pulse raises the temperature of the illuminated region within the hot electron penetration depth, ~10–100 nm in the case of metals and semiconductors that are opaque at the pump light wavelength. This abrupt and localized temperature rise, typically ~100 K, generates a spatiotemporal variation of the thermal stress field, which launches acoustic waves propagating along the sample surface as well as bulk acoustic waves propagating away from the surface. The frequency spectrum of the generated SAWs is limited by the lateral size of the pump light spot: the spot size, typically of micron order, governs the shortest wavelength of the generated SAWs. For a solid with a SAW velocity of 3 km s^{-1}, for example, a wavelength of 2 μm generated by a spot diameter of ~ 1 μm corresponds to a period of 0.6 ns or a frequency of 1.5 GHz. The frequency spectrum of the generated acoustic waves is in theory also limited by the optical pulse duration, but in fact a pulse duration < 1 ps is short enough to generate acoustic frequency components well beyond 1 GHz, and so the high frequency limit for SAW generation is in practice only governed by the pump light spot size. The low frequency limit is normally imposed by the repetition rate of the laser, that is, 80 MHz. The strain amplitude of the SAWs is typically 10^{-6}, corresponding to a surface displacement of ~ 10 pm for a 10 μm wavelength.

The probe light pulses are also focused to a spot of diameter ~1 μm on the sample surface through the same microscope objective. Using an optical interferometer, described in detail in Sect. 7.2.2, the surface displacement (or, more precisely, the difference in out-of-plane surface displacement at two close instants in time) at the moment when the probe light pulse hits the sample surface is converted to an intensity variation of the reflected probe light. By varying the probe light spot position with the scanning system, a two-dimensional SAW field image can be obtained. The spatial resolution of the measurement is governed by the spot size of the probe light spot, and is typically $\sim 1\,\mu$m.

The delay time between the pump and probe light pulse arrival times at the sample surface is controlled by a motorized optical delay line. By varying the optical path length up to 4 m, the delay time can be varied from 0 to 13 ns, which adequately covers a single period (12.5 ns) of the laser pulse repetition.

The reflected probe light at the output of the optical interferometer is fed to a photodetector equipped with Si photodiodes of bandwidth $\sim$ 5 MHz. The relative variation of the intensity of the reflected probe light $\Delta I/I_0$, where ΔI is the intensity variation and I_0 is the detected steady-state intensity, is of the order of 10^{-5}. To observe such relatively small intensity variations, a lock-in detection technique is used. The pump light pulse train is modulated using an acousto-optic modulator operating at $\sim$1 MHz. The photodetector output is fed to a lock-in amplifier for synchronous amplification. Any stray pump light is eliminated with a color filter just before the photodetector. The resolution of this detection system can be honed to achieve near optical-shot-noise performance, limited only by the power of the probe light reaching the photodetectors [24].

The optical pump-probe technique can be extended to the case of pulsed electrical excitation and optical ultrafast imaging of acoustic fields [35]. In this case electrical pulses are arranged to be synchronous with the optical probe pulses. This method is useful for application to piezoelectric devices based on bulk or surface acoustic waves. The advantage of this method is that the acoustic wave amplitude is not limited by the optical pump damage threshold. This method could also be applied to imaging phononic crystals.

7.2.2 *Interferometer*

Several types of interferometer can be used to detect the out-of-plane surface motion. So far Mach-Zender [36], Sagnac [37, 38], and Michelson interferometers have been proposed [5, 24, 30, 39, 40]. Here we describe an example [24] based on a modification of a previously proposed Sagnac interferometer [37]. The design is in fact better termed a Michelson interferometer. Figure 7.2 shows the details of the optics, to be in incorporated into the pump-probe setup of Fig. 7.1. This interferometer is classed as being of time-division common-path type.

The operation is as follows: After passing through a half-wave plate HWP_1, the probe light is linearly polarized with its polarization plane at 45° to the horizontal. At this point it has in-phase horizontal (x) and vertical (y) polarization components. This beam is then directed to a polarizing beam splitter PBS_1 via a non-polarizing

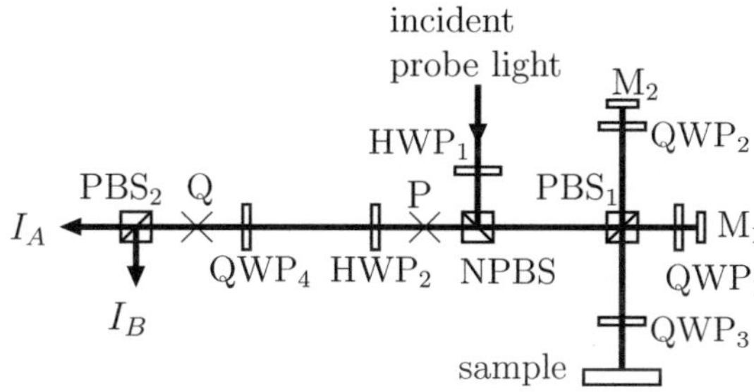

Fig. 7.2 Schematic diagram of a time-division common-path Michelson interferometer. *HWP*: half wave plate, *QWP*: quarter wave plate, *PBS*: polarizing beam splitter, *NPBS*: non-polarizing beam splitter, and *M*: mirror

beam splitter NPBS, and split into two, the resulting beams heading for mirrors M_1 or M_2. The optical axes of the quarter wave plates QWP_1, QWP_2, and QWP_3 are set at 45° to the horizontal so that a combination of a QWP and a mirror (or the sample) flips x-polarized incident light to y- polarized reflected light and vice versa. The light passing through PBS_1 thus reaches to the sample via M_1 and PBS_1, whereas the light reflected at PBS_1 reaches to the sample via M_2 and PBS_1. The distance between PBS_1 and M_2 is set to be longer than that between PBS_1 and M_1, so that probe light pulses travelling via M_2 reach the sample after the corresponding ones travelling via M_1. The associated time delay is $\Delta\tau$, which is typically chosen to be ~300 ps. Due to the combination of QWP_3 and the sample surface, probe light pulses reflected off the sample take opposite paths compared to those they took before reaching the sample. The two consecutive probe light pulses that travel in different paths are finally reunited to a single pulse after passing through (or being reflected at) PBS_1. In other words, the time difference between the arrival of these two probe light pulses at the sample is compensated when the beams are reunited. The y-polarized component of the light in the final beam contains information about the sample position at a time $\Delta\tau$ prior to that contained in the x-polarized component.

When the sample surface is dynamically displaced by the propagating acoustic waves, a finite phase difference $\phi_x - \phi_y$ between the x and y polarization components of the unified probe beam is induced. For the small displacements involved, this phase difference is proportional to the difference in out-of-plane surface displacements at the moments in time when the relevant probe pulses reach the sample surface. Since $\Delta\tau$ is chosen to be much smaller than the period of the SAWs to be measured, this translates as a measurement of the out-of-plane surface velocity of the sample motion. The phase difference $\phi_x - \phi_y$ is converted to a light intensity change by interfering the x and y polarization components using HWP_2, QWP_4, and PBS_2.

A brief description of the optical theory involved follows. Let the electric field amplitude and phase of the probe light at point P in Fig. 7.2 be expressed as $E_{x,y}$ and $\phi_{x,y}$. The corresponding Jones vector is

$$\mathbf{E}_\mathrm{P} = \begin{pmatrix} E_x \exp(i\phi_x) \\ E_y \exp(i\phi_y) \end{pmatrix}. \tag{7.1}$$

HWP_2 is set so that its optical axis is at 22.5° to the horizontal plane. The corresponding Jones matrix is

$$M_{\mathrm{HWP2}} = \frac{\sqrt{2}}{2}\begin{pmatrix}1 & 1\\ 1 & -1\end{pmatrix}. \tag{7.2}$$

QWP_4 is set so that its optical axis is at 45° to the horizontal plane. The corresponding Jones matrix is

$$M_{\mathrm{QWP4}} = \frac{1}{2}\begin{pmatrix}1+i & 1-i\\ 1-i & 1+i\end{pmatrix}. \tag{7.3}$$

The Jones vector at point Q is given as

$$\mathbf{E}_{\mathrm{Q}} = M_{\mathrm{QWP4}} M_{\mathrm{HWP2}} \mathbf{E}_{\mathrm{P}} = \frac{1}{\sqrt{2}}\begin{pmatrix}E_x \exp(i\phi_x) + iE_y \exp(i\phi_y)\\ E_x \exp(i\phi_x) - iE_y \exp(i\phi_y)\end{pmatrix}. \tag{7.4}$$

The x and y components of the electric field are separated using PBS_2, and are detected by a two-channel photodetector characterized by channels A and B, respectively. The corresponding intensities I_A and I_B are given by

$$I_A \propto \frac{1}{2}|E_{\mathrm{Q}x}|^2 = \frac{1}{2}\{E_x^2 + E_y^2 + 2E_xE_y \sin(\phi_x - \phi_y)\}, \tag{7.5}$$

$$I_B \propto \frac{1}{2}|E_{\mathrm{Q}y}|^2 = \frac{1}{2}\{E_x^2 + E_y^2 - 2E_xE_y \sin(\phi_x - \phi_y)\}. \tag{7.6}$$

By taking the difference $I_A - I_B$, one obtains an output proportional to the optical phase difference:

$$I_A - I_B \propto 2E_xE_y \sin(\phi_x - \phi_y). \tag{7.7}$$

If the phase difference is predominantly caused only by the surface displacement u, as is often the case for metal-coated samples, then, to a good approximation,

$$\phi_x - \phi_y = -\frac{4\pi\Delta u}{\lambda}, \tag{7.8}$$

where λ is the wavelength of the probe light and Δu is the displacement over the time interval $\Delta\tau$. Here, $\Delta u > 0$ corresponds to outward displacements. Since $|\Delta u|$ is typically $\sim$10 pm and much smaller than λ, the sine function can be expanded:

$$I_A - I_B \propto -2E_xE_y \frac{4\pi\Delta u}{\lambda}. \tag{7.9}$$

Here the velocity v of the out-of-plane surface motion is given by

$$v \simeq \frac{\Delta u}{\Delta \tau} \propto I_B - I_A, \tag{7.10}$$

where $v > 0$ corresponds to an outward motion of the surface.

Since the modulation in E_x or E_y, like that in phase, is also very small (typically $\Delta E_x/E_x \lesssim 10^{-2}$), one can calculate the numerical value of the velocity of the out-of-plane surface motion from the relation

$$\frac{I_A - I_B}{I_A + I_B} \simeq \sin(\phi_x - \phi_y) \simeq= -\frac{4\pi v \Delta\tau}{\lambda}. \tag{7.11}$$

This common-path interferometer has a significant advantage when applied to the measurement of phononic crystals: Since active stabilization is not required, samples which exhibit a wide variation of optical reflectivity across their surface can be measured.

7.2.3 Scanning System

To obtain two-dimensional images of the propagating acoustic waves, the probe light spot position is scanned across the sample surface for a fixed pump light spot position. One of the simplest ways to achieve this is by use of two independent objective lenses for the pump and probe beams, and to scan the probe beam objective in the plane of the sample surface. This was demonstrated for thin-film metallic samples on transparent substrates [20, 41]. In these experiments the pump light was focused from the substrate side of the sample, whereas the probe light was focused from the front side. For opaque substrates the pump and probe light can be focused from the same side of the sample again with independent objective lenses, with the pump light incident at oblique incidence [42]. However, this setup is difficult to implement because of the close proximity of the two objectives.

To remedy this problem a setup using a single objective lens for both pump and probe light was proposed [24]. This uses two lenses arranged in a so-called 4f configuration which allows the incident angle of the probe beam at the microscope objective to be scanned. Figure 7.3 shows a schematic diagram of this scanning system. By means of a mirror M_3 mounted on an automated biaxial rotation stage, the incident probe beam is directed to a lens L_1. The probe light beam position on M_3 is set at the center of rotation of M_3. The positions of the light spot on M_3, the center of the lenses L_1 and L_2 (both of focal length f), and the center of the aperture of the objective are all arranged on a straight line (labeled by A). L_1 and L_2 are separated by 2f, and these are in turn separated by f from M_3 and the objective input aperture. When M_3 is set so that the probe beam passes along A, the probe beam is focused at the point O on the sample at normal incidence. When M_3 is tilted

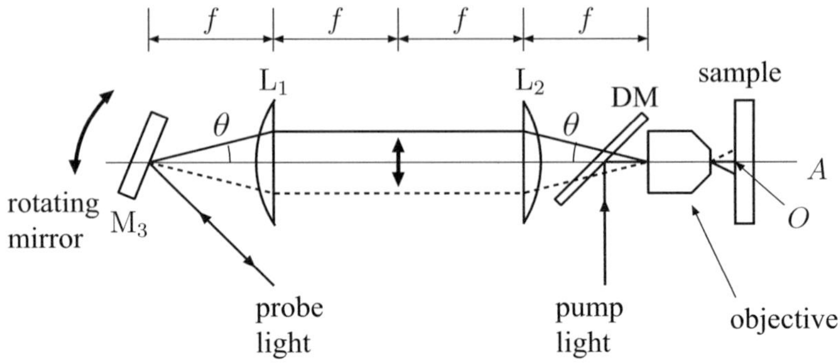

Fig. 7.3 Schematic diagram of the scanning system for the probe light. *DM*: dichroic mirror and *M*: mirror

so that the reflected beam is deflected by an angle θ from A, the probe beam enters the aperture of the objective with a non-zero incident angle θ as a collimated beam, and is focused on the sample at a position $f'\theta$ away from O, where f' is the focal length of the objective. By scanning θ, the probe light spot moves across the sample surface over a region typically up to a maximum size of 500 µm × 500 µm. The probe beam reflected from the sample retraces the path of the incident beam but with a parallel displacement by $2f'\theta$, which is usually much smaller than the probe beam diameter and has no measurable effect on the measurement process. An image of ~ 200 × 200 pixels is typically obtained in 5–10 min. A series of ~ 40 images is recorded at different delay times, taking a total of 3.5–7 h. The limitations of the measurement technique are discussed in more detail in Sect. 7.3.4.

The pump light beam is directed by a dichroic mirror to the objective, and is focused to a fixed point on the sample. To compensate chromatic aberration and any difference in the beam divergence between the probe and pump beams, a pair of additional lenses may be used in the pump or probe beam paths (not shown).

A similar design of the scanning system used with the $4f$ confocal setup makes use of two motorized linear stages instead of the biaxial rotation stage [43]. This system is somewhat easier to align than the setup described above. However, its use leads to small variations, $<$ 100 ps in the pump probe delay time across the scanned area. This variation is, however, negligible in sub-GHz SAW or other nanosecond imaging experiments.

7.3 Applications to Phononic Crystals

Here we review the application of this ultrafast time-domain interferometric imaging technique to surface phononic crystals and describe how the results can be interpreted theoretically in terms of the phononic band structure. We first describe how the spatiotemporal acoustic displacement field can be Fourier analyzed to yield the acoustic dispersion relations, in this case the phononic band structure. We then review experimental results for one- and two-dimensional phononic crystals.

7.3.1 From Time-Resolved Data to the Acoustic Dispersion Relation

The experiment yields the velocity field representing the out-of-plane surface motion $v(\mathbf{r}, t)$ as a function of two-dimensional position $\mathbf{r}$ and time t. One can relate this to the surface displacement field $u(\mathbf{r}, t)$ using Fourier analysis. The corresponding Fourier amplitudes $V(\mathbf{k}, \omega)$ and $U(\mathbf{k}, \omega)$ are obtained as follows:

$$V(\mathbf{k}, \omega) = \frac{1}{(2\pi)^3} \int_{-\infty}^{\infty} v(\mathbf{r}, t) \exp\{-i(\mathbf{k} \cdot \mathbf{r} - \omega t)\} d^2\mathbf{r}\, dt, \tag{7.12}$$

$$v(\mathbf{r}, t) = \int_{-\infty}^{\infty} V(\mathbf{k}, \omega) \exp\{i(\mathbf{k} \cdot \mathbf{r} - \omega t)\} d^2\mathbf{k}\, d\omega, \tag{7.13}$$

$$U(\mathbf{k}, \omega) = \frac{1}{(2\pi)^3} \int_{-\infty}^{\infty} u(\mathbf{r}, t) \exp\{-i(\mathbf{k} \cdot \mathbf{r} - \omega t)\} d^2\mathbf{r}\, dt, \tag{7.14}$$

$$u(\mathbf{r}, t) = \int_{-\infty}^{\infty} U(\mathbf{k}, \omega) \exp\{i(\mathbf{k} \cdot \mathbf{r} - \omega t)\} d^2\mathbf{k}\, d\omega. \tag{7.15}$$

Comparing time derivative of Eq. (7.15) with Eq. (7.13), one obtains

$$U(\mathbf{k}, \omega) = \frac{i}{\omega} V(\mathbf{k}, \omega). \tag{7.16}$$

In general the displacement field $u(\mathbf{r}, t)$ can be expressed as a superposition of normal modes of the medium. For a sample exhibiting periodicity, the displacement (more rigorously, a component of the displacement vector) corresponding to a normal mode, which can be specified by branch index j and wave vector $\mathbf{k}$ in the first Brillouin zone (1st BZ), can be expressed according to Bloch's theorem as

$$u_{\mathbf{k},j}(\mathbf{r}, t) = \mathrm{Re}\left[\sum_{\mathbf{G}} C_j(\mathbf{k} + \mathbf{G}) \exp\{i((\mathbf{k} + \mathbf{G}) \cdot \mathbf{r} - \omega_j(\mathbf{k})t)\}\right], \tag{7.17}$$

where $\mathbf{G}$ is a reciprocal lattice vector (consisting of a linear combination of integer multiples of unit reciprocal lattice vectors), and $C_j(\mathbf{k} + \mathbf{G})$ is the amplitude of the corresponding Bloch harmonic specified by $\mathbf{G}$. The summation covers all possible $\mathbf{G}$'s. The angular frequency $\omega_j(\mathbf{k})$ refers to the mode specified by j and $\mathbf{k}$, and represents part of the dispersion relation. The spatiotemporal displacement field can therefore be expanded as follows:

$$u(\mathbf{r},t)=\frac{1}{2}\sum_j\int_{1\mathrm{stBZ}}\Big\{A_j(\mathbf{k})\sum_{\mathbf{G}}C_j(\mathbf{k}+\mathbf{G})\exp\{i((\mathbf{k}+\mathbf{G})\cdot\mathbf{r}-\omega_j(\mathbf{k})t)\}$$
$$+A_j^*(\mathbf{k})\sum_{\mathbf{G}}C_j^*(\mathbf{k}+\mathbf{G})\exp\{-i((\mathbf{k}+\mathbf{G})\cdot\mathbf{r}-\omega_j^*(\mathbf{k})t)\}\Big\}\,d^2\mathbf{k}. \quad (7.18)$$

$A_j(\mathbf{k})$, termed a source function, is the amplitude for a mode with branch j and wave vector $\mathbf{k}$. The integration over the 1st BZ and the summation over all $\mathbf{G}$ can be extended to an integral over all $\mathbf{k}$-space:

$$\begin{aligned}u(\mathbf{r},t)&=\frac{1}{2}\sum_j\int_{\mathrm{all}\mathbf{k}}\{A_j(\mathbf{k}')C_j(\mathbf{k})\exp\{i(\mathbf{k}\cdot\mathbf{r}-\omega_j(\mathbf{k}')t)\}\\&\quad+A_j^*(\mathbf{k}')C_j^*(\mathbf{k})\exp\{-i(\mathbf{k}\cdot\mathbf{r}-\omega_j^*(\mathbf{k}')t)\}\}\,d^2\mathbf{k}\\&=\frac{1}{2}\int_{\mathrm{all}\mathbf{k}}\sum_j\{A_j(\mathbf{k}')C_j(\mathbf{k})\exp\{-i\omega_j(\mathbf{k}')t\}\\&\quad+A_j^*(-\mathbf{k}')C_j^*(-\mathbf{k})\exp\{i\omega_j^*(-\mathbf{k}')t\}\}\exp(i\mathbf{k}\cdot\mathbf{r})d^2\mathbf{k}\end{aligned}\quad (7.19)$$

where $\mathbf{k}'$ is the wave vector reduced to the 1st BZ zone using the relation $\mathbf{k}=\mathbf{k}'+\mathbf{G}$ with an appropriate $\mathbf{G}$.

Rewriting Eq. (7.15) as

$$u(\mathbf{r},t)=\int_{\mathrm{all}\mathbf{k}}\left\{\int U(\mathbf{k},\omega)\exp(-i\omega t)d\omega\right\}\exp(i\mathbf{k}\cdot\mathbf{r})d^2\mathbf{k}, \quad (7.20)$$

and comparing it with Eq. (7.19), one obtains

$$\begin{aligned}&\int U(\mathbf{k},\omega)\exp(-i\omega t)d\omega\\&=\frac{1}{2}\sum_j\{A_j(\mathbf{k}')C_j(\mathbf{k})\exp\{-i\omega_j(\mathbf{k}')t\}+A_j^*(-\mathbf{k}')C_j^*(-\mathbf{k})\exp\{i\omega_j^*(-\mathbf{k}')t\}\}.\end{aligned}\quad (7.21)$$

When the absorption of the acoustic waves is negligible over the imaged region, $\omega_j(\mathbf{k}')$ is real, that is $\omega_j(\mathbf{k}')=\omega_j^*(\mathbf{k}')$. Equation (7.21) can then be further simplified by multiplying both sides by $\exp(i\omega' t)/(2\pi)$ and integrating with respect to t:

$$U(\mathbf{k},\omega')=\frac{1}{2}\sum_j\{A_j(\mathbf{k}')C_j(\mathbf{k})\delta(\omega'-\omega_j(\mathbf{k}'))+A_j^*(-\mathbf{k}')C_j^*(-\mathbf{k})\delta(\omega'+\omega_j(-\mathbf{k}'))\}. \quad (7.22)$$

Equation (7.22) satisfies the relation

$$U(\mathbf{k}, \omega') = U^*(-\mathbf{k}, -\omega'), \tag{7.23}$$

as expected for the Fourier transform of the real function $u(\mathbf{r}, t)$ [see Eqs. (7.14) and (7.15)].

The time-reversal symmetry of the acoustic wave equation implies $\omega_j(\mathbf{k}') = \omega_j(-\mathbf{k}')$, and also ensures that the amplitudes of the Bloch harmonics can be chosen as $C_j(\mathbf{k}) = C_j^*(-\mathbf{k})$. This can be seen by letting $t \to -t$ in Eq. (7.17) and comparing it with $u_{-\mathbf{k},j}(\mathbf{r}, t)$. Equation (7.22) can therefore be written as

$$U(\mathbf{k}, \omega') = \frac{1}{2} \sum_j \left\{ A_j(\mathbf{k}') C_j(\mathbf{k}) \delta(\omega' - \omega_j(\mathbf{k}')) + A_j^*(-\mathbf{k}') C_j(\mathbf{k}) \delta(\omega' + \omega_j(\mathbf{k}')) \right\}. \tag{7.24}$$

Equations (7.22) or (7.24) indicate that the Fourier amplitude $U(\mathbf{k}, \omega')$ only takes a finite value when the combination $(\mathbf{k}', \omega')$ satisfies the relation $\omega' = \omega_j(\mathbf{k}')$ for the branch j. This shows that the dispersion relation can be determined from a temporal series of time-resolved two-dimensional images of the acoustic field.

If $\omega_j(\mathbf{k}')$ is chosen to be positive, the first term in the sum of Eq. (7.22) or (7.24) corresponds to $\omega' > 0$, whereas the second term corresponds to $\omega' < 0$. Since the Fourier amplitudes for positive and negative ω' are related by Eq. (7.23), it suffices to consider only the first term:

$$U(\mathbf{k}, \omega') = \frac{1}{2} \sum_j A_j(\mathbf{k}') C_j(\mathbf{k}) \delta(\omega' - \omega_j(\mathbf{k}')). \qquad (\omega' > 0) \tag{7.25}$$

For $(\mathbf{k}', \omega')$ satisfying the dispersion relation, that is $\omega' = \omega_j(\mathbf{k}')$, $U(\mathbf{k}' + \mathbf{G}, \omega')$ with any $\mathbf{G}$ can take a finite value, and the ratios among $C_j(\mathbf{k}' + \mathbf{G})$ with different $\mathbf{G}$ are equal to the corresponding ratios among $U(\mathbf{k}' + \mathbf{G}, \omega')$.

7.3.2 One-Dimensional Phononic Crystals

The simplest example of a phononic crystal is one exhibiting one-dimensional periodicity. Such a sample has been investigated using the experimental setup described above [28, 32]. It consists of alternating stripes of copper and silicon oxide formed on a silicon (001) substrate of thickness 0.74 mm (Fig. 7.4a, b). Each stripe has a width of 2 μm and a thickness of 800 nm. The stripes are oriented along the $[\bar{1}10]$ direction of the substrate, and form a one-dimensional periodic structure with a period $a = 4$ μm along the [110] direction. The sample fabrication is based on a damascene process, and the top surface is prepared by chemical-mechanical polishing (CMP) to a flatness less than 10 nm. A 25 nm tantalum layer is formed beneath the Cu stripes as a diffusion barrier. The top surface is then covered by a 30 nm gold film by radio frequency sputtering to produce a sample with uniform optical reflectivity and also to increase the reflectivity of the probe light.

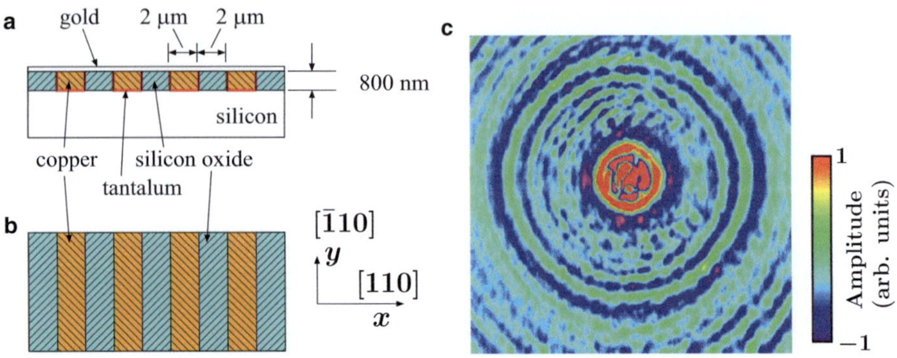

Fig. 7.4 (**a**) Cross section of a one-dimensional phononic crystal sample. (**b**) *Top view* of the phononic crystal sample showing the substrate orientation and the definition of the coordinate system. (**c**) Experimentally obtained SAW snapshot image for the one-dimensional phononic crystal at 2.62 ns after optical excitation. The center of the image corresponds to the pump light spot, and the imaged area is 150 μm × 150 μm

Time-resolved two-dimensional images of the velocity of the out-of-plane surface motion are obtained using the apparatus described in Sect. 7.2. The pump light wavelength is 400 nm and the probe light wavelength is 800 nm. The pump light pulses of energy 0.2 nJ/pulse, duration 200 fs, and repetition frequency 76.3 MHz are focused at normal incidence to a $\sim$2 μm diameter spot on the sample. The probe light pulses are focused to a spot of the same diameter, and the imaged area is 150 μm × 150 μm. Images are recorded at different delay times in steps of $13.1/40 = 0.33$ ns within the laser repetition period of 13.1 ns. Figure 7.4c shows a typical image at delay time 2.62 ns. The center of the image corresponds to the pump light spot. Concentric rings corresponding to SAWs emerging from the excitation point are visible, as well as complicated wave patterns inside the rings.

The origin of this pattern can be better understood after a spatiotemporal Fourier transform. The upper row of Fig. 7.5 shows the modulus of the Fourier amplitude, $|V(\mathbf{k}, \omega)|$ plotted in two-dimensional $\mathbf{k}$-space at some representative frequencies. As described in Sect. 7.3.1, $|V(\mathbf{k}, \omega)|$ only takes a finite value if $(\mathbf{k}, \omega)$ satisfies the dispersion relation $\omega = \omega_j(\mathbf{k})$.

The 458 MHz image (Fig. 7.5a) exhibits two concentric rings (shown in red) with their centers at the Γ point. The near circular rings indicate that these waves propagate almost isotropically at this frequency. The outer ring has a phase velocity of $\simeq 4,000\,\mathrm{ms}^{-1}$, and corresponds to Rayleigh-like waves (RW), whereas the inner ring has a phase velocity of $\simeq 7,400\,\mathrm{ms}^{-1}$, and corresponds to Sezawa waves (SW—higher-order modes of the film/substrate system) [44]. Both rings lie within the 1st BZ ($|k_x| < \pi/a = 0.79\,\mu\mathrm{m}^{-1}$, where $a = 4\,\mu$m is the period of the phononic crystal structure). The RW ring is accompanied by two faint rings on the right and left. These are Bloch harmonics, and have exactly the same shape as the central RW ring. They are shifted by reciprocal vectors $\pm\mathbf{G}_0$, where $\mathbf{G}_0$ is a unit reciprocal

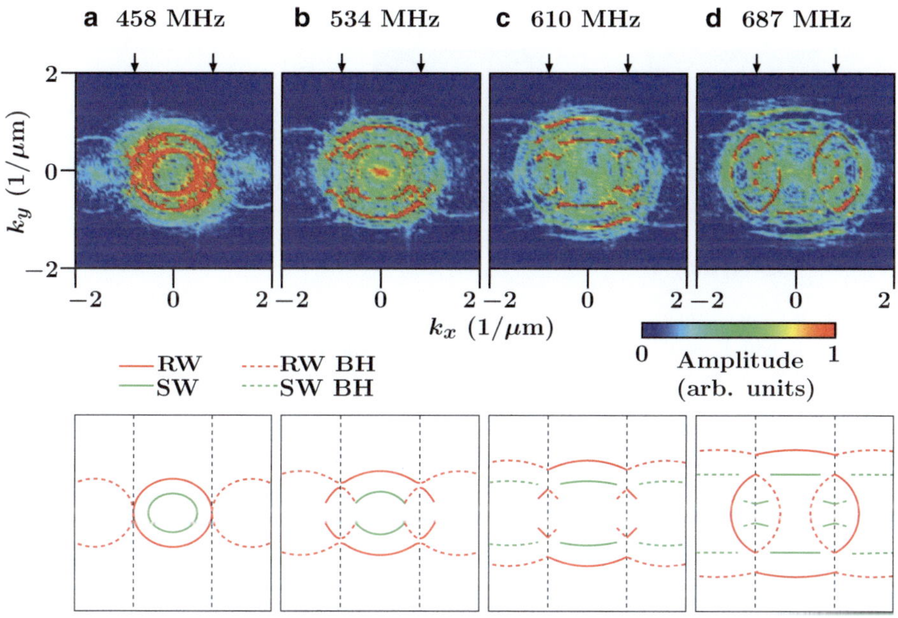

Fig. 7.5 *Upper row*: Amplitude (modulus) images of the spatiotemporal Fourier transform obtained from a set of time-resolved two-dimensional SAW images for a one-dimensional phononic crystal sample at several representative frequencies. (**a**) 458 MHz, (**b**) 534 MHz, (**c**) 610 MHz, and (**d**) 687 MHz. *Arrows* indicate the 1st BZ boundaries. *Lower row*: Mode identification diagram. The *solid lines* represent dominant components in the acoustic dispersion relation, whereas the *dotted lines* correspond to Bloch harmonics (BH). *Red lines*: Rayleigh-like waves (RW). *Green lines*: Sezawa waves (SW). The *black dotted lines* indicate the 1st BZ

vector $2\pi\mathbf{i}/a$ ($\mathbf{i}$ being the unit vector in the x direction), with respect to the dominant rings inside the 1st BZ. The Bloch harmonics for the SW modes are less prominent. Mode identification is shown schematically in the bottom part of Fig. 7.5.

In the 534 MHz image (Fig. 7.5b) the radius of the RW and SW rings becomes larger, but some part of the rings are missing. By defining a propagation angle θ with respect to the k_x axis, one can see that there are the openings for $|\theta| < 15°$ for RW and for $|\theta| < 20°$ for SW, providing evidence for a directional phononic stop band. The shape of the curves indicates that the formation of the phononic stop bands is caused by the interaction between the SW mode and the RW mode [44]. This gap owes its existence to a crossing (i.e., a degeneracy) of the modes in the dispersion relation for an analogous non-interacting system (i.e., an empty lattice). When the interacting system is considered, there is a repulsion of the modes to produce bonding and anti-bonding modes [45]. This is also referred to as an avoided crossing.

In the 610 MHz image (Fig. 7.5c) each segment of the RW and SW branches is pushed further out, whereas the openings in the RW and SW branches around $k_y = 0$ persist. On further increase of the frequency to 687 MHz (Fig. 7.5d), new features

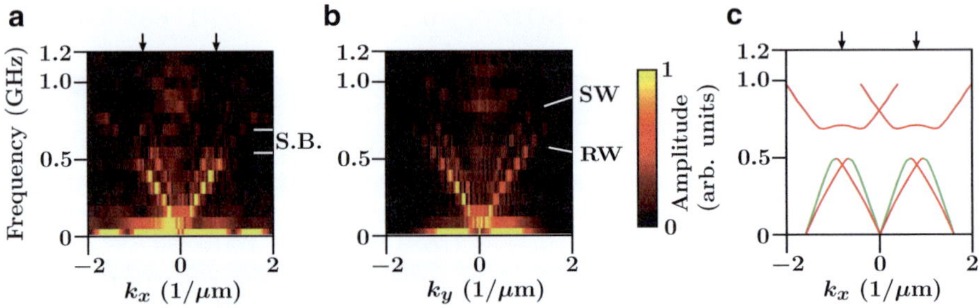

Fig. 7.6 Experimentally obtained dispersion relations for the one-dimensional phononic crystal (**a**) along k_x with $k_y = 0$ and (**b**) along k_y with $k_x = 0$. The *downward-pointing arrows* indicate the 1st BZ boundary. *RW*: Rayleigh waves, *SW*: Sezawa waves, and *S.B.*: stop band (phononic band gap). (**c**) Schematic diagram to help with mode identification. *Red lines*: Rayleigh-like waves (RW). *Green lines*: Sezawa waves (SW)

appear in the region where the opening is located at 610 MHz. This indicates that the directional phononic stop band persists at 534 MHz and 610 MHz but is closed at 458 MHz and 687 MHz.

By stacking constant frequency curves such as those shown in Fig. 7.5, one may construct a three-dimensional dispersion surface in (k_x, k_y, ω) space. Figure 7.6a shows a cross section of such a dispersion surface for a (k_x, ω) plane that represents the dispersion relation along k_x with $k_y = 0$, whereas Fig. 7.6b shows a cross section for a (k_y, ω) plane that represents the dispersion relation along k_y with $k_x = 0$. Each figure contains lines that originate at $k_{x,y} = 0$ with different slopes. The steeper one corresponds to the SW branch whereas the shallower one to the RW branch. As expected, for waves propagating along k_x (Fig. 7.6a), a directional stop band is observed around 0.5–0.7 GHz as a region with missing bright parts. Figure 7.6c shows the origin of the formation of the first stop band. It is formed where the Rayleigh and Sezawa branches cross, a position slightly removed from the 1st BZ boundary [44].

It is also useful to plot real-space images of the temporal Fourier transform $F(\mathbf{r}, \omega)$ of the experimentally obtained spatiotemporal data at constant frequencies. The first row of Fig. 7.7 shows images of the Fourier amplitude $A = |F(\mathbf{r}, \omega)|$ and the second row shows images of the acoustic phase $\psi = \arg F(\mathbf{r}, \omega)$. The spatiotemporal evolution of the acoustic field at a single frequency ω is given by

$$\mathrm{Re}[F(\mathbf{r}, \omega) \exp(-i\omega t)] = \mathrm{Re}[A \exp\{i(\psi - \omega t)\}]. \tag{7.26}$$

Thus, a snapshot of the acoustic field at $t = 0$ is given by $A(\mathbf{r}, \omega) \cos \psi(\mathbf{r}, \omega)$. Images representing this function are shown in the third row of Fig. 7.7. The amplitude images are useful to see the spatial extent of the modes at a given frequency, whereas the phase images are useful to see the wavelength of the modes. The snapshot images are related to both the phase and amplitude images, and show the instantaneous acoustic field at a moment in time during one cycle when exciting at a single frequency.

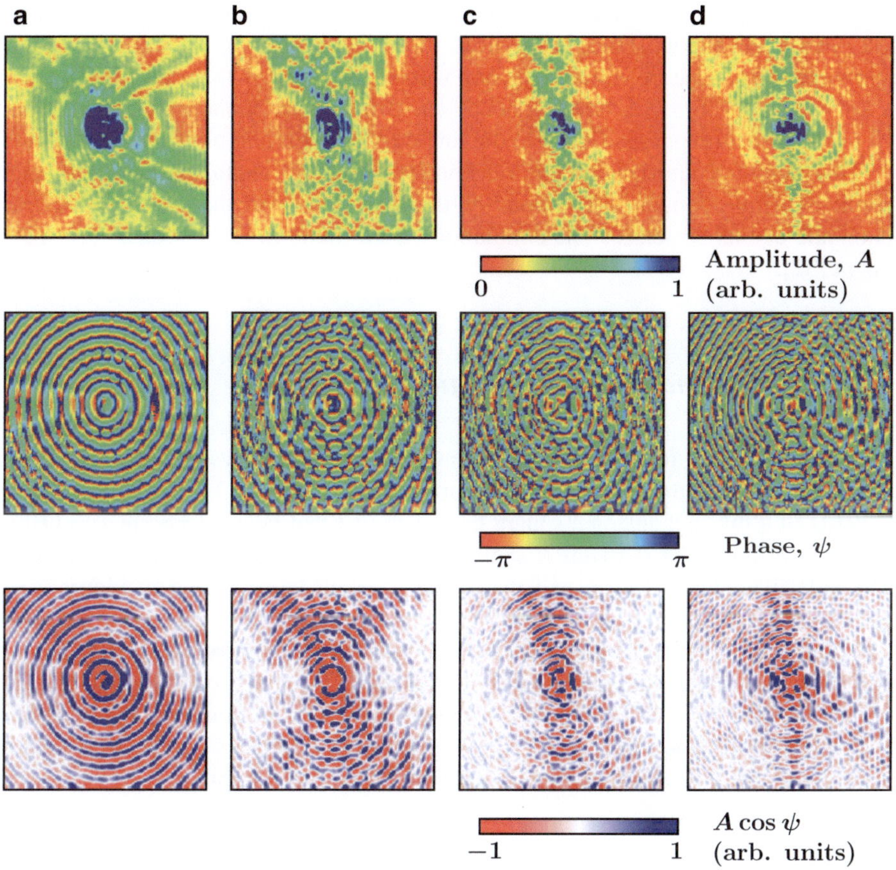

Fig. 7.7 Images of the temporal Fourier transform obtained from a set of time-resolved SAW images for the two-dimensional phononic crystal sample at several representative frequencies: (**a**) 458 MHz, (**b**) 534 MHz, (**c**) 610 MHz, and (**d**) 687 MHz. The *first row* shows images of the modulus of the Fourier amplitude. The *second row* shows the argument of the Fourier amplitude. The *third row* shows snapshot images of the out-of-plane surface velocity for a single frequency component. The imaged area is 150 μm × 150 μm

At 458 MHz (Fig. 7.7a) the wavelength for the RW branch for x-directed propagation is slightly longer than twice the structure periodicity ($2a = 8\,\mu$m), and the phase image displays an approximately circular pattern. At this frequency the excitation of equal-amplitude counter-propagating waves in the x direction would result in adjacent unit cells vibrating in antiphase [46]. For such x-directed propagation this mode is a pure standing wave that satisfies the Bragg scattering condition. At 534 MHz and above (Figs. 7.7b–d) the high amplitude region is more localized along the y direction. This is the result of a directional stop band around the k_x axis and the existence of portions of the constant frequency curves with a relatively low-curvature; since the group velocity is given by the gradient of $\omega_j(\mathbf{k})$, a low-curvature in a constant-frequency curve results in phonon focusing effects along the direction perpendicular to the low-curvature lines in plots like Fig. 7.5.

This temporal and spatiotemporal Fourier transform technique has also been applied to time-domain simulations of the wave fields in this one-dimensional phononic crystal structure. The simulations were found to agree well with experiment [32].

7.3.3 Two-Dimensional Phononic Crystals

The acoustic properties of a two-dimensional phononic crystal have also been investigated using the experimental setup described above [29, 31]. The sample consists of a two-dimensional square lattice of air-filled holes etched in a Si (100) substrate of thickness 0.46 mm using a deep reactive ion etching (DRIE) process. The hole diameter is 12 μm and the lattice constant is $a = 15$ μm. Figure 7.8a shows an optical micrograph of the sample over an area of 60 μm × 60 μm. The hole depth is 10.5 μm, as measured by scanning white light interferometry. The x and y axes of the crystal are in parallel with the [110] and [$\bar{1}$10] directions of the underlying Si substrate. The surface of the phononic crystal sample is covered with a 40 nm gold film to enhance the SAW generation efficiency by the pump light absorption and also to increase the reflectivity of the probe light.

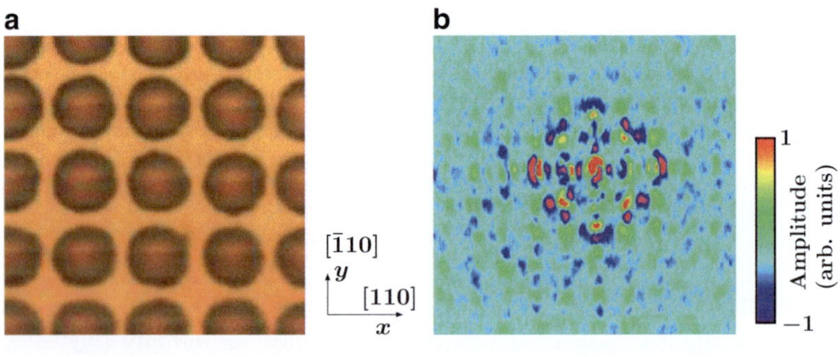

Fig. 7.8 (**a**) Optical micrograph of the two-dimensional phononic crystal fabricated in a Si substrate. The imaging area is 60 μm × 60 μm. The hole diameter is $D = 12$ μm, and the lattice constant is $a = 15$ μm. The substrate orientation and the definition of the coordinate system is also shown. (**b**) Experimentally obtained SAW image of the two-dimensional phononic crystal at 7.4 ns after excitation. The pump light is focused at the center of the image, and the imaged area is 150 μm × 150 μm

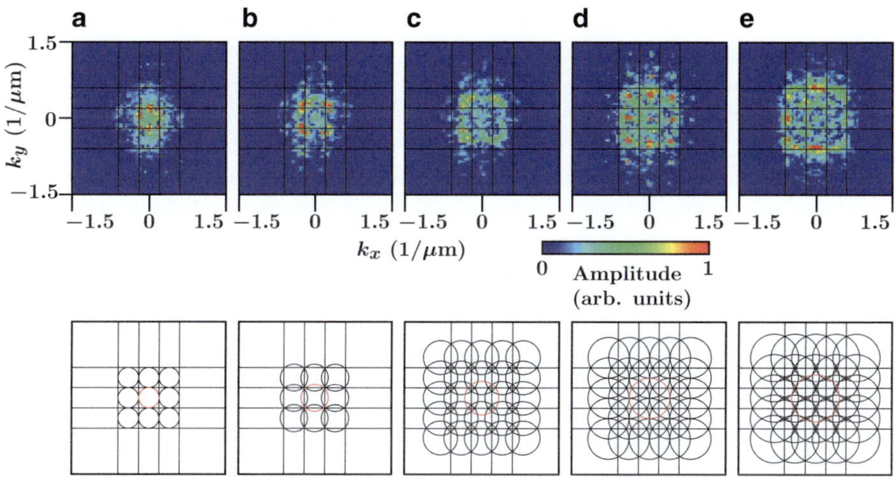

Fig. 7.9 *Upper row*: Amplitude (modulus) images of the spatiotemporal Fourier transform obtained from a set of time-resolved SAW images for a two-dimensional phononic crystal sample at several representative frequencies. (**a**) 153 MHz, (**b**) 229 MHz, (**c**) 305 MHz, (**d**) 382 MHz, and (**e**) 458 MHz. The *thin lines* represent the 1st BZ boundaries at $k_x, k_y = \pm 0.29\,\mu\mathrm{m}^{-1}$ and also at $k_x, k_y = \pm 3 \times 0.29\,\mu\mathrm{m}^{-1}$. *Lower row*: Mode identification diagram based on an empty lattice having the same periodicity as the two-dimensional phononic crystal at the frequencies corresponding to the figures in the *upper row*. *Circles in red* correspond to the modes in a non-periodic uniform medium whose SAW velocity is chosen so that the construction mimics the images in the *upper row*

Time-resolved two-dimensional images of the velocity of the out-of-plane surface motion are obtained using the apparatus described in Sect. 7.2 under similar conditions to those used for the one-dimensional phononic crystal. Figure 7.8b shows a SAW image at delay time 7.4 ns obtained over an area of $150\,\mu\mathrm{m} \times 150\,\mu\mathrm{m}$. The pump light is focused at the center of the image. The SAWs are strongly scattered by the array of holes.

The SAW modes can again be extracted by spatiotemporal Fourier transforms. The upper row of images of Fig. 7.9 shows the Fourier amplitude $|V(\mathbf{k}, \omega)|$ at some representative frequencies. Below each amplitude image a constant frequency circle to match the data for the x- and y-directed velocities at the corresponding frequency is depicted together with replicas shifted by reciprocal lattice vectors of the sample structure. These circles represent the dispersion relation of an empty lattice having the same periodicity as the phononic crystal, and help to understand the origin of the experimental Fourier amplitude images. The 1st BZ corresponds to the central square defined by the thin straight lines. In the periodic zone scheme for representation of the phononic band structure, any other such displaced square can be reduced to the 1st BZ by shifting an appropriate reciprocal lattice vector **G**.

The 153 MHz image (Fig. 7.9a) shows a constant frequency curve approximately circular in shape that is just touching the edges of the 1st BZ. This circle corresponds to Rayleigh-like modes with a velocity $\sim 4,800\,\mathrm{ms}^{-1}$ similar to that of Rayleigh waves ($\sim 5,100\,\mathrm{ms}^{-1}$) observed on bare Si (100) substrates [47]. Bloch harmonics are visible in the squares displaced from the 1st BZ (the central square) by $\mathbf{G} = (\pm 2\pi/a, 0)$ and $(0, \pm 2\pi/a)$.

In the 229 MHz image (Fig. 7.9b) the circle corresponding to the Rayleigh-like mode is larger, and parts of the circle touch the BZ boundary. One can see that the modes survive near the corner of the 1st BZ, but in the other parts of the zone there is an avoided crossing, as noted in the case of the one-dimensional phononic crystal, and a directional phononic stop band opens for propagation angles near the k_x and k_y directions.

In the 305 MHz image (Fig. 7.9c) the dominant high amplitude region lies outside the 1st BZ. This indicates that the modes at this frequency exist in the second or higher-order BZs in the extended zone scheme. The Fourier amplitude diminishes for the directions around k_x and k_y again owing to the persistence of the directional phononic stop band in these directions.

At 382 MHz (Fig. 7.9d) a large square array of dots are superimposed on a rounded-square shaped constant-frequency curve. These dots are located near the points where four adjacent circles centered at $\mathbf{k} = (\pm 2\pi/a, \pm 2\pi/a)$ meet in the empty lattice dispersion image. For this reason the dots have an enhanced intensity.

At 458 MHz (Fig. 7.9e) the constant-frequency curve forms a large square having significant amplitude in higher order BZs. As explained before in the context of the one-dimensional phononic crystal, the low-curvature of these constant-frequency curves again results in phonon focusing effects, in this case manifested by self-collimation along the directions ($\pm x$ and $\pm y$) perpendicular to the low-curvature lines in this plot, as will become evident later.

Figure 7.10a shows a cross section of the three-dimensional dispersion surface corresponding to a (k_x, ω) plane representing the dispersion relation along k_x with $k_y = 0$. It contains lines that originate at $k_{x,y} = 0$, which correspond to the RW

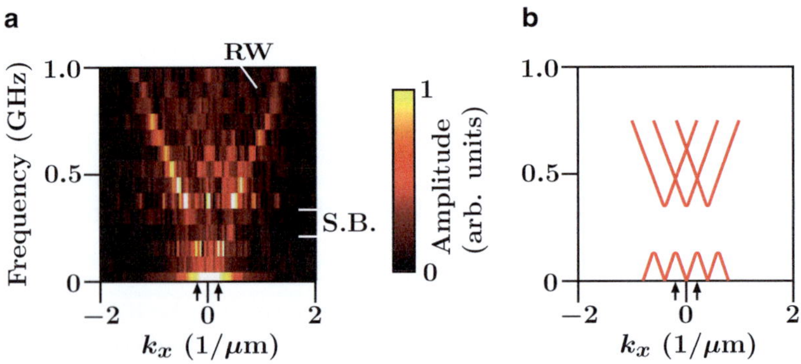

Fig. 7.10 (**a**) Experimentally obtained dispersion relations for the two-dimensional phononic crystal along k_x with $k_y = 0$. The *black arrows* indicate the 1st BZ boundary along k_x. *RW*: Rayleigh waves, and *S.B.*: stop band (phononic band gap). (**b**) Schematic diagram to help with mode identification

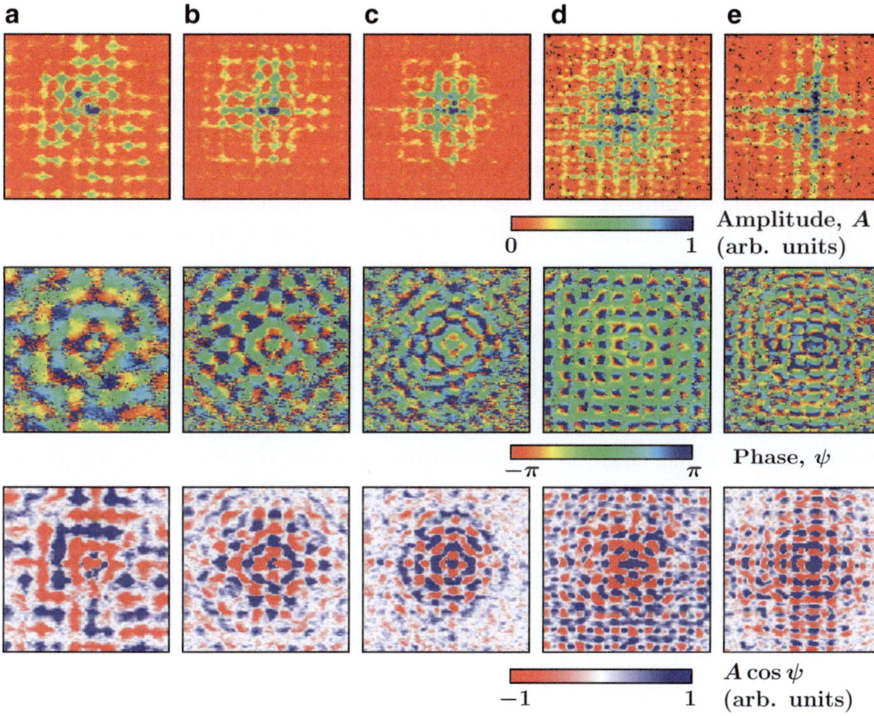

Fig. 7.11 Images of the temporal Fourier transform obtained from a set of time-resolved SAW images for the two-dimensional phononic crystal sample at several representative frequencies: (**a**) 153 MHz, (**b**) 229 MHz, (**c**) 305 MHz, (**d**) 382 MHz, and (**e**) 458 MHz. The *first row* shows images of the modulus of the Fourier amplitude. The *second row* shows the argument of the Fourier amplitude. The *third row* shows snapshot images of the out-of-plane surface velocity for a single frequency component. The imaged area is 150 μm × 150 μm

branch. The directional stop band is observed around 0.3 GHz, as expected, as a region with missing bright parts. Figure 7.10b shows a possible interpretation for the origin of the stop band formation. The lower limit of the first stop band is located where the acoustic branch crosses the 1st BZ boundary. Above the upper limit of this stop band Bloch harmonics are particularly evident near the frequency of 382 MHz where we noted a square array of dots in the constant frequency image.

Further information is again obtainable from the real-space images of the temporal Fourier transform $F(\mathbf{r},\omega)$ of the experimentally obtained spatiotemporal data. The first row of Fig. 7.11 shows images of the Fourier amplitude $A = |F(\mathbf{r},\omega)|$ and the second row shows the images of the phase $\psi = \arg F(\mathbf{r},\omega)$. The third row shows a snapshot of the acoustic field at $t = 0$ as calculated from $A(\mathbf{r},\omega)\cos\psi(\mathbf{r},\omega)$.

At 153 MHz (Fig. 7.11a) the phase images display an approximately circular pattern, as expected from the results in **k**-space. At 229 MHz the acoustic wavelength for x- or y-directed propagation is approximately twice the lattice constant of the sample ($2a = 30\,\mu\text{m}$). The maxima in amplitude occur at points approximately equidistant between four holes, and adjacent unit cells vibrate in antiphase. For the x and y directions this mode is a pure standing wave that satisfies the Bragg scattering condition. At both 229 MHz and 305 MHz (Figs. 7.11b, c) the high amplitude region is more localized near the excitation point. This is the result of either a low group velocity or a high damping rate, both characteristics being expected for modes in the vicinity of the phononic stop band. At 382 MHz, just above the stop band, the wavelength is approximately equal to the period of the lattice ($a = 15\,\mu\text{m}$), resulting in a characteristic square pattern in the phase image. Because of the nearly square shape of the constant-frequency surface at this frequency, to a good approximation this mode is a pure standing wave that satisfies the Bragg scattering condition for all propagation directions. At 458 MHz (Figs. 7.11e) the amplitude image shows a broad cross oriented parallel to the x and y axes, and the phase image again shows a square shape. This is caused by the self-collimation effects discussed in relation to Fig. 7.9e.

This temporal and spatiotemporal Fourier transform technique has also been applied to time-domain simulations of the wave fields in this two-dimensional Si structure. The simulations were found to agree well with experiment [31].

7.3.4 Summary: Capabilities and Limitations

With the experimental technique described in Sect. 7.2, the spatiotemporal evolution of the acoustic displacement field can be measured with a lateral resolution of $\sim 1\,\mu\text{m}$ and a temporal resolution of <1 ps. The sensitivity of the interferometer for vertical displacements is ~ 0.1 pm for a bandwidth of ~ 100 Hz. This sensitivity typically corresponds to a strain of 10^{-8} at $10\,\mu\text{m}$ SAW wavelength. The maximum area which can be scanned with the $4f$ system described in Sect. 7.2.3 is $\sim 500 \times 500\,\mu\text{m}^2$. A larger imaging area would be possible if the sample could be displaced as well as the pump spot position. The time constant of the lock-in amplifier is typically set to be ~ 10 ms, and the sampling rate of the lock-in amplifier output is ~ 100 Hz. So for a typical image of 200×200 pixels, the measurement of a single image requires 7 min. If a full set of experimental data consists of 40 images, the measurement time is ~ 5 h. The stability of the interferometer is sufficient to make such measurements fully automatic.

We have seen how the spatiotemporal Fourier transform of a set of time-resolved two-dimensional images for SAW propagation provides the dispersion relation. Inbuilt in this process is a fundamental limitation: The frequency step obtainable is equal to the laser repetition frequency. This restriction arises because periodic trains of pulses only contain frequency components at integer multiples of the pulse repetition frequency. The typical value of ~ 80 MHz is fairly coarse, and this is problematic if one needs to know the precise frequency range of the phononic stop band, for example.

The number of images covering the period of the laser repetition provides one limit on the maximum frequency obtainable. With 40 images covering a 12.5 ns repetition period (i.e., an 80 MHz repetition frequency), for example, the maximum frequency is 80 MHz $\times$ 40/2 = 1.6 GHz. A larger maximum frequency is advantageous to avoid aliasing effects from waves *excited* at acoustic frequencies beyond this maximum frequency but which are detected at a lower apparent frequency. Likewise the minimum achievable step Δk_x (in m^{-1}) along the k_x axis is determined by the edge length L_x (in m) of the imaged area as $\Delta k_x = 2\pi/L_x$ (and the same applies for Δk_y), leading, for a typical imaged region of $150 \times 150\,\mu m^2$ to $\Delta k_x = \Delta k_y = 4 \times 10^4\,m^{-1}$. The maximum value of k_x or k_y is defined by the dimensions of a single pixel. If these are chosen to be equal to the lateral spatial resolution $\sim$1 μm, one obtains $k_{max} \sim 2\pi \times 10^6\ m^{-1}$ (i.e., a minimum acoustic wavelength of $\sim$ 1 μm). Obviously setting the pixel size smaller than the spatial resolution is not advantageous.

Another limitation of the technique concerns the maximum frequency of the acoustic modes generated. Although the temporal resolution available for the generation of acoustic waves is in the sub-picosecond region, corresponding to a frequency bandwidth >1 THz, the minimum acoustic wavelength for surface wave generation is governed by the diffraction limit of the optical pump system. For a typical solid with a surface wave velocity $\sim$3 km s^{-1}, the maximum frequency generated for such a $\sim$1 μm spot size is $\sim$1 GHz. The above limitation is, however, only applicable to modes on acoustic branches: for the modes on optical branches around the Γ point in **k**-space, the frequency limit for generation increases to $\sim$1 THz or more. These modes, however, are non-propagating or have low group velocities. Comparable limitations apply to the probe beam spot size, again limiting the maximum detectable frequency to $\sim$1 GHz for the above-mentioned sound velocity.

The frequency of detection is also effectively limited by the choice of the time delay $\Delta\tau$ between the two consecutive probe pulses used for measurement. To avoid significant aliasing effects when sampling at two points in time and to access higher frequencies, $\Delta\tau$ should be chosen as small as possible. However, the lock-in amplifier output signal level decreases as $\Delta\tau$ decreases, and so a compromise is usually made to limit the value of $\Delta\tau$ to $\sim 1/(3f_{max})$, where f_{max} is the maximum required frequency for SAW detection. For f_{max} = 1 GHz one therefore chooses $\Delta\tau \sim$ 300 ps, for example. This compromise inevitably results in some distortion of the detected SAW field at the higher end of the required frequency spectrum.

To solve the limitation on spatial resolution or acoustic generation frequency, one could use near field optics, but the relatively low throughput of such optics would be a significant problem. To solve the limitation on frequency resolution, one could use a lower repetition frequency. This is feasible by use of a pulse picker or by varying the laser cavity length, for example. To solve the limitation set by the time delay $\Delta\tau$, one could use a different scanning system in which the two probe pulses are incident at different positions on the sample.

We have also seen how a single temporal Fourier transform of a set of time-resolved SAW images proved useful for the investigation of the mode patterns of the acoustic field at different frequencies. For samples such as phononic crystal waveguides or cavities which have a more complicated structure than uniform phononic crystals, this temporal Fourier transform analysis would allow the visualization of energy storage and leakage.

Of course simulation by the finite element methods (FEM) in the time domain or other numerical methods can circumvent a variety of experimental limitations [31, 32].

7.4 Conclusions

In conclusion, we have discussed how the acoustic properties of phononic crystals can be characterized using time-domain optical imaging of surface motion. In particular we have described the ultrafast optical pump-probe technique combined with interferometric spatial scanning as a means to generate and image gigahertz-order surface acoustic waves with a broad wavelength spectrum. Experimental results for two generic samples are presented corresponding to phononic crystals with one- and two-dimensional periodicity. We showed how the temporal Fourier transforms can be used to access the acoustic mode patterns at specific frequencies, and how spatiotemporal Fourier transforms can be used to access the dispersion relation and phononic stop bands. The role of Bloch harmonics, related to the scattering potential, in the spatiotemporal Fourier transforms was also analyzed mathematically, and we showed how they contribute to the experimental results in **k**-space.

Time domain optical techniques reveal a wealth of physics because both real space and **k**-space are easily accessible through Fourier analysis. We look forward to their continued use in a wide variety of studies on phononic crystals and phononic crystal devices over a broad range of frequencies.

References

1. R. Adler, A. Korpel, P. Desmares, IEEE Trans. Ultrason. Ferroelectr. Freq. Control **15**, 157 (1968)
2. G. Sölkner, A. Ginter, H.P. Graßl, Mat. Sci. Eng. A **122**, 43 (1989)
3. K.L. Telschow, V.A. Deason, R.S. Schley, S.M. Watson, J. Acoust. Soc. Am. **106**, 2578 (1999)
4. J.V. Knuuttila, P.T. Tikka, M.M. Salomaa, Opt. Lett. **25**, 613 (2000)
5. J.E. Graebner, B.P. Barber, P.L. Gammel, D.S. Greywall, Appl. Phys. Lett. **78**, 159 (2001)
6. A. Miyamoto, S. Matsuda, S. Wakana, A. Ito, Electron. Commun. Jpn. Pt. 2 **87**, 1295 (2004)
7. W.C. Wang, Y.H. Tsai, J. Vib. Control **12**, 927 (2006)
8. K. Kokkonen, M. Kaivola, Appl. Phys. Lett. **92**, 063502 (2008)
9. N. Wu, K. Hashimoto, K. Kashiwa, T. Omori, M. Yamaguchi, Jpn. J. Appl. Phys. **48**, 07GG01 (2009)
10. W.F. Riley, J.W. Dally, Geophysics **31**, 881 (1966)
11. J.W. Dally, Exp. Mech. **20**, 409 (1980)
12. Y.H. Nam, S.S. Lee, J. Sound Vib. **259**, 1199 (2003)

13. T. Saito, O. Matsuda, M. Tomoda, O.B. Wright, J. Opt. Soc. Am. B **27**, 2632 (2010)
14. M. Clark, S.D. Sharples, M.G. Somekh, J. Acoust. Soc. Am. **107**, 3179 (2000)
15. A.A. Maznev, A.M. Lomonosov, P. Hess, A.A. Kolomenskii, Eur. Phys. J. B **35**, 429 (2003)
16. J.L. Blackshire, S. Sathish, B.D. Duncan, M. Millard, Opt. Lett. **27**, 1025 (2002)
17. A. Neubrand, P. Hess, J. Appl. Phys. **71**, 227 (1992)
18. K. Nakano, K. Hane, S. Okuma, T. Eguchi, Opt. Rev. **4**, 265 (1997)
19. S.R. Greenfield, J.L. Casson, A.C. Koskelo, Proc. SPIE **4065**, 557 (2000)
20. Y. Sugawara, O.B. Wright, O. Matsuda, M. Takigahira, Y. Tanaka, S. Tamura, V.E. Gusev, Phys. Rev. Lett. **88**, 185504 (2002)
21. C. Glorieux, K. Van de Rostyne, J.D. Beers, W. Gao, S. Petillion, N.V. Riet, K.A. Nelson, J.F. sAllard, V.E. Gusev, W. Lauriks, J. Thoen, Rev. Sci. Instrum. **74**, 465 (2003)
22. J.A. Scales, A.E. Malcolm, Phys. Rev. E **67**, 046618 (2003)
23. D.H. Hurley, K.L. Telschow, Phys. Rev. B **71**, 241410(R) (2005)
24. T. Tachizaki, T. Muroya, O. Matsuda, Y. Sugawara, D.H. Hurley, O.B. Wright., Rev. Sci. Instrum. **77**, 043713 (2006)
25. R.E. Vines, J.P. Wolfe, A.V. Every, Phys. Rev. B **60**, 11871 (1999)
26. A. Sukhovich, L. Jing, J.H. Page, Phys. Rev. B **77**, 014301 (2008)
27. S. Peng, X. Mei, P. Pang, M. Ke, Z. Liu, Solid State Commun. **149**, 667 (2009)
28. D.M. Profunser, O.B. Wright, O. Matsuda, Phys. Rev. Lett. **97**, 055502 (2006)
29. D.M. Profunser, E. Muramoto, O. Matsuda, O.B. Wright, U. Lang, Phys. Rev. B **80**, 014301 (2009)
30. B. Bonello, L. Belliard, J. Pierre, J.O. Vasseur, B. Perrin, O. Boyko, Phys. Rev. B **82**, 104109 (2010)
31. O.B. Wright, I.A. Veres, D.M. Profunser, O. Matsuda, B. Culshaw, U. Lang, Chin. J. Phys. **49**, 16 (2011)
32. I.A. Veres, D.M. Profunser, O.B. Wright, O. Matsuda, B. Culshaw, Chin. J. Phys. **49**, 534 (2011)
33. C. Thomsen, J. Strait, Z. Vardeny, H.J. Maris, J. Tauc, J.J. Hauser, Phys. Rev. Lett. **53**, 989 (1984)
34. C. Thomsen, H.T. Grahn, H.J. Maris, J. Tauc, Phys. Rev. B **34**, 4129 (1986)
35. T. Fujikura, O. Matsuda, D.M. Profunser, O.B. Wright, J. Masson, S. Ballandras, Appl. Phys. Lett. **93**, 261101 (2008)
36. B. Perrin, B. Bonello, J.C. Jeannet, E. Romatet, Prog. Nat. Sci. **S6**, S444 (1996)
37. D.H. Hurley, O.B. Wright, Opt. Lett. **24**, 1305 (1999)
38. M. Nikoonahad, S. Lee, H. Wang, Appl. Phys. Lett. **76**, 514 (2000)
39. C.J.K. Richardson, M.J. Ehrlich, J.W. Wagner, J. Opt. Soc. Am. B **16**, 1007 (1999)
40. T. Dehoux, M. Perton, N. Chigarev, C. Rossignol, J.M. Rampnoux, B. Audoin, J. Appl. Phys. **100**, 064318 (2006)
41. Y. Sugawara, O.B. Wright, O. Matsuda, Appl. Phys. Lett. **83**, 1340 (2003)
42. Y. Sugawara, O.B. Wright, O. Matsuda, Rev. Sci. Instrum. **74**, 519 (2003)
43. D.H. Hurley, O.B. Wright, O. Matsuda, S.L. Shinde, J. Appl. Phys. **107**, 023521 (2010)
44. A.A. Maznev, O.B. Wright, O. Matsuda, New J. Phys. **13**, 013037 (2011)
45. C. Kittel, *Introduction to Solid State Physics*, 6th edn. (Wiley, New York, 1986)
46. A.A. Maznev, O.B. Wright, J. Appl. Phys. **105**, 123530 (2009)
47. R.E. Vines, M.R. Hauser, J.P. Wolfe, Z. Phys. B **98**, 255 (1995)

Chapter 8
Optical Characterization of Phononic Crystals in the Frequency Domain

Kimmo Kokkonen

8.1 Introduction

Need to understand the physics of wave motion and propagation in materials as well as the ongoing interest in developing microacoustic components for, e.g., signal processing applications has created a demand for versatile characterization methods.

Electroacoustic components in which an electrical signal is transformed into mechanical domain (into acoustic wave), for example, via piezoelectric effect, such as surface acoustic and bulk acoustic wave (SAW and BAW) devices, as well as microelectromechanical (MEMS) resonators, are typically studied in electrically excited test structures or in real device configurations. Research and development of electroacoustic devices has benefitted substantially from the advances in the electrical measurement technology, such as vector network analyzers. Much of the characterization therefore relies on measuring the electrical response of the structure and comparing that to simulations. However, even though the electrical behavior of a component can be thoroughly characterized, it still often leaves one wondering "what is really going on" in the device. Although the electrical measurement methods are well established, and elaborate and ingenious test geometries have been developed for various purposes, in the end they provide only secondary information of the wave motion and the underlying device physics.

Optical probing, on the other hand, enables direct, non-contact measurement of the vibration fields within the sample structure. The ability to measure and visualize the actual vibration fields yields valuable information on the wave physics and on the device performance. Due to the extensive history and successful application of laser interferometers for characterization of microacoustic devices, there are currently

---

K. Kokkonen (✉)
Department of Applied Physics, Aalto University School of Science and Technology,
P.O.Box 13500, 00076 Aalto, Finland
e-mail: kimmo.kokkonen@aalto.fi

A. Khelif, A. Adibi (eds.), *Phononic Crystals*, DOI 10.1007/978-1-4614-9393-8_8

a number of research groups that utilize an interferometer as an important part of their research facilities. Frequency-domain laser interferometric techniques are well suited for research of electroacoustic devices. In SAW research, for example, they can be used for imaging wave field amplitude distributions and for revealing possible sources of losses and unwanted responses, such as excitation of transverse modes in resonators, escaping acoustic beams and acoustic crosstalk in filters. In the case of BAW research, wave dispersion properties derived from vibration field measurements play an important role in device design and characterization. A good example is energy escaping from a resonator. Source of energy loss may be difficult to identify using electrical measurements alone (there seem to be unexpected losses, but why) whereas via optical probing the acoustic beams escaping from the structure can be readily seen. For best results, optical probing should be combined with electrical measurements to acquire the most complete picture of device behavior.

Many interferometric techniques developed for SAW, BAW, and MEMS devices are directly applicable to the research of phononic crystals (acoustic metamaterials, PnCs). It is a relatively new and rapidly evolving field, which is expected to benefit from laser interferometric measurements. Interferometric frequency-domain measurements allow to select the excitation frequencies of interest and to image the resulting wave fields. The technique is well suited to studying wave interaction with complex PnC structures.

This chapter gives an overview on the scanning laser interferometry (frequency domain technique) of surface vibrations. The detection principle and experimental setup are introduced in Sect. 8.2 and the capabilities and limitations of the technique discussed in Sect. 8.3. Application examples and measurement results are provided for SAW, BAW, and PnC samples in Sect. 8.4. Time domain techniques are discussed elsewhere in this book.

8.2 Scanning Laser Interferometry in Frequency Domain

In physical acoustics, it is important to be able to detect and characterize minute surface vibrations. Vibration amplitudes vary greatly depending on the type of the device and application, but in the case of SAW and BAW research for ultrasonic applications (f > MHz), the typical maximum amplitudes are of the order of a few nanometers. Therefore, in order to ensure a good signal-to-noise ratio (SNR) and to be able to study weak effects, the minimum detectable vibration amplitude should be around 1 pm, which is roughly 1/100 of the size of an atom. Furthermore, as operating frequencies range from MHz to several GHz, the detection method should have a flat frequency response to facilitate quantitative vibration analysis. In addition, since the devices often are sensitive to perturbations from the surroundings and have high Q-values, a non-contact method is called for.

Laser interferometry offers a non-contact optical method which can be utilized to detect such vibrations. In the conceptually simple case of the homodyne Michelson interferometer (see Fig. 8.1), the interferometer consists of a measurement arm

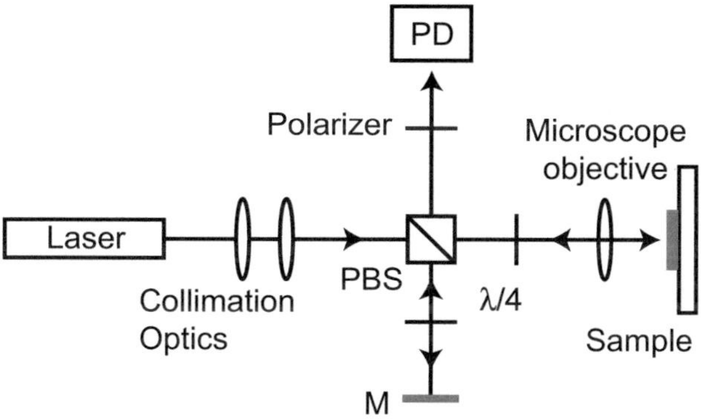

Fig. 8.1 Schematic presentation of a scanning homodyne Michelson laser interferometer. The laser beam is collimated and then split into the two interferometer arms by a polarizing beam splitter (PBS). The beam going straight through the PBS is the measurement arm and the deflected beam is the reference arm of the interferometer. The polarization state of the measurement beam is transformed from linear to circular by the quarter-wave plate ($\lambda/4$). The beam is focused to a spot on the sample surface, from which it is reflected back. The back reflected beam propagates through the quarter-wave plate again, to facilitate correct beam steering at the PBS. The reference beam is also transformed into circularly polarized, before being reflected back from a reference mirror (M). The two beams are combined at the PBS, with orthogonal linear polarization states, after which they propagate through a polarizer to a photodetector (PD). The photodetector detects the resulting varying light intensity due to interference (see Fig. 8.2)

and a reference arm. The sample with the vibrating surface is placed in the measurement arm. The surface vibration leads to a variation of the optical path length of the measurement arm, resulting in an optical phase difference with respect to the reference beam. The laser interferometer converts the minute optical phase modulation between the two laser beams into a variation of light intensity via interference of the beams. The resulting intensity signal can then be measured with a photodetector. In this way, the length reference for the measurement is derived from the well-defined laser transition, in the case of red HeNe laser, $\lambda = 632.8$ nm.

There are many ways to take advantage of the interference phenomenon in order to enable extremely sensitive measurements, and hence, there are a number of different laser interferometer setups. For a review of optical detection of ultrasound, including not only interferometric methods, see, e.g., [1–3].

In the following, we concentrate on scanning laser interferometry with sinusoidal electrical excitation of the vibrations in the sample and their detection in the frequency domain. This concept has been very successful in SAW and BAW device research, where the samples are excited electrically and the resulting wave fields within the samples can be measured with an interferometer. As the excitation frequency can be precisely controlled, the frequency domain operation allows detailed study of mechanical resonances and modes arising in the test structures. Furthermore, by limiting the detection bandwidth, the SNR can be enhanced, allowing the detection of smaller vibration amplitudes, even less than 1 pm.

8.2.1 Homodyne Detection

Homodyne interferometer setups used in physical acoustics research are typically based on either Michelson or Mach-Zender interferometers. A Michelson type homodyne interferometer is depicted in Fig. 8.1. A simplified schematic presentation of the homodyne interferometer signal due to a change in optical path length, together with the response to a surface vibration, is presented in Fig. 8.2. Here the signal is considered to be devoid of noise or any other unidealities.

For simplicity, consider a one-dimensional case, where two coherent monochromatic waves of equal optical frequencies (with the same polarization state) are superimposed. The waves have intensities I_1 and I_2 and optical phases θ_1 and θ_2. Denoting $\theta = \theta_2 - \theta_1$, we can write

$$I(\theta) = I_1 + I_2 + 2\sqrt{I_1 I_2}\,\cos(\theta). \tag{8.1}$$

This interference equation shows that the resulting light intensity $I(\theta)$ depends not only on the two intensities I_1 and I_2, but also on the phase difference θ between the two constituent waves. The phase difference in turn depends on the optical path lengths traveled by the two beams. The last part of Eq. (8.1), the interference term,

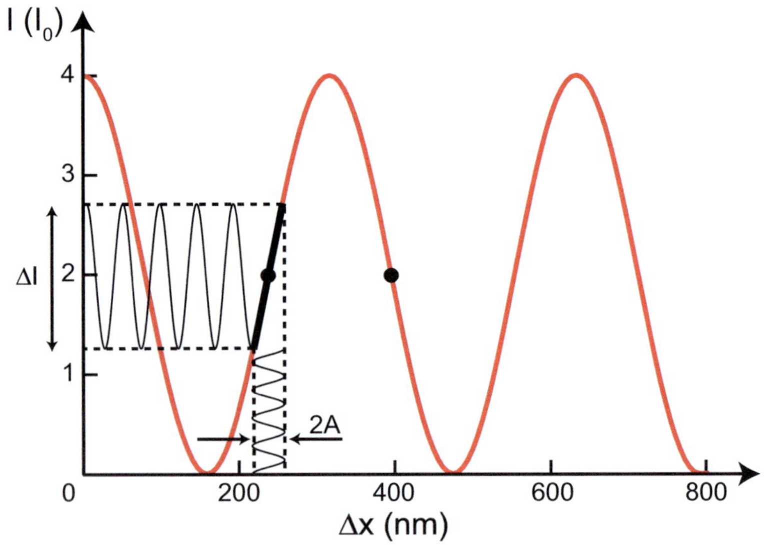

Fig. 8.2 Schematic presentation of a homodyne laser interferometer signal due to a moving object (either sample, or reference mirror), assuming ideal conditions (equal optical powers, perfect interference). In ideal case, the optimum operation point of the interferometer offering maximum sensitivity and linearity is found at the so-called quadrature-point, marked with a *black dot*. In this operation point, the smallest object movement with an amplitude A results in a largest change in the detected light intensity I

may be either positive or negative, corresponding to a constructive or destructive interference of the two waves. In the special case of interference between two waves with equal intensities ($I_1 = I_2 = I_0$), the resulting intensity I varies between 0 and $4I_0$, depending on the optical phase difference between the two waves, see Fig. 8.2. The phase difference θ carries the information of the varying optical path length of the interferometer and, hence, ultimately that of the surface vibration to be probed. Therefore, the interference of the two waves serves to convert a small optical path length difference between the two beams into an intensity variation measurable with a photodetector.

Assuming a sinusoidal surface displacement $A\sin(2\pi f_{\text{vib}}t + \psi_{\text{vib}})$, where A is the amplitude and ψ_{vib} is the phase of the surface vibration under study, the resulting time-dependent variation of the optical phase difference can be written as

$$\varphi(t) = \phi + \frac{4\pi A}{\lambda}\sin(2\pi f_{\text{vib}}t + \psi_{\text{vib}}), \tag{8.2}$$

where f_{vib} is the frequency at which the sample is electrically excited (assuming a linear mechanical response), λ is the wavelength of the laser light, and the term ϕ is a slowly varying (compared to f_{vib}) phase term that represents the optical phase difference between the two beams due to changes in ambient conditions. Substituting Eq. (8.2) into θ in Eq. (8.1), and in the limit of small vibrations (for linear detection, typically $A < 10$ nm), the homodyne detection equation can be approximated as

$$\begin{aligned} I(t,\phi) &= I_1 + I_2 + 2\sqrt{I_1 I_2} \\ &\quad \times \left[\cos(\phi) + \frac{4\pi A}{\lambda}\sin(2\pi f_{\text{vib}}t + \psi_{\text{vib}})\sin(\phi)\right]. \end{aligned} \tag{8.3}$$

From Eq. (8.3), it is evident that in homodyne detection the optical phase difference ϕ between the two beams has to be controlled in order to detect the signal due to the surface vibration. This can be realized by actively stabilizing the interferometer (see, e.g., [4]) into a so-called quadrature point, where $\sin(\phi) = 1$, see Fig. 8.2. The quadrature point offers both the highest sensitivity and the best linearity in the detection of small vibrations. The role of stabilization is to compensate for any slow drifts between the optical path lengths of the two interferometer arms, typically caused by air flows, thermal expansion, or simply by differences in the heights of the features on the sample surface. The stabilization counteracts this change, for example by displacing slightly the reference mirror, in order to ensure operation at the quadrature point. Stabilized setups enable phase sensitive detection of the vibrations, e.g., via sample excitation and photo detector signal measurement with a vector network analyzer. Acquisition of absolute amplitude data requires an additional measurement to quantify and to compensate for gain drift due to differences in optical reflectivity of the sample surface, drift in the quality of interference, etc.

8.2.2 Heterodyne Detection

In heterodyne interferometers one of the laser beams is optically frequency shifted with respect to the other using, e.g., an acousto-optical modulator (AOM). A schematic view of a possible realization is depicted in Fig. 8.3.

When the surface-vibration amplitude (A) is small compared to the laser wavelength λ, the interference term of the heterodyne detection equation may be expanded as [5]

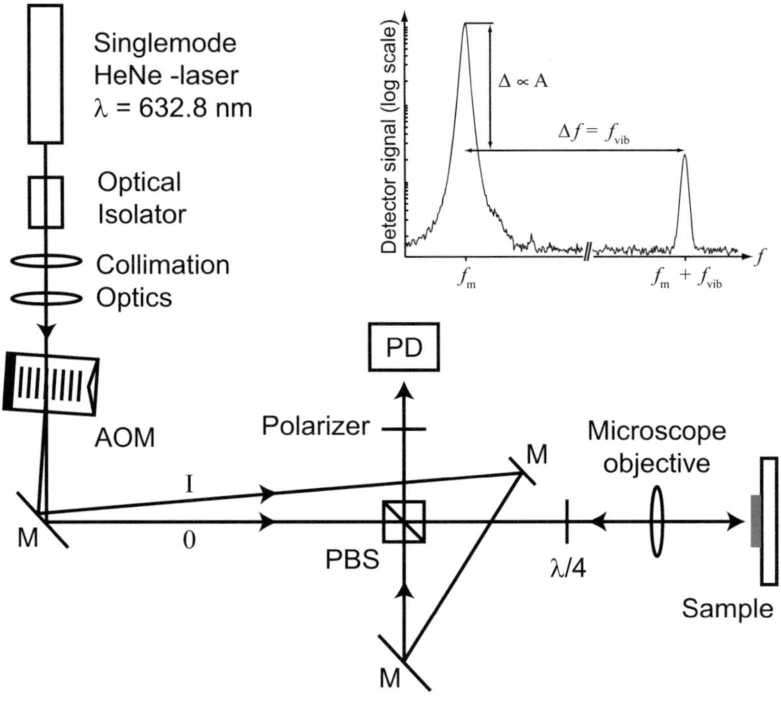

Fig. 8.3 Schematic presentation of a scanning heterodyne laser interferometer [6]. The heterodyne signal detected from the photodetector is shown schematically as an *inset*. The acousto-optical modulator (AOM) is used to split the laser beam into a measurement beam (0) and a reference beam (I). The reference beam is optically frequency-shifted, and the frequency difference of the two beams will result in the modulation peak in the detected frequency spectrum. Note that the signal detected due to the surface vibration is also frequency shifted, hence diminishing the electromagnetic feedthrough. Furthermore, the use of a single-mode HeNe laser eliminates longitudinal mode beating of the laser light, thereby ensuring a clean frequency spectrum

$$
\begin{aligned}
I_{12}(t) = 2\sqrt{I_1 I_2}\big[&\cos\left(2\pi f_{\mathrm{m}} t + \varphi_0\right) \\
&+ \frac{2\pi A}{\lambda}\{\cos\left(2\pi\left(f_{\mathrm{m}} + f_{\mathrm{vib}}\right)t + \phi + \varphi_0\right) \\
&- \cos\left(2\pi\left(f_{\mathrm{m}} - f_{\mathrm{vib}}\right)t - \phi + \varphi_0\right)\}\big].
\end{aligned} \tag{8.4}
$$

The error in the calculated amplitudes due to this approximation is less than 1 %, when $A \leq 10\,\mathrm{nm}$ and $\lambda = 632.8\,\mathrm{nm}$. In Eq. (8.4), ϕ is the phase of the surface vibration, and φ_0 denotes the slowly varying phase difference between the two laser beams due to ambient conditions.

When observed in the frequency domain, the detection signal consists of a modulation peak (f_{m}) and two satellite peaks ($f_{\mathrm{m}} \pm f_{\mathrm{vib}}$). Typically, when measuring actual samples, only the modulation peak and the upper ($f_{\mathrm{m}} + f_{\mathrm{vib}}$) satellite peak (the signal peak) are detected (for frequency domain presentation, see the inset in Fig. 8.3). The modulation and signal peaks (two frequencies) are detected simultaneously and the absolute amplitude of the surface vibration can be acquired from their amplitude ratio. Also, by comparing the phases of the two signals, the phase of the surface vibration is acquired, and any slow variations in the optical path lengths cancel out. The ability to measure the absolute amplitude of the surface vibration also provides immunity to variations in the local optical surface reflectivity of the sample. Furthermore, the heterodyne detection diminishes radio frequency (RF) leakage problems as there is a frequency offset between the detected frequency ($f_{\mathrm{m}} + f_{\mathrm{vib}}$) and the frequency at which the sample is driven (f_{vib}).

8.3 Capabilities and Limitations of Scanning Laser Interferometers

Optical probing provides a means for a non-contact measurement with lateral resolution determined by the laser beam spot size on the sample surface (in the case of scanning laser interferometers), ultimately limited by laser physics. Typical minimum spot size on the sample is $\sim 1\,\mu\mathrm{m}$ (with lasers operating in the visible range $\lambda \sim 400$–$700\,\mathrm{nm}$). The small spot size (high lateral resolution) results in short depth of focus, creating a need for accurate focusing onto the sample structure.

In the interferometric detection of vibration amplitudes, the relatively short wave length of the laser light itself is used as a reference for the measurement. However, characterization of surface vibrations in microacoustic components and in PnC structures is demanding, because typical vibration amplitudes range from less than a picometer to a few nanometers. For a lossless homodyne interferometer with perfect interference, optical power of 1 mW, laser (HeNe) wavelength of 632.8 nm and a detector quantum efficiency of 10 %, the theoretical detection limit according to [7] is $A_{\mathrm{min}} \sim 6 \times 10^{-6}\mathrm{nm}/\sqrt{\mathrm{Hz}}$. Therefore, narrow-band frequency domain detection allows to measure sub-picometer vibration amplitudes in setups in which the sample is excited harmonically with a single frequency. Time-independent device operation is assumed.

Laser interferometric probes can be roughly divided to scanning and full-field systems. Both of these methods have advantages and disadvantages. In general, the full-field method is significantly faster (even $\sim 100\times$) than scanning, but typically the minimum detectable amplitude is not nearly as good as what can be achieved with a scanning probe. The measurement speed of a frequency domain scanning probe utilizing a frequency sweep is a compromise between a fast sweep and minimum detectable vibration amplitude. The overall measurement speed can be increased by choosing a larger detection bandwidth, but at the expense of SNR. Furthermore, scanning probes allow to choose freely the scanning steps (x and y), as well as the scan area. In contrast, the full-field probe is typically limited to a set of magnifications resulting from changing microscope objectives, that is, the lateral resolution is the same in x and y directions, and there is a connection between the lateral resolution and the scan area due to the size of the camera's chip (also pixel limitation). However, the full-field systems can, at best, achieve almost video-rate measurement speed, and hence, almost instant characterization of the wave motion.

Several authors have published interferometer systems with minimum detectable amplitudes in the picometer-range, see, e.g., [5, 6, 8–14]. Of these, two extremes can be found: The early work of Monchalin et al. [5], a quantum noise limited heterodyne interferometer, is to be noted for a particularly small minimum detectable amplitude limit of 0.06 pm, with a laser power $P_{\mathrm{opt}} = 5\,\mathrm{mW}$ and a detection bandwidth of 1 Hz. On the other hand, for great measurement speed, the work by Telschow et al. [13], a full-field interferometer utilizing dynamic photorefractive holography, is to be noted. Due to the full-field imaging, the interferometer has a image acquisition speed of 18 fps, leading to a measurement speed of 275,000 pnt/s, with a minimum detectable amplitude of 100 pm.

The non-contact nature and flat frequency response of laser interferometers facilitate surface vibration measurements from DC to several tens of GHz, with the frequency range typically limited by the detection electronics used. Typically, only the vibration component along the surface normal is detected, and therefore the detection of horizontal movement such as shear-SAWs is possible indirectly, through their out-of-plane vibration component.

Detection of the smallest vibration amplitudes involves low-level RF-signals and signal processing. When characterizing modern micro-acoustic devices with operating frequencies ranging from hundreds of MHz to several GHz, electromagnetic (EM) feed through has to be minimized, since it can severely mask the actual measurement signal. All interferometric characterization systems are more or less vulnerable to EM feed through, especially when detecting at frequencies beyond 1 GHz. Homodyne systems, in which the sample is excited at the same frequency at which the extremely small signal due to the surface vibration is to be detected, are naturally the most vulnerable to this kind of interference. In contrast, heterodyne systems, when designed and implemented correctly, alleviate much of this problem, as the detected signals are offset in frequency from the signal that the sample is driven at.

8.4 Applications to Phononic Crystals

Laser interferometric frequency domain techniques are especially suitable for studying wave behavior in electrically excited micro- and nanoacoustic structures, such as SAW, BAW, and MEMS devices. The possibility to realize extremely sensitive setups, with detection limits in the sub-picometer range, opens up a wealth of possibilities both for pure science and for applied research. Typically, in SAW and BAW research, the maximum vibration amplitudes are below a few nanometers. In order to characterize small effects, like weak escaping SAW beams, or to detect weak vibration modes in BAW devices, detection limit on the order of picometers suffices. The sub-micron lateral resolution obtainable with laser interferometry is also sufficient for most applications. The ability to select the (single) excitation frequency at will allows to study in detail complicated device frequency responses featuring rapid changes, such as resonances or transition from passband to stopband.

Electrically excited micro- and nanoacoustic devices are widely used for signal processing and sensing functions, with one major application being (for SAW and BAW) RF-filters utilized in mobile communications systems. Interferometric characterization of acoustic wave fields has proven to be useful technique in the research and development of SAW and BAW devices. Research on phononic crystals in connection with microacoustics is an emerging field, and is expected to benefit from the interferometric techniques developed for SAW, BAW, and MEMS devices. Electrical measurement of complex micro-structures, such as PnCs, is not always enough to give a complete picture of the acoustic operation of the device. Characterization of the actual wave interaction with the PnC is central in order to increase understanding of the physics involved and to develop better modeling tools.

In the following, some wave field characterization possibilities are presented with measurement examples. The ability to extract detailed information on waves propagating into different directions is illustrated with a measurement example of a random SAW field unveiling material anisotropy. Measurement example of SAW interaction with a 2D PnC at $f \sim 200\,\text{MHz}$ showcases the detailed information available from wave field measurements. Two measurement examples of BAW interaction with a thin-film 1D PnC are presented, illustrating the characterization of plate-wave dispersion, resonances, and transfer function of an acoustic mirror in detail.

8.4.1 Surface Acoustic Waves in Micro-Structured Acoustic Metamaterials

Scanning laser interferometry is an established technique to characterize wave fields in SAW devices. Most notably, laser interferometry has led to a discovery of a new significant SAW loss mechanism and has contributed to advancement of the theory to explain it [15]. In addition to detecting the SAWs at the device surface, the technique can also be used to study BAW radiation from SAW devices. As an example, BAWs radiated by a low-loss SAW resonator were characterized at ~1 GHz [16]. In this case, the BAWs reflecting from the back-surface of the substrate were detected (on the back-side) and the measured radiation angles were compared to simulations to identify wave modes. Laser interferometric imaging of SAW wave fields suits also well to characterization of escaping acoustic energy because acoustic beams leaving, e.g., a resonator are directly visible, see [17]. In addition to being able to observe acoustic losses, theoretical predictions of radiation angles can be validated. Interferometric measurements can often be used not only to characterize a device but also to confirm its operation principle. Furthermore, correct modeling of the device behavior can be validated, as an example, characterization of a double-resonance SAW filter is presented in [18].

All of the above-mentioned capabilities are expected to apply to the characterization of not only conventional SAW devices, but to SAW interaction with micro-structured acoustic metamaterials as well.

8.4.1.1 Extracting Propagation Information from Random SAW Field

Laser interferometric phase-sensitive measurements enable the separation of waves traveling into different directions, analysis of transmission and reflection, study of wave dispersion, and characterization of wave propagation in anisotropic materials. As an example, consider SAW propagation in an anisotropic medium [19]. This example illustrates the possibilities of deriving quantities from high-quality interferometer data. In the experiment, SAW beams are scattered from micro-structures to generate a random wave field in the measurement area (see Fig. 8.4d for a schematic presentation). The measured wave field (see Fig. 8.4a, b for amplitude and phase fields, respectively) features very low amplitudes (< 20 pm) and seems random, almost like noise. When the wave content is further analyzed in the wavevector space, utilizing Fourier transform (FT) methods, the SAWs traveling to all directions form a closed contour in the wavevector, or slowness, space, see Fig. 8.4c. Substrate anisotropy is manifested in the shape of this contour deviating from a circle. The measurement may therefore be utilized in the acquisition of the elastic tensor of the material. Furthermore, the "filled" interior of the contour implies that the structure also scatters SAWs into BAWs in a relatively uniform way. For further information, see [19].

8.4.1.2 Two-Dimensional Phononic Crystal for 200 MHz SAWs

Recently, there has been a growing research interest to design PnCs for SAWs with characteristic feature sizes in the micrometer scale. As an example, consider a 2D PnC for SAWs, with a bandgap at around 200 MHz [20]. The sample is fabricated onto a standard, piezoelectric, 500 μm thick Y-cut $LiNbO_3$ substrate and it features a PnC in a delay line configuration, similar to that presented in [21]. The PnC structure

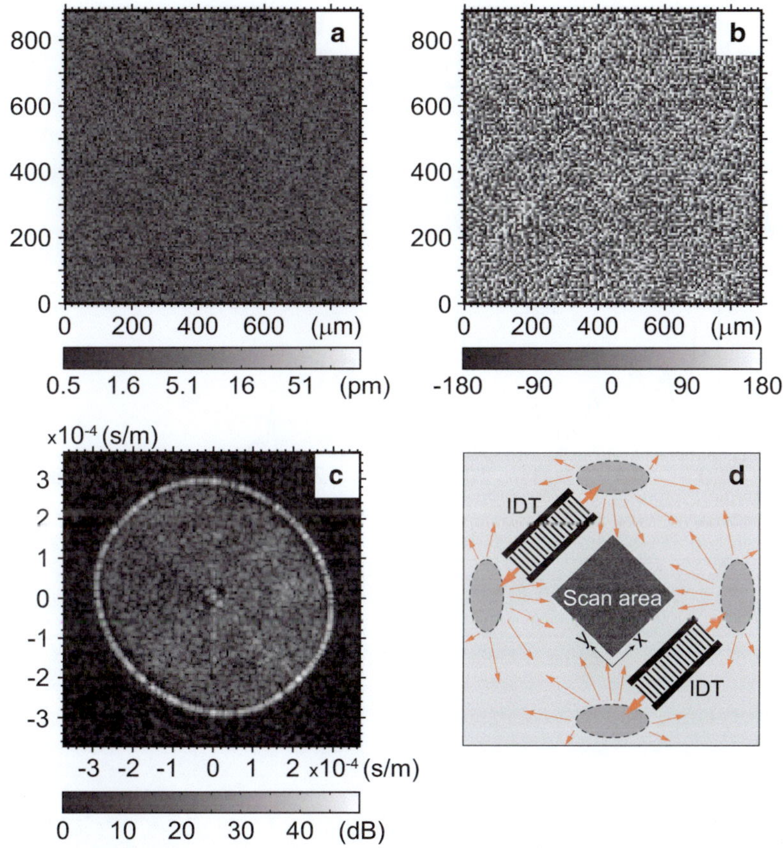

Fig. 8.4 Random scattering of SAWs unveils substrate anisotropy. Measured absolute amplitude and phase (in degrees) of the surface vibration field at 223 MHz are presented in (**a**) and (**b**), respectively. The Fourier transform of the measured wave field presents the wave content in the slowness (inverse phase velocity) space. The SAW slowness curve is seen as the continuous outer boundary and the waves scattered into the bulk as the filling of the disk interior

between the two interdigital transducers (IDTs) is a square lattice of 10 μm deep holes with a diameter of 9.4 μm and a pitch of 10 μm, resulting in a filling fraction of 69 %. Schematic picture and an SEM image of the sample are presented in Fig. 8.5a, b, respectively. Electrical measurements determined that the full band gap extends from 200 up to 230 MHz.

Here we consider experiments in the ΓM direction. For electrical characterization of wave transmission through the PnC, see Fig. 8.5c. The electrical measurements show the existence of the theoretically predicted band gap (BG), some beginning of re-transmission after the BG, but most of the frequency range above the BG shows no wave transmission. This is contrary to the theoretical predictions.

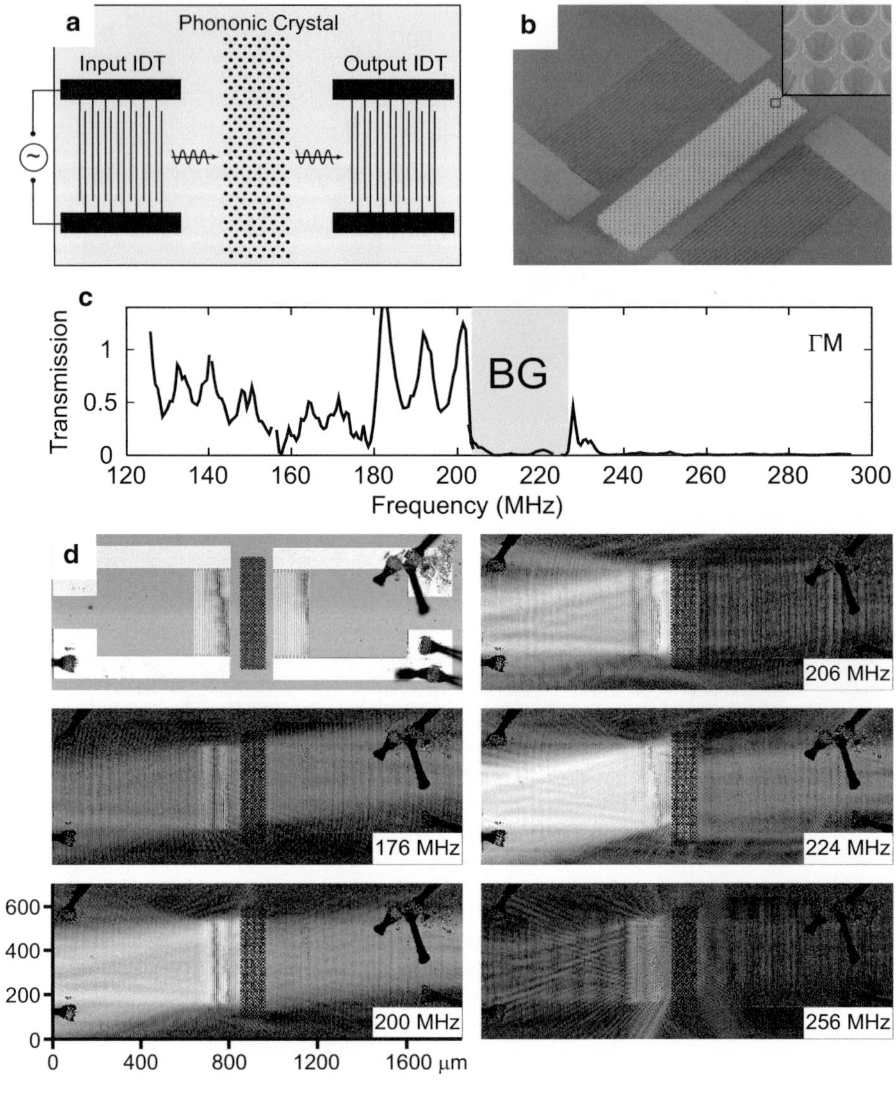

Fig. 8.5 Selected results from a PnC structure (similar to that presented in [21]). (**a**) schematic presentation of a SAW delay line with a PnC structure in between the two IDTs, intended to allow for electrical characterization. (**b**) SEM image of the component showing both IDTs and the PnC hole structure in between. The *inset* provides a close-up view of the hole grid etched into the lithium niobate single crystal substrate. (**c**) SAW transmission characteristics determined via electrical measurements. The band gap (BG) frequency range is indicated with *gray shading*. (d) Light power image and selected wave field amplitude scans to illustrate the acoustic behavior in the device. Below the BG, the wave amplitudes, on both sides of the PnC structure, are similar, indicating a good transmission. In contrast, within the BG frequency range, the PnC is very reflective, resulting in a strong standing wave pattern on the *left side* of the PnC. This behavior is accompanied by a low transmission leading nearly to an absence of wave amplitude on the other side of the PnC structure. Above the BG, re-transmission is expected, but instead, the PnC acts as an anisotropic diffraction grating, and the incoming waves are scattered by the PnC [20]

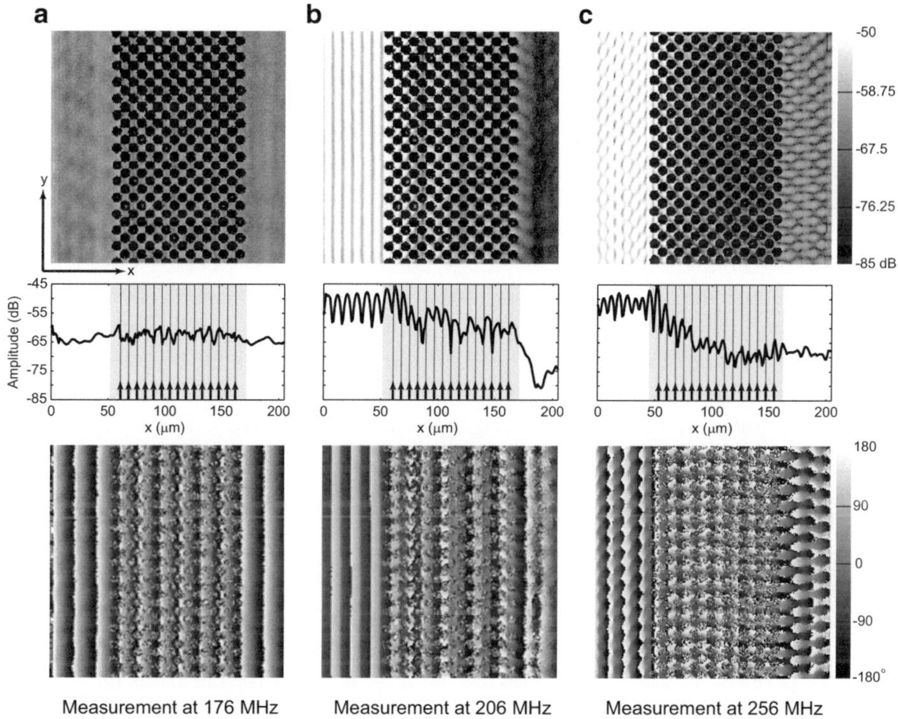

Fig. 8.6 Selected detailed wave field measurements for the same PnC structure as in Fig. 8.5, with IDTs operating (**a**) below the band gap, at 176 MHz (**b**) within the band gap, at 206 MHz, and (**c**) above the band gap, at 256 MHz. The acoustic band gap exists for frequencies between 200 and 230 MHz. In each case, the *left-hand side* IDT is emitting while the *right-hand side* IDT is receiving. The PnC structure is overlayed on the amplitude images to indicate the locations of the holes. The amplitude data of the *first row* are averaged in the y-direction and the resulting line profiles of the averaged amplitude along the wave propagation direction (x) are presented as graphs in the *second row*. The location of the PnC structure is marked with *gray area*. Due to the PnC geometry, filling fraction, and scan step used, there are x-coordinate values at which only few good data points are available for the averaging. These locations are marked on the graphs by *arrows* and *lines*. The phase data is presented on the *third row*. For more information and color figures, see [20]

Interferometric measurements can be used to study the mechanical behavior to find the origin of the contradiction. A large area scan ($1,850 \times 700\,\mu\text{m}^2$) was performed to provide an overview of the wave generation by the input IDT and the wave interaction with the PnC structure, for selected scans, see Fig. 8.5d. Detailed measurements were performed with a lateral scanning step of $1\,\mu\text{m}$, covering the space between the two IDTs, corresponding to an area of $205 \times 205\,\mu\text{m}^2$. These scans reveal the SAW interaction with the PnC-structure, see Fig. 8.6. The amplitude data has been averaged at each frequency, orthogonally to the wave propagation direction, to provide a line profile of the wave amplitude along the propagation path (x).

At frequencies below the band gap frequency range ($f < 200$ MHz), the SAWs pass through the PnC lattice with relatively undisturbed phase fronts, characteristic for a traveling wave, and without significant attenuation. For an example, see the data at $f = 176$ MHz in Figs. 8.5d and 8.6a. Furthermore, no significant reflection, scattering, or other losses are observed, indicating that the PnC does not significantly interfere with the wave motion. The SAW beam is, however, displaced slightly after propagating through the PnC, and therefore part of the wave misses the output IDT (see the measurement at 176 MHz, Fig. 8.5d).

At 200, 206, and 224 MHz (Fig. 8.5d), the input IDT is working closer to its center frequency, resulting in stronger emission of SAWs. In the band gap frequency range, the PnC structure is very reflective, resulting in a strong standing wave pattern seen on the left side of the PnC in Figs. 8.5d and 8.6b, between the transmitting IDT and the PnC. This behavior is accompanied by a low transmission leading nearly to an absence of wave amplitude on the right side of the PnC structure. The line profile in Fig. 8.6b shows the strong standing wave behavior on the left side of the PnC and the amplitude diminishing to the right side of the PnC (an attenuation of ~ 20 dB). The phase fronts of the wave field still correspond to a plane wave.

Above the band gap frequency range, re-transmission of SAWs is expected. The PnC is observed to start transmitting the SAWs again weakly at 224 MHz (see Fig. 8.5d). Note that the SAW beam is displaced after passing through the PnC similarly, but in opposite direction than at 176 MHz. At higher frequencies, however, the PnC is observed to scatter the wave field at angles different from the normal incidence. This results in scattered SAW beams (see the X-pattern at 256 MHz in Fig. 8.5d) and in a lobe structure visible both in the measured amplitude and phase fields (see Fig. 8.6c). Despite the scattering, the PnC does not provide as significant an attenuation to the wave field as at 206 MHz within the band gap. The PnC acts as an anisotropic diffraction grating above a certain threshold frequency, which explains the observed scattering. The scattering leads to the low transmission value observed in the electrical characterization, since most of the SAWs miss the output IDT altogether, and furthermore, there is a mismatch between the IDT geometry and the SAW wavefronts.

The PnC structure might also scatter some waves into the bulk, e.g., due to the conicity and finite depth of the etch holes. With proper sample preparation, the backside of the sample wafer can also be interferometrically imaged and the wave scattered into bulk impinging onto the back surface would be visible.

Besides confirming the existence of the theoretically predicted band gap for SAWs (extremely reflective in 200 MHz $< f <$ 224 MHz), the measurements reveal SAWs passing through the PnC almost unaffected below the band gap (at 176 MHz) and strongly reflected within it (at 206 MHz). Above the band gap, scattering to higher diffraction orders is observed, an effect difficult to characterize and quantify by electrical measurements alone.

8.4.2 *Bulk Acoustic Waves in Acoustic Metamaterials*

Bulk acoustic wave devices are utilized to provide high-performance filters needed in modern wireless communication systems, such as in mobile phones. Solidly mounted thin-film BAW resonators and filters utilize an acoustic mirror to isolate the resonator from the substrate and to provide desired plate wave dispersion behavior. In solidly mounted BAW devices, the acoustic mirror is a thin-film layer stack, a one-dimensional PnC, with an engineered frequency response. In the development of high-performance thin-film BAW resonators and filters, the acoustic mirror and the dispersion properties of the resonator play an important role.

Laser interferometric measurements are well suited, for example, to characterize the wave fields excited in BAW devices, to measure the plate-wave dispersion properties, and to study losses (leakage laterally and through the mirror). In the following, two examples are provided to illustrate the interferometer capabilities. Excited wave fields are measured, the data is used to extract the plate-wave dispersion properties, and these are further utilized to analyze the mirror transmission properties.

8.4.2.1 Thin-Film Mirror Transfer Properties

A thin-film acoustic mirror with alternating layers having high and low acoustic impedances can be regarded as a 1D PnC. Interferometric imaging is an excellent method for characterizing the plate wave dispersion properties in such structures and the ability to measure the wave fields on top of the resonator as well as on the bottom of the mirror stack enables experimental characterization of the mirror transmission properties.

The sample here is a solidly mounted 932 MHz thin-film BAW resonator with ZnO as the piezoelectric material and an acoustic mirror consisting of two $W - SiO_2$ layer pairs [22]. The substrate material is glass, in order to avoid capacitive shunting (the W-layers in the mirror were patterned for the same reason). A photograph of the resonator sample, a schematic presentation of the thin-film layer stack and measured input impedance (Z_{11}), calculated from wafer-level network analyzer data, are presented in Fig. 8.7. The main figures of merit (series and parallel resonance frequencies, f_s and f_p, and their Q-values Q_s and Q_p) are provided as an inset in Fig. 8.7c.

In order to thoroughly characterize the wave behavior in the sample, and the performance of the acoustic mirror, interferometric wave field measurements were carried out on both sides of the sample (at the resonator surface and at the boundary between the bottom of the acoustic mirror and the glass substrate). To study the frequency response in detail, the measurements were carried out at frequencies ranging from 350 to 1200 MHz.

Measured wave field amplitudes, on both sides, at selected frequencies are presented in Fig. 8.8. The fundamental, longitudinal, resonance is observed at 933 MHz

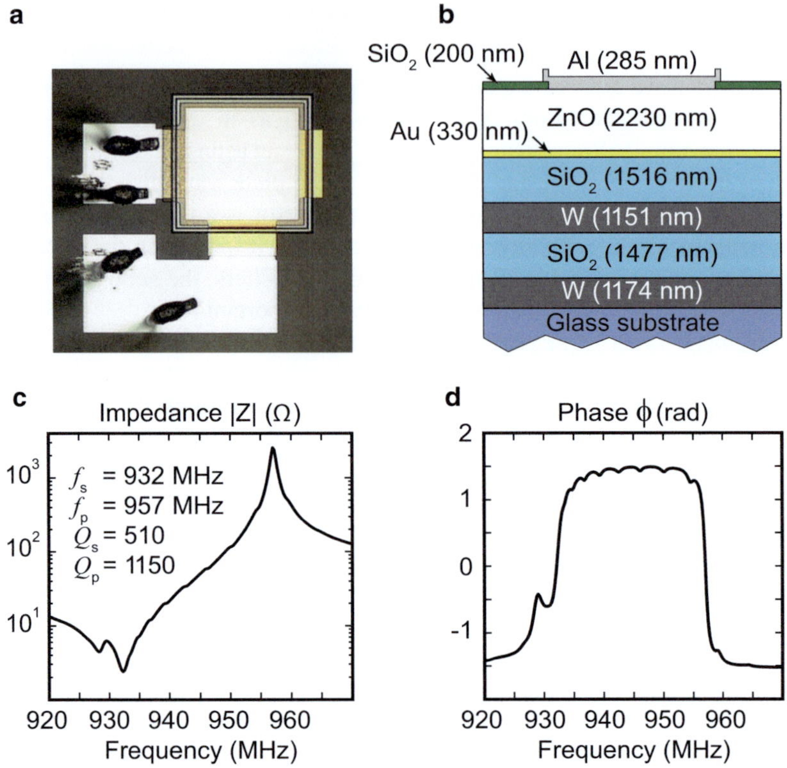

Fig. 8.7 (**a**) Microscope image of the square-shaped BAW resonator with bond wires to provide electrical contacts. (**b**) Schematic presentation of the thin-film layer stack and sample structure. (**c**) Impedance and (**d**) phase response obtained from a wafer-level electrical measurement, with figures of merit provided as an *inset* in (**c**). The electrical response features spurious resonances, seen as *dips* in the curves in (**c**) and (**d**)

as a single, strong, rather uniform amplitude distribution over the resonator area. It is also seen that the corresponding distribution on the bottom of the mirror shows very small amplitudes, indicating a good performance by the mirror as the wave mode is confined within the resonator. In contrast, at 416 MHz, the amplitudes on the top and at the bottom have similar distributions and values, indicating a wave mode not localized within the resonator, but rather extending throughout the layer stack. Furthermore, at 448 MHz, the amplitudes on the bottom are actually larger than those on the top. The measurement at 936 MHz features a lobe pattern due to the standing lateral resonances of the fundamental thickness extensional mode (TE_1). This lateral standing wave pattern reflects the square resonator geometry. In addition, there is ripple on all of the amplitude images, due to simultaneous excitation of wave modes with shorter acoustic wave lengths (higher lateral wave vectors).

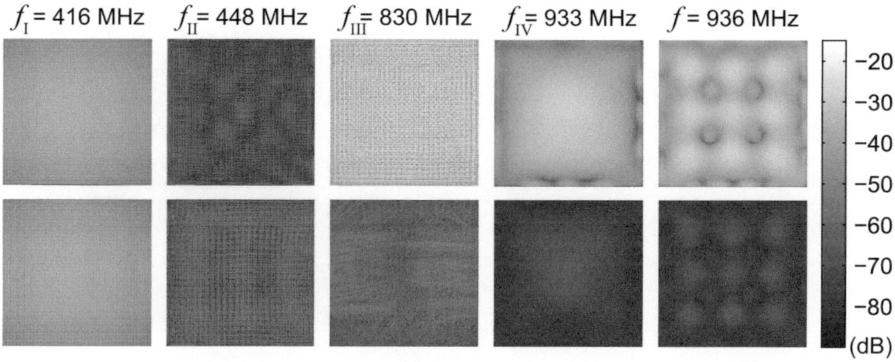

Fig. 8.8 Selected measured amplitude distributions both on *top* of the resonator (*first row*) and at the *bottom* of the acoustic mirror (*second row*). All the images share a common color scale to facilitate comparison. The first four frequencies are labeled (I–IV) for further reference

Let us now proceed to characterization of plate-wave dispersion properties and wave transmission through the mirror by using the dispersion diagrams. The wave dispersion can be experimentally measured utilizing laser interferometry. The measurement principle is presented here in connection with the characterization of the plate-wave dispersion in thin-film BAW devices, but can be applied to other wave types, such as SAWs, as well.

The process of obtaining the dispersion diagram from the interferometric wave field measurements is illustrated in Fig. 8.9. The measured wave field data, at a single frequency, is a superposition of all the wave components excited at that frequency. Fourier transform methods are used to provide a modal decomposition of the wave field data. In the FT result (Fig. 8.9, upper right), different wave modes have different lateral wave vectors (e.g., several concentric circles) and different excitation strengths (the intensity of the color in the figure). Note that the TE_1 mode contribution peaks on the axes (lateral resonances due to sample symmetry) and that the scattered waves produce circles in the wave vector space, indicating a good in-plane isotropy of the sputtered piezo-electric ZnO thin-film. Measuring the wave field distributions at several frequencies allows to obtain a dispersion diagram for the waves. Note that the frequency domain technique allows to excite the sample with frequencies chosen at will and thus detailed characterization of wave behavior at frequencies of interest, e.g., near resonances is possible.

The dispersion diagram for this sample is presented in Fig. 8.10. The dispersion diagram contains a wealth of information about wave behavior in the thin-film layer stack and enables to identify the different wave modes and their relative excitation strengths. Four frequencies of special interest have been labeled with Roman numerals (I-IV). I corresponds to a longitudinal resonance, not confined within the resonator, but extending throughout the whole layer stack. Frequencies II to IV correspond to the starting frequencies of dispersion curves, first and second thickness shear modes (TS_1 and TS_2) and to the TE_1, respectively. It is not surprising

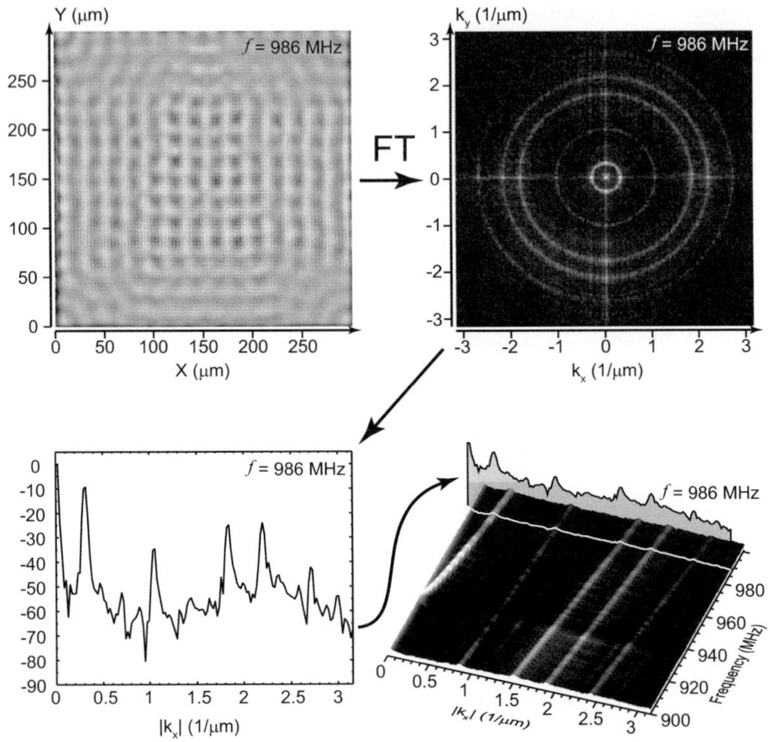

Fig. 8.9 An example illustrating the acquisition of the BAW dispersion curves from laser interferometric measurement. The wave field data at a single frequency ($f = 986\,\text{MHz}$) (*upper left*) is Fourier transformed to wave-vector space (*upper right*). The FT result displays a modal decomposition of all the wave content in the measurement data in wave-vector space (in order to show most of the dynamic range of the measurement, the figures use logarithmic scale). In this case, the sample is a square ZnO thin film BAW resonator, and the symmetry due to the resonator geometry is clearly seen as, e.g., the TE_1 wave mode content is mostly localized to the x- and y-axes (*bright white spots on the innermost circle*). The purely longitudinal wave component is seen as a *bright spot* in the center, $k_x = k_y = 0$. Due to the symmetry, it is feasible to extract the dispersion curves along one of the principal axes, dictated by the device and eigenmode geometry, e.g., along the x-axis ($k_y = 0$). This will result in slightly better SNR for the dispersion curves, when compared to circular averaging of the FT result. The extracted k_x-line is further processed to yield just one line, a $|k_x|$-slice (*bottom left*), which contributes one (*frequency*) line to the final dispersion diagram (the line obtained indicated in *bottom right*)

that in the beginning of TS_1 (II), at the intersection with the f-axis, the measured amplitude is very weak, because at that limit the mode is characterized by a pure horizontal displacement, which the interferometer does not detect.

In addition to comparing the measured wave fields (Fig. 8.8) on the top of the resonator and at the bottom of the mirror, the data at both sides can be utilized to calculate dispersion diagrams (on top and at the bottom), which can be compared to extract the mirror transfer properties in the wave-vector frequency space, see

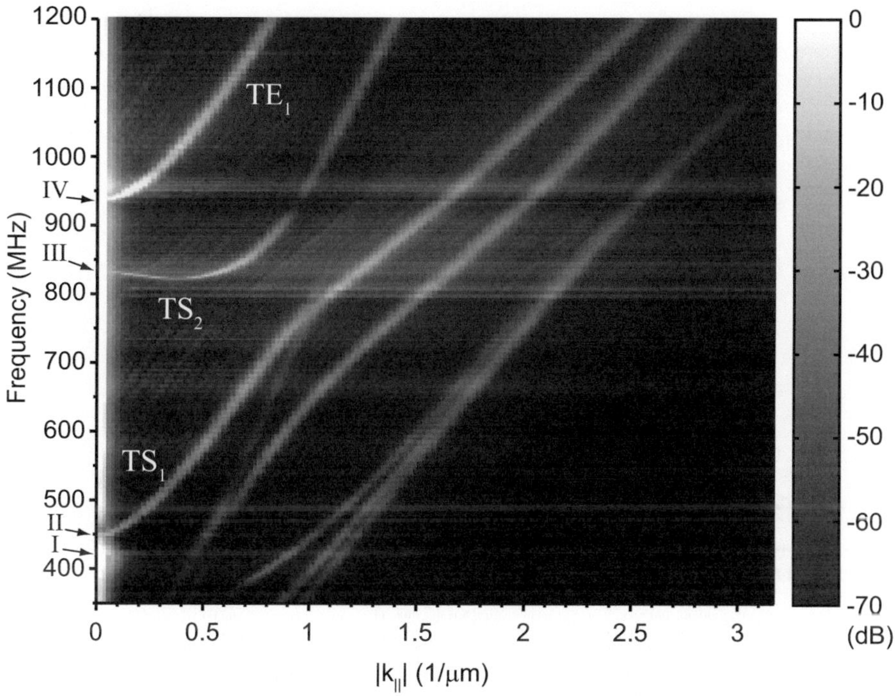

Fig. 8.10 Plate wave dispersion diagram for the 932 MHz thin-film BAW resonator. A longitudinal wave resonance, marked with I at 416 MHz, is observed as a maximum localized on $|k_{||}| = 0$. The curves starting at II and III correspond to the first and second thickness shear modes (TS_1 and TS_2). The TE_1 wave mode starts from the series resonance frequency $f_s \sim 932$ MHz (IV) with monotonically increasing lateral wave vector

Fig. 8.11. The results provide first direct experimental confirmation that the thin-film acoustic mirror is performing almost according to a 1D simulation, in the limit of pure longitudinal waves, $|k_{||}| = 0$, (see the attenuation line graph in Fig. 8.11). The thin-film mirror, with its engineered frequency response, serves to confine the wave energy within the resonator, isolating it from the substrate near the series and parallel resonance frequencies, by providing an attenuation of ~ 37 dB for the longitudinal waves. In other words, the structure does not support longitudinal wave propagation at this frequency range, and can be regarded to produce a band gap.

For a more complete description of the extensive characterization of dispersion and mirror transmission characteristics of this 932 MHz BAW sample, see [22]. For further examples of recent results on frequency-domain dispersion measurements, see, e.g., [23–30].

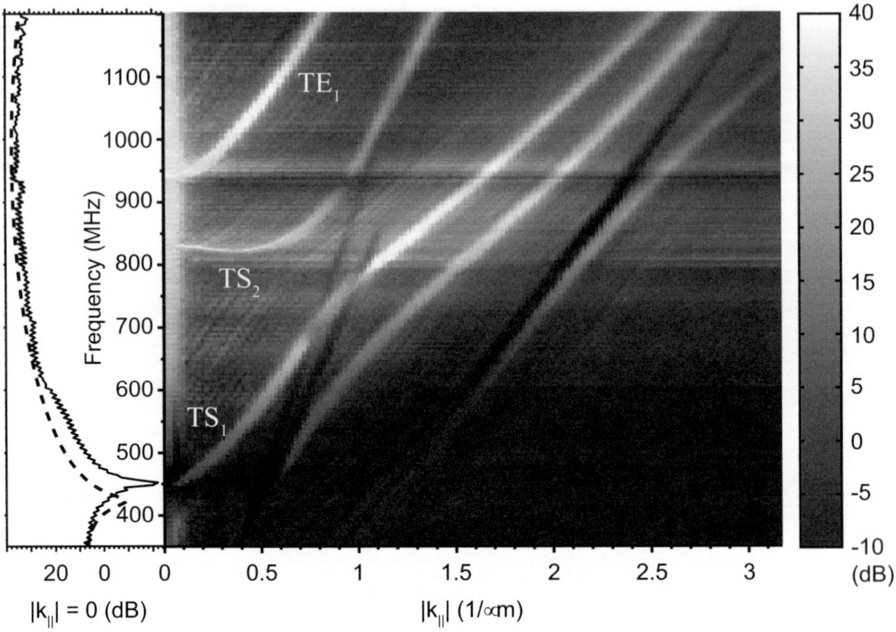

Fig. 8.11 Attenuation of vibrations due to the acoustic mirror obtained by comparing dispersion diagrams calculated from the data acquired both from the *top* of the resonator and from the *bottom* of the mirror. The attenuation of the purely longitudinal wave component, $|k_{||}| = 0$, is also provided as a function of the frequency (*solid line*) together with theoretical prediction (1-D simulation, *dashed line*)

8.4.2.2 Further Use of Dispersion Diagrams in Data Analysis

To illustrate how dispersion diagrams can be used to delve deeper into device physics, we consider a solidly mounted 1,820 MHz thin film BAW resonator [31]. The resonator is composed of an AlN piezoelectric thin-film sandwiched between metal electrodes, and of a thin-film acoustic mirror. Photograph of the resonator, schematic of the thin-film layer stack and wafer-level electrical measurements are presented in Fig. 8.12. The resonator operates in the lowest order thickness extensional TE_1 mode, in which about half an acoustic wavelength is contained within the thickness formed by the piezoelectric AlN-film and the metal electrodes. Since the TE_1 mode can also propagate horizontally as a plate wave, standing wave resonances are formed within the laterally finite-sized resonator. These lateral eigenmodes induce ripple in the electrical response near the main resonance, see Fig. 8.12c, d.

Despite the measured wave fields being superpositions of all the excited wave modes, the properties of the individual lateral eigenresonances can be studied by further utilizing the dispersion diagram, see Fig. 8.13. The lateral high-Q resonances show in the dispersion diagram as a chain of maxima instead of a uniform mode branch, see the inset in Fig. 8.13. A single wave mode (dispersion curve) can

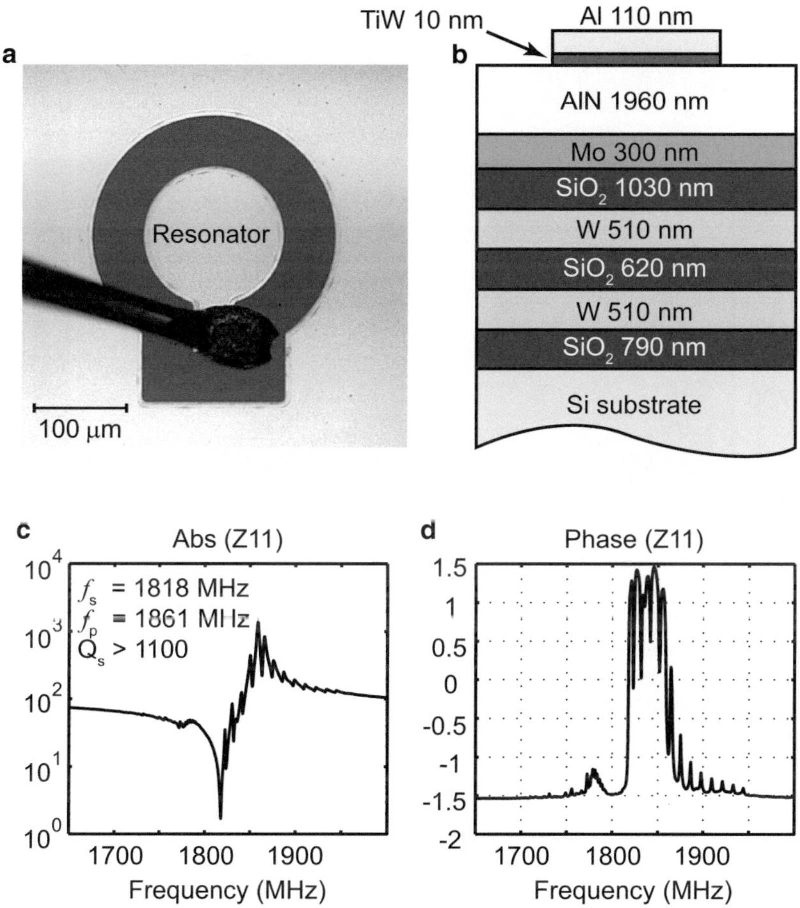

Fig. 8.12 (**a**) Microscope image of the BAW resonator with bond wire to provide electrical contact. (**b**) Schematic presentation of the thin-film layer stack and sample structure. (**c**) Impedance and (**d**) phase response obtained from a wafer-level electrical measurement, with figures of merit provided as an *inset* in (**c**). The electrical response features strong spurious resonances due to lateral eigenmode resonances, seen as *dips* in the curves in (**c**) and (**d**)

be followed and the amplitude values (or wave vector values, if desired) can be extracted as a function of frequency. This allows to single out a wave mode and, for example, to determine the Q-values of the individual eigen resonances. For further details, see [31].

Tailoring a desired plate-wave dispersion and ensuring both proper lateral and longitudinal (mirror) energy trapping is central for successful design of modern high-performance thin-film BAW resonators and devices. This requires designing appropriate dispersion properties not only in the resonator region but also in the regions surrounding it. Interferometric measurements are a valuable tool for

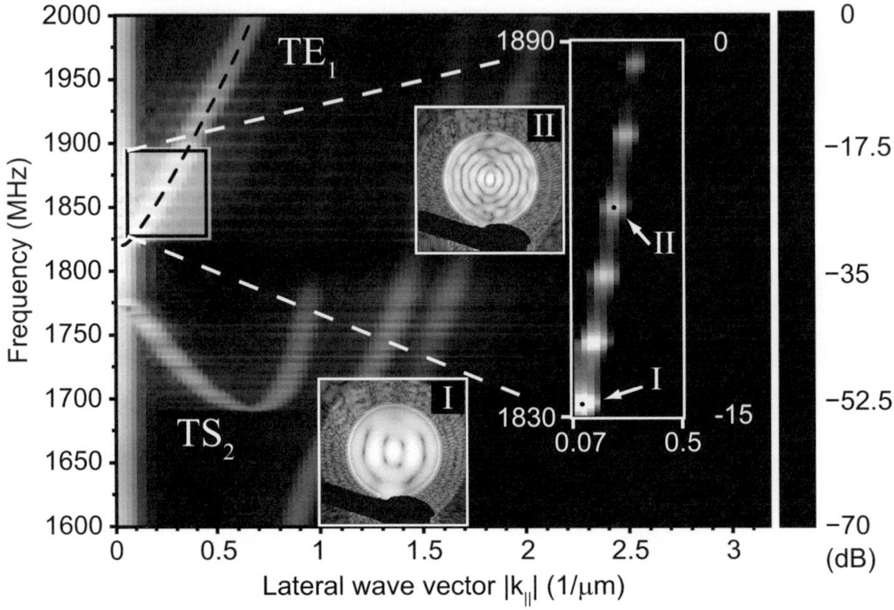

Fig. 8.13 Plate wave dispersion diagram from the measured wave field data for the 1,820 MHz thin-film BAW resonator. Due to the finite lateral dimensions of the sample and high-*Q* resonances, the TE_1 dispersion curve (marked with *dashed line*) is seen in the dispersion diagram as a chain of discrete maxima. Part of the TE_1 curve is further enlarged and provided as an *inset* to enable better visualization of the maxima. The amplitude distributions corresponding to two of the eigen frequencies, 1,831.5 and 1,863.5 MHz, are marked with I and II, respectively

verifying acoustic behavior of produced samples. Lately it has been shown that dispersion measurements can be performed also in the regions outside the active resonator area, which extends the applicability of these measurements [32].

Furthermore, measured wave fields can be visualized selectively, by utilizing filtering in the Fourier domain, i.e., selecting wave modes of interest in the wave vector space, propagating with certain wave vector $\boldsymbol{k}$ and then converting back to a wave field image via inverse Fourier transform. This method can also be used to gain extra SNR, for example, when visualizing very weak wave motion at the detection limit of the system, but one with a known wave vector or of a known range of wave vectors.

8.5 Conclusion

Frequency domain scanning laser interferometry has become an established characterization method in physical acoustics research. It has the potential to reveal significant new effects and to provide valuable information on the wave behavior and on device physics. Interferometry should not be considered as a substitute for electrical measurements, but rather it provides complementary information on the research subject. Modern scanning laser interferometer setups provide better than 1 μm spatial resolution together with minimum detectable vibration amplitudes down to sub-picometer range. Furthermore, measurement speed can reach millions of points per hour, rendering the technique useful not only in the research but also in product development. Optical probing excels in revealing effects that can be directly "seen" from the wave field images, but whose characterization via electrical measurements may be very difficult and time consuming, such as escaping acoustic beams. Phase sensitive measurements enable to separate different wave modes and waves traveling to different directions via Fourier transform methods, and hence allows to analyze wave transmission, reflection, and scattering, as well as to extract plate-wave dispersion properties.

References

1. R.L. Whitman, A. Korpel, Appl. Opt. **8**, 1567 (1969)
2. G.I. Stegeman, IEEE Trans. Sonics Ultrason. **SU-23**(1), 33 (1976)
3. J.P. Monchalin, IEEE Trans. Ultrason. Ferroelectr. Freq. Control **33**(5), 485 (1986)
4. J.E. Graebner, in *Proc. IEEE Ultrasonics Symposium*, vol. 1 (2000), pp. 733–736
5. J.P. Monchalin, Rev. Sci. Instrum. **56**(4), 543 (1985)
6. K. Kokkonen, M. Kaivola, Appl. Phys. Lett. **92**, 063502 (2008)
7. J.W. Wagner, J.B. Spicer, J. Opt. Soc. Am. B **4**(8), 1316 (1987)
8. R.L. Whitman, L.J. Laub, W.J. Bates, IEEE Trans. Sonics Ultrason. **SU-15**(3), 186 (1968)
9. J.V. Knuuttila, P.T. Tikka, M.M. Salomaa, Opt. Lett. **25**, 613 (2000)
10. J.E. Graebner, B.P. Barber, P.L. Gammel, D.S. Greywall, S. Gopani, Appl. Phys. Lett. **78**(2), 159 (2001)
11. G.G. Fattinger, P.T. Tikka, Appl. Phys. Lett. **79**(3), 290 (2001)
12. H. Yatsuda, S. Kamiseki, T. Chiba, in *Proc. IEEE Ultrasonics Symposium* (2001), pp. 13–17
13. K.L. Telschow, V.A. Deason, D.L. Cottle, I.J.D. Larson, IEEE Trans. Ultrason. Ferroelectr. Freq. Control **50**(10), 1279 (2003)
14. H. Martinussen, A. Aksnes, H.E. Engan, Opt. Express **15**(18), 11370 (2007)
15. J. Koskela, J.V. Knuuttila, T. Makkonen, V.P. Plessky, M.M. Salomaa, IEEE Trans. Ultrason. Ferroelectr. Freq. Control **48**, 1517 (2001)
16. J.V. Knuuttila, J.J. Vartiainen, J. Koskela, V.P. Plessky, C.S. Hartmann, M.M. Salomaa, Appl. Phys. Lett. **84**, 1579 (2004)
17. O. Holmgren, T. Makkonen, J.V. Knuuttila, M. Kalo, V.P. Plessky, W. Steichen, IEEE Trans. Ultrason. Ferroelectr. Freq. Control **54**(4), 861 (2007)
18. J. Meltaus, S.S. Hong, O. Holmgren, K. Kokkonen, V.P. Plessky, IEEE Trans. Ultrason. Ferroelectr. Freq. Control **54**(3), 659 (2007)
19. V. Laude, K. Kokkonen, S. Benchabane, M. Kaivola, Appl. Phys. Lett. **98**(6), 063506 (2011)
20. K. Kokkonen, S. Benchabane, A. Khelif, V. Laude, M. Kaivola, Appl. Phys. Lett. **91**, 083517 (2007)
21. S. Benchabane, A. Khelif, J.Y. Rauch, L. Robert, V. Laude, Phys. Rev. E **73**, 065601(R) (2006)
22. K. Kokkonen, T. Pensala, M. Kaivola. Dispersion and mirror transmission characteristics of bulk acoustic wave resonators. IEEE Trans. Ultrason. Ferroelectr. Freq. Control **58**(1), 215–225 (2011)

23. J.E. Graebner, H.F. Safar, B. Barber, P.L. Gammel, J. Herbsommer, L.A. Fetter, J. Pastalan, H.A. Huggins, R.E. Miller, in *Proc. IEEE Ultrasonics Symposium*, vol. 1 (2000), pp. 635–638
24. G.G. Fattinger, P.T. Tikka, in *Proc. IEEE MTT-S Int. Microwave Symp.*, vol. 1 (2001), pp. 371–374
25. K.L. Telschow, J.D. Larson, in *Proc. IEEE Ultrasonics Symposium* (2003), pp. 280–283
26. T. Makkonen, T. Pensala, J. Vartiainen, J.V. Knuuttila, J. Kaitila, M.M. Salomaa, IEEE Trans. Ultrason. Ferroelectr. Freq. Control **51**(1), 42 (2004)
27. G.G. Fattinger, S. Marksteiner, J. Kaitila, R. Aigner, in *Proc. IEEE Ultrasonics Symposium* (2005), pp. 1175–1178
28. K.L. Telschow, J.D. Larson, in *Proc. IEEE Ultrasonics Symposium* (2006), pp. 448–451
29. K. Kokkonen, T. Pensala, in *Proc. IEEE Ultrasonics Symposium* (2006), pp. 460–463
30. K. Kokkonen, T. Pensala, M. Kaivola, in *Proc. IEEE MTT-S Int. Microwave Symp.* (2007), pp. 2071–2074
31. K. Kokkonen, T. Pensala, J. Meltaus, M. Kaivola, Appl. Phys. Lett. **96**(17), 173502 (2010)
32. K. Kokkonen, J. Meltaus, T. Pensala, M. Kaivola. Characterization of energy trapping in a bulk acoustic wave resonator. Appl. Phys. Lett. **97**, 233507 (2010). http://dx.doi.org/10.1063/1.3521263

Chapter 9
Future Prospects of Phononic Crystals and Phononic Metamaterials

Saeed Mohammadi, Abdelkrim Khelif, and Ali Adibi

9.1 Introduction

The principles of operation of phononic crystal structures hold in a variety of frequency ranges starting from a portion of a Hertz to a few Tera Hertz. Therefore, phononic crystal structures can be beneficial for a variety of applications at different ranges of frequencies using various configurations and regimes of operation. In this chapter, we briefly outline the areas with great potential for applications in case further developments and research are pursued on phononic crystals and metamaterials.

9.1.1 Timing Elements Using Phononic Crystals

The ability of phononic crystal structures to form high quality resonators can be extended towards the development of better timing elements in combination with electronic circuitry. Involvement of more than two materials to form the phononic

S. Mohammadi (✉)
Multimodality Molecular Imaging Lab, Stanford University, 318 Campus Drive, East Wing, 1st Floor, Stanford, CA 94305-5427, USA
e-mail: saeedm@gmail.com; saeedm@gatech.edu

A. Khelif
Institut FEMTO-ST, Centre National de la Recherche Scientifique, Besançon Cedex, France
e-mail: abdelkrim.khelif@femto-st.fr

A. Adibi
School of Electrical Engineering, Georgia Institute of Technology, Ferst Dr NW 266, Atlanta, GA 30332, USA
e-mail: adibi@ee.gatech.edu

A. Khelif, A. Adibi (eds.), *Phononic Crystals*, DOI 10.1007/978-1-4614-9393-8_9

crystals can lead to materials with temperature compensation in addition to the possibility of obtaining phononic band gaps.

9.1.2 Signal Processing Functions for Communications

One of the major envisioned applications of phononic crystal structures is the development of high-frequency signal processing devices for wireless communications. Two major already established platforms for implementing such functions are bulk acoustic wave (BAW) platform (e.g., FBARs) and surface acoustic wave (SAW) platform. Each of these platforms have their own advantages and applications and it is worth to investigate the implementation of phononic crystals based on them. The basic components demonstrated in this book, i.e. resonators and waveguides, can serve as the building blocks of larger and more complex integrated acoustic systems. Wireless communications devices such as filters, multiplexers, and demultiplexers can be realized and optimized by proper coupling of such fundamental components. Further, the possibility of engineering the dispersion in phononic crystal structures allows for much needed functionalities such as delay lines. Possible advantages of phononic crystal structures with band gaps are the elimination of certain types of losses (see Chap. 7) and better control over the coupling between various components in the system. The reader is advised to refer to Chaps. 6 and 7 of this book for more details.

9.1.3 Phononic Crystal Sensors

It is also worth exploring the phononic crystal structures for sensing applications. Along with gaseous environments, liquid environments are very desirable for biosensing applications. Therefore, design and analysis of structures that can sense the targeted species in gaseous and liquid environments would be of great interest. The loss according to attenuation in liquids can be minimized by using the appropriate modes of the structure. This area of research has just launched and bears a great potential for further research [1].

9.1.4 Negative Refraction and Superlensing

Supporting phononic band gaps is not the only advantage of composite structures. Since such structures greatly modify the dispersion characteristics of the propagating waves, they can be used to attain effects such as negative refraction [2], efficient and compact lensing [3], and cloaking capabilities. Such phenomena may be used for various applications such as earthquake and tsunami protection, compact lenses for medical and non-destructive testing applications, and radar and sonar evasion. This area of research has become one of the most active areas in phononic crystal and metamaterial research.

9.1.5 Phononic Crystals for Opto-Mechanics

Although photonics and phononics have their own separate applications, the interaction between the two particles and the two types of waves can form unprecedented hybrid devices. Some of these devices utilize acousto-optic and optomechanical interactions [4, 5]. Controlling and harnessing such interactions using the concept of band gaps can be extremely useful in semiconductors and can lead to novel and more efficient devices. Such simultaneous band gaps have been proven to exist in photonic crystal fibers [6], in two-dimensional structures with a very large extension into the third dimension [7, 8], in three dimensional opals [9], and in more promising integrable "phoxonic" crystal slabs [10–12].

Since the optomechanical interaction between photons and phonons is inherently nonlinear, using a medium that can provide small-area confinement and large mode volumes can increase the effectiveness of such interactions. Photonic band gap structures are well known for their very large mode volumes not obtainable using other cavity types; therefore, photonic band gap structures are one of the best candidates to host strong optomechanical interactions. Of course, if such photonic band gap structures can also hold a phononic band gap, it will be possible to have simultaneous confinement of photons and phonons in the small confinement of cavities and waveguides to enhance their interactions. This can lead to novel acousto-optic and optomechanical devices for communications, sensing, and many other applications. In recent years, the possibility of simultaneous existence of photonic and phononic band gaps has been demonstrated. It was first theoretically discussed for in-plane propagation in two dimensional structures with very large third dimensions [13]. Its possibility in practical slab structures was first revealed in 2008 [10] and then published in more extensive forms in 2010 [11, 12]. This trend was followed by the experimental demonstration of high-Q simultaneous cavities [14] and more recently with a proposal for a photon/phonon translator [15].

9.1.6 Phononic Crystals for Liquid Control and Handling

Handling of microfluidics is essential for various applications including the development of a laboratory-like environment for disease or toxin diagnosis and for drug delivery. Often various functions over liquids (such as mixing, centrifuging, or delivering) are needed to develop the proper procedure for such applications. These functions call for a high concentration of power to be applied to the fluids and their droplets. Ultrasound is a favorable method for handling and manipulation of liquids to provide functionalities such as mixing and nebulization. On chip generation of such power densities that can selectively manipulate liquids on chip, either requires high levels of input power with various transducers or appropriate handling and guiding of elastic waves to the locations they are needed the most. Phononic crystals with their unique characteristics can be one of the best candidates for such applications. By their natural frequency selective characteristics and guiding capabilities they can play a major role in developing lab on chip applications. Reboud et al. have demonstrated the capability of phononic crystal structures in handling liquids [16–18]. Various functions such as microcentrifugation of particles and blood cells and selective nebulization of liquid micro-droplets have been demonstrated to be very effective using phononic crystal structures. This seems to be only the tip of the iceberg appearing and a more pronounced use of phonic crystal structures for handling on chip microfluids is expected.

9.1.7 Nonlinearity Effects in Phononic Crystals

Most of the discussions, investigations, and analyses performed so far on the phononic area have been on the linear propagation of acoustic waves, where the level of vibrations and tensions are relatively small. However, a great deal of physics is underlying the nonlinear nature of acoustic wave propagation in periodic structures. The nonlinearity arises when the displacement of the vibrations reaches a level that takes the media out of the linear elasticity region. Besides the great potential for the linear phononic crystals, including the nonlinearity adds a whole new dimension to their potential.

9.1.8 Energy Scavenging Using Phononic Crystals

Energy scavenging and harvesting is becoming more and more important as the need for clean and renewable energy for various applications is on the rise. Phononic crystals may find their role in this domain. However except for a few sparse efforts [19], there have not been many significant developments in this area so far. Therefore, this area of research has a great potential for further exploration.

9.1.9 Thermal Phonon Control Through the Use of Phononic Crystals

Controlling the flow of thermal phonons has a lot of potential for the development of novel devices for various applications. Exemplarily, thermoelectric materials are of interest for applications such as energy generation out of heat and the development of heat pumps. The performance of thermoelectric materials for such applications is pinpointed by the figure of merit of *ZT*, in which *Z* is a measure of the material thermoelectric properties and *T* is the absolute temperature. The quest for obtaining figure of merits much higher than unity at room temperature has been continued since nearly five decades ago. Despite the efforts the progress towards such goal has been sluggish so far. However, phononic crystal structures with nano-scale dimensions have shown initial promise for such improvements [20] and the next breakthrough in thermal control may just be around the corner.

9.1.10 Noise Control and Sound Proofing

From the observation of sound attenuation in a periodic sculpture published in Nature a few years ago [21], the efforts for using phononic crystal structures for noise control and isolation have been conducted [22]. Major applications besides sound and acoustic control may include the vibration isolation in airplanes, and vibration-free (e.g., optical) tables. While periodic structures were experimentally known to isolate vibrations, more systematic way of looking at them did not reveal itself before the appearance of phononic crystals analyses.

9.1.11 Phononic (Acoustic or Thermal) Diodes and Transistors

The analogy of the concept of periodic structures and band gaps in electrical and acoustic domains suggests the development of the revolutionary electronic devices such as diodes and transistors in the acoustic domain. However, even with phononic band gaps such devices are not trivial to form. The directive guidance of acoustic or thermal phonons is not straightforward and usually requires nonlinear effects [23]. However, there are efforts to emulate such behaviors in linear media as well [24]. Although such structures will be extremely useful, more rigorous research is required in this area before a practical solution can be achieved.

9.1.12 Immunization to Environmental Variables

Many of the applications of MM structures, such as signal processing components, timing elements, and many types of sensors require accurate frequency control. The frequencies of operation of such structures should not be drastically altered by the unwanted environmental variables such as temperature. Therefore, designing phononic crystal structures in more complex substrates (such as multilayer substrates), that can lead to more stability, would be of great benefit. Further, hermetically packaging of these structures can protect them from many environmental adverse effects. This type of packaging can also lead to higher quality factors and better devices with improved characteristics if vacuum packaging is used.

9.1.13 Note on Acoustic Metamaterials

Although phononic crystals and acoustic metamaterials are similar in some properties and the terms are sometimes used interchangeably, there is a possible distinction that can be made to refer to separate structures that can reduce the clutter in the terminology. Phononic crystals are usually referred to structures in which the main mechanism of dispersion is Bragg reflection, therefore, the periodicity and position of scatterers is a very important element in forming the desired properties. On the other hand, metamaterials are usually referred to structures that involve local resonators, usually resonating at lower frequencies compared to what Bragg-related frequencies can provide. Phononic crystals and acoustic metamaterials are therefore analogs of photonic crystal and Veselago medium in electromagnetism, respectively [25].

9.2 Conclusion

In this chapter, a brief overview of the possible applications and the areas of potential further investigations of phononic crystals and metamaterials were presented. The above-mentioned topics though, are only a subset of the possible emerging applications of the phononic crystals and phononic metamaterial structures. These devices can find their niche uses in many other areas of acoustic/mechanical systems.

References

1. R. Lucklum, J. Li, Phononic crystals for liquid sensor. Meas. Sci. Technol. **20**, 124014-1-12 (2009)
2. X. Zhang, Z. Liu, Negative refraction of acoustic waves in two-dimensional phononic crystals. Appl. Phys. Lett. **85**(2), 341 (2004)
3. T.-T. Wu, Y.-T. Chen, J.-H. Sun, S.-C.S. Lin, T.J. Huang, Focusing of the lowest antisymmetric Lamb wave in a gradient-index phononic crystal plate. Appl. Phys. Lett. **98**(17) 171911 (2011)
4. T. Carmon, H. Rokhsari, L. Yang, T.J. Kippenberg, K.J. Vahala, Temporal behavior of radiation-pressure-induced vibrations of an optical microcavity phonon mode. Phys. Rev. Lett. **94**, 1–4 (2005)
5. M. Eichenfield, R. Camacho, J. Chan, K.J. Vahala, O. Painter, A picogram- and nanometre-scale photonic-crystal optomechanical cavity. Nature **459**, 550–555 (2009)
6. P. Dainese, P.S.J. Russell, N. Joly, J.C. Knight, G.S. Wiederhecker, H.L. Fragnito, V. Laude, A. Khelif, Stimulated Brillouin scattering from multi-GHz-guided acoustic phonons in nanostructured photonic crystal fibres. Nat. Phys. **2**, 388–392 (2006)

7. M. Maldovan, E.L. Thomas, Simultaneous complete elastic and electromagnetic band gaps in periodic structures. Appl. Phys. B **83**(4), 595–600 (2006)
8. M. Maldovan, E.L. Thomas, Simultaneous localization of photons and phonons in two-dimensional periodic structures. Appl. Phys. Lett. **88**(25), 251907 (2006)
9. A.V. Akimov, Y. Tanaka, A.B. Pevtsov, S.F. Kaplan, V.G. Golubev, S. Tamura, D.R. Yakovlev, M. Bayer, Hypersonic modulation of light in three-dimensional photonic and phononic band-gap materials. Phys. Rev. Lett. **101**, 033902–033905 (2008)
10. S. Mohammadi, A.A. Eftekhar, A. Adibi, Large simultaneous band gaps for photonic and phononic crystal slabs, in *2008 Conference on Lasers and Electro-Optics*, Paper CFY1, OSA Publishing, 1–2, May 2008. https://www.osapublishing.org/abstract.cfm?uri=CLEO-2008-CFY1 (Also published by IEEE: http://ieeexplore.ieee.org/xpl/login.jsp?tp=&arnumber=4571339&url=http%3A%2F%2Fieeexplore.ieee.org%2Fiel5%2F4560187%2F4571172%2F04571339.pdf%3Farnumber%3D4571339)
11. S. Mohammadi, A.A. Eftekhar, A. Khelif, A. Adibi, Simultaneous two-dimensional phononic and photonic band gaps in opto-mechanical crystal slabs. Opt. Express **18**(9), 9164–9172 (2010)
12. Y. Pennec et al., Simultaneous existence of phononic and photonic band gaps in periodic crystal slabs. Opt. Express **18**(13), 14301–14310 (2010)
13. M. Maldovan, E.L. Thomas, Simultaneous localization of photons and phonons in two-dimensional periodic structures. Appl. Phys. Lett. **88**(25), 251907 (2006)
14. A.H. Safavi-Naeini, O. Painter, Design of optomechanical cavities and waveguides on a simultaneous bandgap phononic–photonic crystal slab. Opt. Express **18**(14), 14926–14943 (2010)
15. A.H. Safavi-Naeini, O. Painter, Proposal for an optomechanical traveling wave phonon–photon translator. New J. Phys. **13**(1), 013017 (2011)
16. R. Wilson, J. Reboud, Y. Bourquin, S.L. Neale, Y. Zhang, J.M. Cooper, Phononic crystal structures for acoustically driven microfluidic manipulations. Lab Chip **11**(2), 323–328 (2011)
17. Y. Bourquin, R. Wilson, Y. Zhang, J. Reboud, J.M. Cooper, Phononic crystals for shaping fluids. Adv. Mater. (Deerfield Beach, Fla.) **23**(12), 1458–1462 (2011)
18. J. Reboud, R. Wilson, Y. Zhang, M.H. Ismail, Y. Bourquin, J.M. Cooper, Nebulisation on a disposable array structured with phononic lattices. Lab Chip **12**(7), 1268–1273 (2012)
19. S. Gonella, A.C. To, W.K. Liu, Interplay between phononic bandgaps and piezoelectric microstructures for energy harvesting. J. Mech. Phys. Solids **57**(3), 621–633 (2009)
20. P.E. Hopkins, C.M. Reinke, M.F. Su, R.H. Olsson, E.A. Shaner, Z.C. Leseman, J.R. Serrano, L.M. Phinney, I. El-Kady, Reduction in the thermal conductivity of single crystalline silicon by phononic crystal patterning. Nano Lett. **11**(1), 107–112 (2011)
21. R. Martinez-Sala, J. Sancho, J.V. Sanchez, V. Gomez, J. Llinares, F. Meseguer, Sound attenuation by sculpture. Nature **378**(6554), 241 (1995)
22. J.H. Wen, G. Wang, D.L. Yu, H.G. Zhao, Y.Z. Liu, Theoretical and experimental investigation of flexural wave propagation in straight beams with periodic structures: application to a vibration isolation structure. J. Appl. Phys. **97**(11), 114907 (2005)
23. N. Boechler, G. Theocharis, C. Daraio, Bifurcation-based acoustic switching and rectification. Nat. Mater. **10**(9), 665–668 (2011)
24. Y. Li, J. Tu, B. Liang, X.S. Guo, D. Zhang, J.C. Cheng, Unidirectional acoustic transmission based on source pattern reconstruction. J. Appl. Phys. **112**(6), 064504 (2012)
25. J. Li, C. Chan, Double-negative acoustic metamaterial. Phys. Rev. E **70**(5), 1–4 (2004)

MIX
Papier aus verantwortungsvollen Quellen
Paper from responsible sources
FSC® C105338

If you have any concerns about our products,
you can contact us on
ProductSafety@springernature.com

In case Publisher is established outside the EU,
the EU authorized representative is:
Springer Nature Customer Service Center GmbH
Europaplatz 3, 69115 Heidelberg, Germany

Printed by Libri Plureos GmbH
in Hamburg, Germany